I0754658

THE BUREAU OF AGRICULTURE—WASHINGTON, D. C

OPENING A FARM IN IOWA TO RAISE CORN AT 16 CENTS A BUSHEL.

HISTORY

OF THE

GRANGE MOVEMENT;

OR, THE

FARMER'S WAR AGAINST MONOPOLIES:

BEING

A FULL AND AUTHENTIC ACCOUNT OF THE STRUGGLES OF THE AMERICAN FARMERS AGAINST THE EXTORTIONS OF THE RAILROAD COMPANIES.

WITH

A HISTORY OF THE RISE AND PROGRESS

OF THE ORDER OF

PATRONS OF HUSBANDRY,

ITS OBJECTS, PRESENT CONDITION AND PROSPECTS.

TO WHICH IS ADDED

SKETCHES OF THE LEADING GRANGERS.

BY

EDWARD WINSLOW MARTIN,
(JAMES D. McCABE, Jr.,)

AUTHOR OF "BEHIND THE SCENES IN WASHINGTON," ETC., ETC.

Illustrated with 60 Fine Engravings and Portraits of Leading Grangers.

Issued by subscription only, and not for sale in the book stores. Residents of any State desiring a copy should address the Publishers, and an Agent will call upon them. See page 541.

NATIONAL PUBLISHING COMPANY,
CHICAGO, ILL.; CINCINNATI, OHIO; ST. LOUIS, MO.;
PHILADELPHIA, PA.; AND ATLANTA, GA.
A. L. BANCROFT & CO., SAN FRANCISCO, CAL.
1874.

TO THE

FARMERS OF THE UNITED STATES,

THE STRONG-ARMED, TRUE-HEARTED HOPE OF THE REPUBLIC, NOW, AS IN THE PAST, THE FIRST TO RISE AGAINST OPPRESSION AND WRONG, THE AUTHOR DEDICATES THIS BOOK AS A TOKEN OF HIS SYMPATHY WITH THEM IN THEIR SUFFERINGS, AND HIS ADMIRATION OF THE HEROIC BATTLE THEY ARE WAGING FOR THE OPPRESSED OF THE WHOLE COUNTRY.

A PRAIRIE HOME.

PREFACE.

AMONG the many remarkable events of the present century, there are none more worthy of patient and careful study than the important movement among the agricultural classes, which has been popularly termed "The Farmer's War Against Monopolies." The rapid and astounding growth of this movement, the formation of Farmers' Granges in every part of the Union, and the remarkable success which has attended every step of the Order of Patrons of Husbandry, have made it the most closely and anxiously observed of any of the movements of the day. The people of the United States are deeply interested in it, and on all sides there is a growing desire to know more of it. Men cannot help regarding with a deep interest an organization which bids fair to embrace the whole agricultural population at an early period, and which proposes to exert the enormous strength and power of this class of our countrymen as a compact and united force for the accomplishment of a definite object. They naturally desire to know if this new and powerful element in our public affairs is to exert its power for good or for evil; whether it is to work for the good of the whole country, with a wise and generous regard for the welfare of all classes, or whether it seeks only the advancement of its own interests regardless of the rights or well-being of others. Even those who laughed at the movement in its infancy, are now forced to confess that the

Patrons of Husbandry are to-day a power which no political party can afford to ignore, and which will soon, perhaps in the next year, be able to decide the majority of the popular elections throughout the Union.

It is but natural, then, that the people of the United States should be very desirous of knowing more about this powerful Order. The whole land is full of rumors regarding it, the majority of which are utterly without foundation. The present volume is offered to the public as a means of satisfying their legitimate curiosity upon this subject. It presents a careful, and, it is believed, impartial account of the wrongs from which the agricultural classes have been suffering; the causes which led to the organization of the Order of Patrons of Husbandry; together with an account of the establishment of the Order, its history, its present condition, its objects, and its plans and prospects for the future.

It has long been evident to earnest thinkers that the farmers of the United States are the most cruelly oppressed class of our community. In these pages the writer has sought to set forth these wrongs, and to plead the cause of the farmer, in the hope of awakening the general public to a realization of the case. We cannot afford to allow the farmer to suffer at the hands of his enemies. Upon his weal or woe depends the prosperity of the entire nation. The farmers' cause is that of the people, and it is the aim of this work to show that in battling for the farmers' rights, the Grange is fighting the cause of the whole people.

For several years past the country has been suffering from evils of which all have been conscious, but which none had the courage to remedy, until the Grange took up the cause of the oppressed. Prominent among these are the burdens that have been fastened upon the people by the reckless and unscrupulous course of the great Railroad Monopolies that have sprung up in our midst. These vast and powerful corporations have inaugurated a series of abuses which have gradually and effectually undermined the solid basis upon which our finances were supposed to rest. They have debauched and demoralized our Courts and Legislatures; have bribed and taken into their pay the high public officials charged with the making and execution of our laws; have robbed the nation of a domain sufficient to constitute an empire; have flooded the land with worthless stocks and other so-called securities; have established a system of gambling at our financial centres that has resulted in a monetary crisis which must cover the whole land with ruin and suffering; have set at defiance the laws of the land, and have trampled upon individual and public rights and liberties, openly boasting that they are too powerful to be made amenable to the law; and not content with all this, not satisfied with the ruin they have wrought, they propose to petition the National Legislature to give them still greater means of robbing and oppressing the people.

The Grange seeks to array the agricultural class—nearly one-half of our whole population—as a compact body against these evils, and by thus opposing a solid front to the monopolists and their selfish and unpatriotic schemes, to awaken the entire nation to a sense of the danger with which it is threatened, and secure its co-operation in the enforcement of measures which will remove the evil and bring about a more healthful state of affairs. The Grange offers to the farmers the most practicable means of bettering their condition, and while it confines its membership strictly to the agricultural class, it appeals powerfully to the general public for sympathy and encouragement. Believing as he does, that the farmer has suffered great and cruel wrongs, the Author has endeavored to tell his story for him, and to show to the reader wherein it is true.

The great and overwhelming interest manifested by the public in the question has made this a fitting time for the appearance of such a book. Evil days are upon us, and it becomes the duty of everyone to inquire the cause of this unhappy state of affairs, which is as remarkable as it is distressing. A more singular phenomenon was never offered for our consideration. There is every reason why trade this season should be abundantly prosperous. Our harvest has been abundant. The markets for our productions are in our favor. We have not only a ready sale for our breadstuffs and provisions, but we have begun to send abroad manufactured goods, for which we have had no foreign purchasers heretofore. Our mills have not been overstocked. Industry has generally shown a healthy and steady activity. Our depressed shipping interest has wonderfully revived. And yet in spite of all this, the country has been driven into one of the most serious and alarming conditions of financial depression it has ever experienced. Everyone is interested in knowing the cause of this evil, and in taking measures to bring about a better state of affairs. The Author has endeavored in these pages to shed some new light upon the matter in consideration, and assist the reader in the intelligent discharge of his duty as a responsible member of the community.

E. W. M.

LIST OF ILLUSTRATIONS.

CONTENTS.

PART I.

RAILROAD MONOPOLIES.

CHAPTER I.

THE RISE AND PROGRESS OF AMERICAN RAILROADS.

CHAPTER II.

HISTORY OF THE "LAND GRAB."

CHAPTER III.

WATERED STOCKS.

CHAPTER IV.

THE CONSOLIDATION PROCESS.

CHAPTER V.

THE TRANSPORTATION TAX SWINDLE.

CHAPTER VI.

RAILROAD TYRANNY.

CHAPTER VII.

THE CAPTURE OF THE COURTS.

CHAPTER VIII.

RAILROAD LEGISLATION.

CHAPTER IX.

RAILROAD STOCK GAMBLING.

CHAPTER X.

THE GREAT RAILROAD PANIC.

CHAPTER XI.

WILD CAT RAILROADS.

CHAPTER XII.

THE CASE OF THE NORTHERN PACIFIC RAILROAD.

CHAPTER XIII.

DANGER AHEAD.

PART II.

THE COAL MONOPOLY.

CHAPTER XIV.

OPERATIONS OF THE COAL RING.

PART III.
THE FARMERS' WRONGS.

CHAPTER XV.
THE AGRICULTURAL CLASSES AND THEIR WRONGS.

CHAPTER XVI.
THE MIDDLE-MEN.

CHAPTER XVII.
THE RAILROADS AND THE FARMERS.

T. R. Allen, Master of Missouri. Col. John Cochrane, Master of Wisc

John Weir, Master of Indiana. Mrs. J. C. Abbott, "Flora."

Wm. Saunders, First Master, Nat'l Grange. D. W. Adan

SOME OF THE LEADING OFFICERS OF T

sconsin. S. H. Ellis, Master of Ohio. F. H. Dumbauld, Master of Kansas.
Mrs. D. W. Adams, "Ceres." C. D. Beeman, Gen'l Nat'l Deputy.
ms, Master, Nat'l Grange. O. H. Kelley, Secretary, Nat'l Grange.

THE ORDER OF PATRONS OF HUSBANDRY.

CHAPTER XVIII.

THE STORY OF FARMER GREEN'S REAPER.

CHAPTER XIX.

FARMER SMITH SPEAKS HIS MIND.

CHAPTER XX.

VIEWS OF A WISCONSIN FARMER.

CHAPTER XXI.

HOW THE GOVERNMENT ROBS THE FARMERS.

CHAPTER XXV.

THE LAWS OF THE ORDER.

CHAPTER XXVI.

THE GRANGE AS A MEANS OF PROTECTION.

CHAPTER XXVII.

SOCIAL ASPECT OF THE GRANGE.

CHAPTER XXVIII.

THE LESSONS OF THE GRANGE.

CHAPTER XXIX.

THE COÖPERATIVE FEATURE.

CHAPTER XXX.

THE FUTURE OF THE GRANGE.

CHAPTER XXXI.

HISTORY
OF THE
GRANGE MOVEMENT;
OR, THE
FARMER'S WAR AGAINST MONOPOLIES.

PART I.

RAILROAD MONOPOLIES.

CHAPTER I.

THE RISE AND PROGRESS OF AMERICAN RAILROADS.

The first Railroad Enterprises—The Pioneer Railroad—A Modest Beginning—The Mauch Chunk Railroad—Inauguration of the Railroad System—Introduction of Steam—The First Locomotive—Opening of the Baltimore and Ohio Railroad to the Potomac—Improvements in the Construction of the Roads—Rapid increase of Railroads—Building of the Great Trunk Lines between the East and West—Efforts of the Eastern Cities to secure the Western Trade—Completion of the great Roads—Commencement of the Railroad System of the West—Its Rapid Growth—Statement of the Annual Growth and Cost of the Railroads of the United States—Their Present Condition.

FEW things have been so remarkable in the wonderful growth of the American Republic as the increase and expansion of its railway system. A comparatively young man finds no difficulty in remembering the time when the only means of communication between the various parts of the Union were the canal boat and the stage coach. Half a century has witnessed the wonderful growth of the American railways. It has also witnessed the gradual change of the system, which was designed to be a lasting benefit to the Republic, into one of the greatest dangers which now threaten its existence.

We propose to trace in these pages the growth of the railway system of the country, and to present to the reader a statement of its present condition, in order that he may the better appreciate the grave danger with which this immense system threatens the land.

It was not until 1826 that capitalists became satisfied of the value of the railway as a means of communication between distant points. The first road of this kind in America was a mere tramway for the transportation of granite from the quarries at Quincy to the Neponsett River, in Massachusetts. The total length of this road was about three miles. It terminated at the quarries in a self-acting inclined plane. It was built upon granite sleepers, seven and a half feet long, laid eight feet apart. The rails were laid five feet apart, were of pine, a foot deep, and covered with an oak plate, and this with flat bars of iron. The cars were drawn by horses.

In January, 1827, a second road was begun, and completed in May of that year, from the coal mines, at Mauch Chunk, Pennsylvania, to the Lehigh River, a distance of nine miles. "From the summit within half a mile of the mines the descent to the river was 982 feet, of which 225 feet were included in a self-acting plane at the river, and twenty-five feet more in a shute by which the coal was discharged into the boats. The remainder was in a continual descending grade, down which the loaded wagons ran by gravity, one of them being appropriated to the mules by which the empty wagons were drawn back. The rails were of timber, laid on wooden sleepers and strapped with flat iron."

These roads demonstrated the fact that heavy loads could be drawn with a slighter expenditure of power than must be used on the ordinary highways. Admirable results had been obtained in Europe by the adoption of this system, and by the close of 1827 it had come to be acknowledged in the United States that the railway was the best method of transporting freight. Accordingly measures were set on foot in many of the States, especially in Massachusetts, New York, New Jersey, Pennsylvania, Maryland, and South Carolina, for the construction of various roads, some of which were designed upon a grand scale.

Steam had now been introduced upon the European roads as a motive power, and the managers of the American roads at once adopted it as the cheapest and the best means of propelling their cars. The Delaware and Hudson Canal Company began, in 1828, the construction of a road from their coal mines to Honesdale, the terminus of the canal. Before the road was fairly begun, they sent Mr. Horatio Allen to England to investigate and report upon the railways of that country, and also to purchase their railroad iron. Later still, he was directed to buy three locomotive engines. These purchases were made, and the first, a locomotive, built by George Stephenson, at his works at Newcastle-upon-Tyne, arrived in New York in the Spring of 1829. It was exhibited for some time in New York, and attracted great attention. The other locomotives were built by other persons, and arrived and were placed on the road in the latter part of 1829. They were the first locomotives used in this country. They were engines on four wheels, furnished with the multitubular boiler and the exhaust blast.

In 1829 work was begun on the South Carolina Railroad, designed to connect Charleston with Hamburg, on the Savannah River, opposite the City of Augusta, Ga. Six miles of this road were completed from Charleston before the close of the year. It is an interesting fact that some time previous to the commencement of work upon this road, the directors determined, under the advice of their engineer, Mr. Horatio Allen, who was convinced that horses could not draw cars over this road with profit to the Company, to make a species of horse locomotive "the motive power, and the road was constructed in such a manner as to be wholly dependent upon them, being built upon piles often at a great height above the ground. The company offered a premium of $500 for the best plan of horse locomotive, and this was awarded to Mr. C. E. Detmold, afterward of New York, who constructed one with the horse working on an endless chain platform. It carried twelve passengers at the rate of twelve miles per hour. The same gentleman, in the winter of 1829–30, made drawings of the first American steam locomotive, called the 'Best Friend,' which was planned by Mr. E. L. Miller, then residing in Charleston, made by the Kembles at their shop in West street, and placed on the road late in the summer of 1830. It was a small four-wheeled engine, with upright boiler and water flues close at bottom, and the flame circulating around them. It worked successfully for about two years, when it exploded, and was rebuilt with a flue boiler. Upon this road, in 1831, was first introduced on any road, either abroad or in the United States, the important arrangement of two four-wheeled trucks for locomotives and long passenger cars. These were built from plans de-

A GRANGERS' PROCESSION AND MASS MEETING.

signed by Mr. Horatio Allen in 1830; and with no essential change his system of double truck running gear, including the application of pedestals to the springs, has ever since been adopted upon all the roads of the country."

In 1828, the Baltimore and Ohio Railroad, the first of the great routes connecting the Eastern and Western States, was begun, the first stone being laid with great rejoicing in the city of Baltimore on the 4th of July of that year. It was originally designed to use horse cars only upon it, and in June, 1830, the road was finished to Ellicott's Mills, a distance of fourteen miles from Baltimore. The cars were drawn by horses. "For a long time," says the historian of the road, "notwithstanding the use of horse power, the railroad was regarded as a great novelty; and the people of Baltimore, with their wives, sisters, or friends, patronized it very extensively. A ride to Ellicott's Mills by railroad was a daily or weekly amusement; and that interesting village became exceedingly popular with all classes of people. The number of cars provided by the company proved entirely inadequate to the trade, both for passengers and merchandise; and although but one track had been finished, the receipts for the first four months showed an aggregate of over twenty thousand dollars." It was at one time proposed to propel the cars on this road by means of sails.

By April, 1832, the road was opened to the Point of Rocks, on the Potomac, a distance of seventy miles. Long before this, it had become evident to the company that horse power was utterly inadequate to the demands upon the road. The successful introduction of steam upon the English roads encouraged the directors to believe that they could use it with equal success in their own undertaking, and rewards were offered for the construction of the best locomotive suited to the needs of the road. A small locomotive, built in Baltimore by Mr. Peter Cooper, was placed on the line in 1830, and gave considerable satisfaction, but it did not fully meet the requirements of the company.

"Agreeably to the invitation of the President and Directors, three locomotive engines were introduced upon the road, in the summer of 1831, of which only one proved to be successful, according to the stipulations of the company. This was the *York*, having been erected in the town of that name, in Pennsylvania, situated fifty-seven miles north of Baltimore. This engine was erected by Phineas Davis (a very ingenious and worthy man, who subsequently met with an accident which proved fatal, while experimenting with his machinery), of the firm of Davis & Gardiner, and, after undergoing some slight modifications, was found capable of conveying fifteen tons, at the rate of fifteen miles per hour, on a level portion of the road. It was employed for a considerable time between Baltimore and Ellicott's Mills, and generally performed the trip in one hour, with four cars, being a gross weight of fourteen tons. The engine, it will be observed, was mounted on four wheels, of thirty inches diameter, like those of common cars; and the velocity was obtained by means of gearing with a spur wheel and pinion, on one of the axles of the road wheels. The entire weight was but three and a half tons, and it not unfrequently attained a speed ranging from twenty to thirty miles per hour. It passed over curves with much facility, overcoming those of four hundred feet radius, the shortest on the road, at the rate of fifteen miles per hour. The fuel used was anthracite coal, which is said to have answered the purpose very well; but the lightness of the machine prevented it from drawing very heavy loads over ascending grades."

The railroads at this time were built of longitudinal rails pinned down to the wooden or stone cross ties, which were imbedded in the ground, and upon the rails flat bars of iron about half an inch thick, and from two and a half to four and a half inches in width, were fastened by spikes, the heads of which were countersunk in the iron. This species of rail was generally adopted as the cheapest, but it was not long before it attained an unenviable notoriety as the most dangerous. The ends of the rails would frequently curl up, and being caught by the wheels would be thrust through the bottoms of the cars, causing sometimes very serious accidents.

The Baltimore and Ohio road was carried steadily forward, accomplishing in its construction feats of engineering which were justly regarded as national triumphs. It climbed to the summit of the Alleghanies, and passed them by a series of grades at which the most accomplished engineers had halted in dismay when told of them by the friends of the road, and it is not too much to assert that the success of this road was one of the greatest encouragements offered to the persons interested in these enterprises.

In the meantime, however, other roads were begun and carried forward with energy. The use of steam as a motive power had overcome the chief obstacle in their way, and the whole country was engaged in schemes for their increase and diffusion throughout its limits. In 1830–31, as many as twelve railway companies were chartered by the Legislature of Pennsylvania. By the close of the year 1832, Pennsylvania is said to have had sixty-seven lines of railway in operation. In the same year, the principal lines in Massachusetts and New Jersey were also begun.

The most important of these lines were designed to connect the cities of the Atlantic seaboard with the States west of the mountains. The Great West was just beginning its wonderful growth in population and prosperity, and each of the principal seaports of the East became eager to secure the rich harvest offered by the trade of the West. By means of the Western Railway of Massachusetts, Boston was brought into direct communication with the Erie Canal and the roads now constituting the New York Central Railway, over which the grain of the West was conveyed to Albany. This induced the capitalists of New York to undertake the construction of the Erie Road, which was begun in 1833, but was not finished until 1857. In order to compete still more advantageously with New York, Boston furnished the means for the construction of the Michigan Central road, and its extension from Detroit to Chicago, thus bringing the products of the West directly to Albany and thence to Boston. "It was also sending out its long arms toward the Northwest, reaching the outlet of the great lakes at Ogdensburg, before this point was connected by railroad with the metropolis of its own State. These enterprises stimulated Pennsylvania to perfect her line of communication between Philadelphia and Pittsburg, which from Harrisburg to Hollidaysburg was by canal, and thence over the Alleghany Mountains by a succession of five inclined planes and intervening levels up the mountain on one side, then by a long level to the five inclined planes and levels which terminated below at Johnstown, where another canal took the boats that had been brought over the mountain in sections, and conveyed them to Pittsburg. The canals and inclined planes were done away with, and a continuous road was opened across the State." This was the now famous Pennsylvania Central Railroad. Connections were pushed out from it to Lake Erie at Cleveland, to Chicago, and by way of Columbus and Cincinnati, with the railroads of Ohio, Indiana and Illinois.

"Baltimore, feeling the effects of these advances, was impelled to push forward the Baltimore and Ohio road, which had long stopped in the coal region of Cumberland, and it was at last completed to Wheeling on the Ohio. Charleston and Savannah early appreciated the importance of connecting their harbors with the productive districts of the interior by railroads; and when these had penetrated their own States, the line of equal importance to both was extended through North Georgia into Tennessee, connecting, in 1849, Chattanooga with those cities.

"All these advances into the valleys of the branches of the Mississippi affected the cities of the Gulf of Mexico, and Mobile and New Orleans hastened forward the lines which in the early history of American railroads they had projected for securing to themselves the trade of these valleys."

In the Western States the growth of the railway system was not less marked than in the East. In 1838 there were but 22 miles of railway in operation in the West, and this in the State of Kentucky. Four years later, in 1842, there were 274 miles in operation in the West, of which Michigan contained 138 miles. In 1850 the Western States contained over 1400 miles of railway; and in 1860 over 13,000 miles out of a total of 31,185 miles in the entire Union. In 1872, out of a total of 62,647 miles in the whole Union, the Western States contained over 34,000 miles of railroad.

The following Table will show the gradual growth of our railways from the commencement to the present day:

Year.	Miles open.	Yearly increase.
1827	3	—
1828	3	—
1829	28	25
1830	41	13
1831	54	13
1832	131	77
1833	576	445
1834	762	186
1835	918	156
1836	1102	184
1837	1431	329
1838	1843	412
1839	2220	377
1840	2797	577
1841	3319	522
1842	3877	558
1843	4174	297
1844	4311	137
1845	4522	211
1846	4870	348
1847	5336	466
1848	5682	346
1849	6350	668
1850	7475	1125
1851	8589	1114
1852	11,027	2438
1853	13,497	2470
1854	15,672	2175
1855	17,398	1726
1856	19,251	1853
1857	22,625	3374

Year.	Miles open.	Yearly increase.
1858	25,090	2465
1859	26,755	1665
1860	28,771	2016
1861	30,593	1822
1862	31,769	1176
1863	32,471	702
1864	33,860	1389
1865	34,442	582
1866	35,351	909
1867	36,896	1545
1868	38,822	1926
1869	42,272	3450
1870	48,860	6588
1871	54,435	5575
1872	62,647	8212

In Poor's "Railroad Manual of the United States" for 1873–74, the following estimate is given of the cost and capital invested in the railways of the United States:

States.	Capital Stock.	Funded and other Debt.	Total Capital Account.	Cost of R.R. per mile.
New England States	$129,012,748	$101,597,046	$230,609,794	$50,418
Middle States	558,838,174	363,862,600	922,700,774	79,427
Western States	724,686,046	747,939,186	1,472,625.232	50,550
Southern States	171,683,155	230,230,112	401,913,267	36,575
Pacific States	63,623,990	67,950,000	131,573,990	98,300
Total	$1,647,844,113	$1,511,578,944	$3,159,423,057	55,116

A still more definite idea of the immense system which has grown from the modest beginning at Quincy in 1826, will be gained from the following extract from the same Manual:

"The total cost of the railroads, the operations of which are given for the past year, is $3,159,423,057, made up of $1,647,844,113 of capital stock, and $1,511,-578,944 of various forms of indebtedness, chiefly of bonds maturing at distant periods. The capital stock amounted to 52·15 per cent., and the debt to 47·85 per cent. of the total cost. The cost of these roads per mile was $55,116. The gross earnings for the year were $473,241,055, of which $132,309,270, or 28 per cent., was received for transportation of passengers, and $340,931,785, or 72 per cent., for the transportation of freight, including under this head the small amount received from 'miscellaneous sources.' The receipts per mile were $8256. The ratio of earnings to population was $11.76 per head. The operating expenses for the year were $307,486,682, or 65 per cent. of the gross receipts, leaving $165,754,373, or 35 per cent., as net earnings."

Thus, in the comparatively short period of forty-six years, over 67,000 miles of railroads have been constructed in the Union, involving an outlay for construction alone of over three thousand millions of dollars. These vast corporations are the possessors of immense quantities of real and personal estate, employ thousands of operatives, and receive and pay out annually sums of hundreds of millions of dollars. It is a wonderful history.

PETER COOPER—BUILDER OF THE FIRST LOCOMOTIVE.

CORNELIUS VANDERBILT—THE GREATEST RAILROAD KING IN THE WORLD.

CHAPTER II.

HISTORY OF THE "LAND GRAB."

How to build Railroads at the Expense of the People—The Public Domain of the Union a rich Field of Operations for Railroad Managers—The first Land Grants—How the Illinois Central Road obtained its Lands—A bad Example—Handsome Profits—Inauguration of the System of Land Grants—The Result—The Nation robbed by Wild Cat Railroad Companies—How Congress aids the Roads in robbing the People—Actual Workings of the Subsidy System—Detailed Statement of the Amount of the Public Lands granted to each Corporation—Greed of the Railroads—Bonds and Money demanded in addition to Lands—The Railroad Ring—Eloquent Denunciation of these Schemes of Plunder by Hon. E. B. Washburne of Illinois.

A RECENT writer, describing the construction of a railway, says: "The first step, after selecting the route, is to purchase the land upon which the proposed road is to be built." Many of our roads are built upon land fairly purchased and paid for, but not all; and it was regarded as a great step gained in scientific railroad financiering when a shrewd railway magnate of the West conceived the happy idea of building a road at the cost of the people of the whole country.

Previous to 1850, the United States possessed vast tracts of lands in the Western States and Territories. These lands were the common property of the States, and were held by the General Government for their benefit. It was believed at one time that the sale of these lands would produce a large revenue for the Republic, which could be expended in various enterprises for the benefit of the country at large.

In 1850, however, it occurred to certain of our public men that the public lands might be advantageously used for the purpose of defraying the cost of the various railways which were then in contemplation. Who first conceived the idea is not known, but it was caught up by the Illinois Central Railroad Company, and they succeeded in interesting the late Stephen A. Douglas, Senator from Illinois, in the scheme. Mr. Douglas was captivated by the idea of the great railway intersecting the entire State, and bringing Northern and Southern Illinois into rapid and direct communication with Chicago and Cairo. He saw the importance of the undertaking, recognized the magnitude of the expense attending it, and, in an evil hour for the country, adopted the opinion that the General Government should aid the construction of the road by bestowing upon the company a portion of the public lands, since the successful accomplishment of the undertaking would result in building up the population and increasing the wealth of Illinois. Mr. Douglas, with all his great genius, did not seem to recognize the fact that he was really asking the people of the United States to build a road for a corporation in his own State, or that he was opening a way for a systematic fleecing of the nation for the benefit of private individuals.

In 1850, the application of the Illinois Central Company for assistance from the Government was presented in Congress. It was hotly opposed, but supported by the persuasive eloquence of Senator Douglas, the petition was granted. An Act of Congress, approved September 20th, 1850, granted to the State of Illinois six sections of land per mile of road in aid of the construction of a railroad from Cairo to Chicago and Dunleith.

This grant was transferred by the State to the Illinois Central Company, in consideration of which, and in lieu of all other taxes, the company agreed to pay to the State an amount equal to seven per cent. of the gross earnings from freight and passengers moved over their lines. The amount of land embraced in this grant was about 2,595,000 acres. This immense property consisted of a broad strip of land lying on each side of the line of the road throughout the entire length of the State, and for a distance of about six miles on each side of the track. It was a magnificent grant.

The company made a good use of the lands thus acquired. They were promptly surveyed and laid off in sections. Liberal offers—for the company could afford to be liberal since the lands had cost them practically nothing—were made to actual settlers. As they comprised some of the best lands in the State, the railway sections were rapidly taken up, and all along the line of the road there sprang up farms and settlements as if by magic. By the first of January, 1873, the sales of the company amounted to 2,250,633 acres, leaving 344,367 acres on hand. The amount received and due for the lands thus sold up to January 1st, 1873, stands as follows:

Principal	$ 23,320,463
Net cash	5,268,557
Advance interest	976,133
Interest notes	710,328
Notes and deferred payments	18,762,243
Sales, including advance interest	24,296,596

The example of the Illinois Central Company was not lost upon other corporations. Each had its champion in Congress, and applications for land grants began to pour in upon that body. Having granted such aid in one case, Congress could not refuse it in others, and the result was that the greater portion of the public domain was given away to railway corporations, the people of the country practically receiving no valuable consideration for the grants. These grants were made to the States and by them conveyed to the respective railways. Congress conveyed to each of the applicants "six alternate sections of public lands of 640 acres each (and equalling 3480 acres to the mile), to be taken by the *odd* numbers within six miles of the line of the road proposed. In case such a number of sections of *odd* numbers of public lands could not be found within six miles of such line (in consequence of previous sale), then the grant was to be enlarged so as to apply to the *odd* sections within fifteen miles of the line on either side, so as to make up the full amount intended to be granted. Many of the grants were subsequently further enlarged so as to apply to sections of odd numbers within twenty miles of the line."

So common has the custom of giving the public land to a railway corporation, to enable it to build its road, become, that at present, the first care of the directors of a new enterprise of this kind is to obtain, from the Government, land enough to defray the cost of the road. In other words, men forming a corporation to build a road for their own profit, are shrewd enough to throw the expense of their enterprise upon the people of the country at large. The people pay for the roads; the stockholders receive the profits. Members of Congress seem to agree thoroughly with the railway directors of the present day in the belief that it is the duty of the General Government to make the tax-payers of the United States build roads for the benefit of the stockholders.

Very many of the railway enterprises of the present day would not be undertaken, but for the hope of receiving Government aid. The men who organize them, although they do so for their own private benefit, rely upon using the property of the whole people rather than their own. Their plan is very simple. If they can secure a grant of land from the General Government, the public property thus placed in their hands will afford them the means of carrying out their schemes. To be plainer—their plan is simply to rob the nation of its possessions, with the aid or connivance of the august body to whose keeping the trust has been confided.

The people of the United States are not averse to the granting of aid by the Government to enterprises which are national in their character, which are for the public good, and which will render, at some time, an equivalent for the aid thus extended. The American people are decidedly, and very sensibly, averse to giving their property away, for the benefit of a private corporation, and are opposed to such a use of it by Congress. Just now they are very sore over the immense sums that have been squandered by Congress in this way. The Honorable members are aware of this, but they appear to entertain a lofty contempt for the will of the people, fancying that they are the masters rather than the servants of the nation. The public feeling has been repeatedly expressed, but the work of "subsidizing" by the Government still goes on.

Not long since, a leading New York journal gave

the following forcible statement of the popular view of this question:

"Let us say that the property of the Government of the United States—meaning thereby of course the common property of the people of the United States—is worth $4,000,000,000, or $100 a head. In the management of this property by the few hundred men who make up what we call the Government, the implied trust is that the property will in all cases be managed for the benefit of the whole people, and that in no case shall one or two, or half a dozen, or a hundred citizens be given any portion to use for their own peculiar personal profit, to the exclusion of the remaining millions. Now if the Government—*i. e.*, the men under this trust, the trustees of the people in other words—give, say, $500,000,000 of this property to a score of men associated together as a railway or other company, to have and to hold and to use it as their own as much as if it were the product of their own toil, the implied trust is broken; the trustees betray the confidence reposed in them. This is not a fashionable view, we know, but still it is a true one. The wrong is the same in the few men called and calling themselves the Government as if they had committed it in their individual capacities and as private citizens. No man in any capacity has any right to betray a trust reposed. And yet, that such betrayal is not only not wrong, but that it is even nobly, gloriously, beautifully right, is the doctrine underlying the subsidy system. The Government, (so the subsidy doctrine runs,) may, and not only may but should, give the Union Pacific Railroad Company, and the rest, hundreds of millions of public acres and scores of millions of public money for the purpose of building up and operating a business for the exclusive profit of the said company, to the utter and eternal exclusion of any and all of the millions of other citizens whom the act of incorporation fails to recite. And as with this one particular donation so with scores of others; it is right and proper for the Government to give away to whom it will so much of the $4,000,000,000 as it deems proper. It develops the country to do this; it is progress; it is in the line of the best patriotic thought; the wilderness is thereby made to bloom and blossom as a rose—there are, in short, an infinite variety of fine phrases to conceal the real nature of the breach of trust. One particularly specious plea is that unless the millions were thus robbed in behalf of the scores, the scores could not provide great and beneficent instrumentalities for the use of the millions. It is forgotten that the scores charge the millions as much for the use of the instrumentalities as if they had not been built with the millions' own means, but had come bodily out of the bank accounts of the scores. If a man steal from me enough to buy him a horse and vehicle, and then insists he is doing me an immense service by charging me $5 for carrying me a mile on my own property, he does that on a small scale which subsidized corporations, railroad or any other, do upon a large. Such then is the morality of the subsidy system, which has been fostered into such magnificent proportions. The natural operation of the system is to generate about it a fine swarm of adventurers, of all grades, from the benevolent looking company president, whose gold-rimmed glasses would shrivel in the heat of his indignation did any one call him an adventurer, down to the professional lobbyist, whom he uses as the huntsman uses his hound, to run down the game. There being millions at stake, these adventurers, each in his sphere, are instant in action. They cajole, they seduce, they ensnare. All the arts of temptation ooze from their tongues in drops of honey, and fall from their hands in streams of gold. What wonder if success only too often rewards their nefarious efforts—if the not over-stubborn normal virtue of the Senator or Congressman succumbs? If the records of the *Crédit Mobilier* investigation reveal anything, they disclose this—that tactics of this kind were employed with exquisite skill and relentless tenacity; and, despite the half-frantic denials of the victims, it is perfectly evident the strategy of the subsidy adventurers won."

The lands granted by the Government to various railway corporations make up a total area of 198,165,794 acres, or about 300,000 square miles—an area larger than the State of Texas, which contains 237,504 square miles. Texas and the two Virginias combined would make an area smaller than that of the lands thus given away. The total area of the United States is 3,578,392 square miles, and the railway subsidies, it will be seen, comprise nearly one-tenth of the entire Union.

The following Table shows the land grants to railways since 1850, and the amount of land actually received or certified by each company:

GRAND CENTRAL RAILWAY DEPOT, NEW YORK CITY—THE FINEST RAILROAD DEPOT IN THE UNITED STATES.

Date of Laws.	Name of Road.	Acres Granted.	Acres Certified.
	ILLINOIS.		
Sept. 30, 1850.....	Illinois Central....................................	2,595,053	2,595,053
	MISSISSIPPI.		
Sept. 20, 1850......	Mobile and Ohio....................................	1,004,640	737,130
Aug. 11, 1856.....	Southern Railroad....................................	404,800	171,550
" "	Gulf and Ship Island Railroad....................................	652,800	....
	ALABAMA.		
Sept. 21, 1850.....	Mobile and Ohio....................................	230,400	419,528
May 27, 1856......	Alabama and Florida....................................	419,520	394,522
" "	Alabama and Tennessee....................................	481,920	440,700
June 3, 1856.......	Northeastern and Southeastern, Alabama and Chattanooga....................................	691,840	289,535
" "	Wills Valley, Alabama and Chattanooga....................................	206,080	171,920
" "	Coosa and Tennessee....................................	132,480	64,784
" "	Mobile and Girard....................................	840,880	504,145
" "	Coosa and Chattanooga....................................	150,000	....
" "	Tennessee and Alabama Central....................................	576,000	....
	FLORIDA.		
May 17, 1856......	Florida Railroad....................................	442,542	281,984
" "	Alabama and Florida....................................	165,688	165,588
" "	Pensacola and Georgia....................................	1,568,729	1,275,012
" "	Florida, Atlantic and Gulf Central....................................	183,183	37,583
	LOUISIANA.		
June 3, 1856.......	Vicksburg and Shreveport....................................	610,880	353,211
" "	New Orleans, Opelousas and Great Western....................................	967,840	719,193
	ARKANSAS.		
Feb. 9, 1853........	Cairo and Fulton....................................	1,100,667	1,115,408
July 28, 1856......	Cairo and Fulton—additional....................................	966,722	....
Feb. 9, 1853........	Memphis and Little Rock—additional....................................	438,646	127,238
July 28, 1866......	Memphis and Little Rock—additional....................................	365,539	....
Feb. 9, 1853........	Little Rock and Fort Smith....................................	550,525	550,520
July 23, 1866......	Little Rock and Fort Smith—additional....................................	458,771	....
July 4, 1866.......	Iron Mountain Railroad....................................	864,000	....
	MISSOURI.		
June 10, 1852......	Hannibal and St. Joseph....................................	791,944	493,812
" "	Pacific and Southwestern Branch....................................	1,161,235	1,158,073
Feb. 9, 1853........	Cairo and Fulton....................................	219,262	63,540
July 28, 1866......	Cairo and Fulton—additional....................................	182,718	....
July 4, 1866.......	Iron Mountain (Pilot Knob to Helena, Arkansas)........	1,400,000	....
	IOWA.		
May 15, 1856......	Burlington and Missouri River....................................	948,643	291,725
June 2, 1864......	Burlington and Missouri River....................................	101,110	95,656
May 15, 1856......	Chicago, Rock Island and Pacific....................................	1,144,904	481,774
June 2, 1864......	Chicago, Rock Island and Pacific....................................	116,276	144,229
May 15, 1856......	Cedar Rapids and Missouri River....................................	1,298,739	778,869
June 2, 1864.......	Cedar Rapids and Missouri River....................................	123,370	342,406
May 15, 1856......	Dubuque and Sioux City....................................	....	
June 2, 1864.......	Change of route (Fort Dodge to Sioux City)....................................	1,226,163	1,226,163
May 12, 1864......	McGregor and Western....................................	1,536,000	....
	Railroad from Sioux City to Minnesota....................................	256,000	....
July 2, 1864.......	Sioux City and Pacific....................................	580,000	....
	MICHIGAN.		
June 3, 1856.......	Port Huron and Milwaukee....................................	312,384	6,468
" "	Detroit and Milwaukee....................................	355,420	30,998
" "	Jackson, Lansing and Saginaw.................................... }	1,052,469	721,469
July 3, 1866.......	Time extended seven years.................................... }		
June 3, 1856.......	Flint and Père Marquette....................................	....	....
July 3, 1866.......	Act to change the western terminus of road....................................	586,828	511,425
June 3, 1856.......	Grand Rapids and Indiana....................................	629,182	629,182
June 7, 1864.......	Grand Rapids, from Fort Wayne to Grand Rapids, etc...	531,200	191,607
June 3, 1856.......	Bay de Noquette and Marquette....................................	218,880	218,881
March 3, 1865....	Bay de Noquette and Marquette....................................	128,000	....
June 3, 1856.....	Marquette and Ontonagon....................................	309,315	216,919
March 3, 1865....	Marquette and Ontonagon....................................	243,200	49,086
June 3, 1856......	Chicago, St. Paul and Fond-du-Lac....................................	208,062	174,020

Date of Laws.	Name of Road.	Acres Granted.	Acres Certified.
	MICHIGAN—*Continued.*		
....	Chicago, St. Paul and Fond-du-Lac (branch) to Marquette	188,507	162,044
July 5, 1862	Chicago and Northwestern,	375,680	....
March 3, 1865	Chicago and Northwestern	188,800	....
	WISCONSIN.		
June 3, 1856	Toma and Lake Superior	894,907	324,943
May 5, 1864	Toma and Lake Superior	675,000	163,263
June 3, 1856	St. Croix and Lake Superior	524,714	524,718
May 5, 1864	St. Croix and Lake Superior	350,000	....
June 3, 1856	Branch to Bayfield	318,737	318,740
May 5, 1864	Branch to Bayfield	215,000	....
June 3, 1856	Chicago and Northwestern }		
April 25, 1862	(Resolution) change of route }	600,000	311,307
May 5, 1864	From Portage City to Bayfield, thence to Superior	1,800,000	....
	MINNESOTA.		
March 3, 1857	St. Paul and Pacific	660,000	466,566
March 3, 1865	St. Paul and Pacific	500,000	....
March 3, 1857	Branch St. Paul and Pacific }	750,000	438,000
March 3, 1865	Branch St. Paul and Pacific }	725,000	....
July 12, 1862	Authorized change of route }	....	....
March 3, 1857	Minnesota Central	353,403	174,578
March 3, 1865	Minnesota Central	290,000	....
March 3, 1857	Winona and St. Peter	720,000	342,376
March 3, 1865	Winona and St. Peter	690 000	....
March 3, 1857	Minnesota Valley }	....	....
May 12, 1864	Minnesota Valley }	860,000	711,442
July 13, 1866	Extends the time for said road seven years }	150,000	1,040
May 5, 1864	Lake Superior and Mississippi }	....	....
July 13, 1866	Authorized to make up deficiency within thirty miles of west line of said road }	800,000	367,424
July 4, 1866	Minnesota Southern	735,000	125,480
July 4, 1866	Hastings and Dakota River	550,000	....
	KANSAS.		
March 3, 1863	Leavenworth, Lawrence and Galveston }		
July 1, 1864	Atchison, Topeka and Santa Fé }	2,550,000	....
July 1, 1864	Union Pacific Southern Branch (M. K. & T.) }		
July 23, 1866	St. Joseph and Denver City	1,700,000	....
July 25, 1866	Kansas and Neosho Valley	2,350,000	....
July 26, 1866	Southern Branch Union Pacific from Fort Riley to Fort Smith, Arkansas	1,203,000	....
	CALIFORNIA.		
July 25, 1866	California and Oregon	1,540,000	....
July 13, 1866	Placerville and Sacramento Valley	200,000	....
March 2, 1867	Stockton and Copperopolis	320,000	....
	OREGON.		
July 25, 1866	Oregon and California	1,660,000	....
May 4, 1870	From Portland to Astoria and McMinnville	1,200,000	....

In addition to the above grants, the General Government has granted to the various Pacific railways immense tracts of land, making these grants direct to the companies. It has granted to the Central and Union Pacific Railroads a total of 35,000,000 acres, of which only 544,759 had been certified up to January, 1872. To the Northern Pacific road it has given 58,000,000 of acres, and to the Atlantic and Pacific road 42,000,000; making a total of 135,000,000 acres to the three Pacific roads, or about 200,000 square miles.

Lands, however, lavishly as they have been given, no longer form the limit of railway expectations. The greed of these corporations has extended to the public funds, and bonds and money are now demanded with as much coolness as lands. The railway incorporators have learned that with a pliant Congress it is easy to draw from the National Treasury the funds they are not willing to provide for their enterprises.

In order to effect this, they maintain at Washington a force of paid lobbyists, whose business it is to influence the legislation of Congress by unpatriotic and illegal means. What these means are was shown by the investigations attending the Crédit Mobilier scandal of the last session. Yet, that the reader may the better understand how these railroad leeches fasten upon the Government, we give the following account of the schemes that were introduced into the Fortieth Congress, which was particularly distinguished for them. Many of these schemes were successful:

"At present," says a correspondent writing from Washington early in the session, "perhaps there is more money in the various railroad schemes than in any other. And this thing is on a scale which the country does not comprehend, notwithstanding the constant talk about it. Thus far, in the Fortieth Congress, there have been *seventy-two* railroad bills introduced into the Senate alone. *Eight* were presented at the first short session, *fifty-two* at the second session, and in the two weeks of the present session *eleven* have been reported and printed. And these last do not include four as gigantic as any which have been passed, yet to come. One is in preparation for which its friends are now gathering power, for the Northern Pacific, one for the Albuquerque line and its several connections; one for Mr. Pomeroy's little private Atchison Pacific—one of the nicest and fattest speculations ever concocted and worked through—having these special qualifications of nice and fat, on account of the small number to divide the spoils; one for two roads south and west from St. Louis, and two or three for Southern Pacific lines from Memphis, New Orleans, and points in Texas.

"In all this there are four lines across the Continent, with connecting roads enough to stretch out into two more; and then such little ventures as the Atchison and Denver lines by the score.

"Of all these bills, fully three-fourths were originated by Republicans. Four Senators brought in nearly half of them. Mr. Pomeroy reported eleven, Mr. Ramsey seven, Mr. Conness five, and Mr. Harlan four.

"Mr. Pomeroy did not confine his attention to any particular part of the country. He proposed one land grant through the rich lands about Port Royal, South Carolina, and another one of his measures was for the benefit of his Wisconsin brethren; but, not desiring to be reckoned as worse than an infidel, he made full provision for his own political household in Kansas. We find his name attached to a land grant for a railroad from Lawrence to the Mexican line; to three bills for roads from Fort Scott to Santa Fé; to a pleasant arrangement for the Southern branch of the Union Pacific Road—whatever that may be—and also to a land grant from Irwing, Kansas, to New Mexico; and all for the national good, of course.

"These, it must be remembered, are such railroads as Northern companies, Northern lobby-men, and Northern Congressmen have concocted. The word concocted is good for most, though a few are meritorious. The Southern States are just beginning to vote, and the scent of Southern men in Congress is now as keen in respect to all material interests as the Northern Congressman's nose. The reason is evident. Southern smelling is now done with Northern noses. Carpet bags have wrought this change for the South; and as a result, among the very first subjects to call out bills from Southern men are the railroad interests.

"And heading the column, comes Mr. Senator Spencer, with a bill making a land-grant, not through the public domain on the plains, but through the States of Tennessee, Georgia, Alabama, Mississippi, and Louisiana, with permission to get all the 'earth, stone, timber, and other materials' for the construction of its roads, off the public lands along its line, and then to receive ten sections of land to the mile, wherever they can find that amount within twenty miles of the line they may see fit to locate, and from Mobile onward to the western boundary of Louisiana; if the land cannot be found within twenty miles of the road, these patriotic gentlemen are to be obliged to hunt it up within forty miles north of their line. As this is the first attempt on the part of a Southern Senator to follow in the paths already worn so smooth by his fellow Republicans from the North, it will be interesting to see what a fine start Senator Spencer, of Alabama, makes. Section second of his bill is in part as follows:

"'SEC. 2. *And be it further enacted*, That, for the purpose of aiding in the construction of the railroad of said company, there be, and is hereby granted to said company, from the public lands of the United States, ten sections of land for each mile of railroad completed and placed in running order by said company, pursuant to its charter. That said lands are granted as follows: On the line of said railroad from the city of Chattanooga, in the State of Tennessee, to the city of Mobile, in the State of Alabama, every alternate section of public land designated by odd numbers, to the amount of ten alternate sections per mile, for each mile of said railroad from the said city of Chattanooga to the said city of Mobile, such alternate section of land to be selected within the limits of ten miles on each side of the centre line of said railroad; and if public lands sufficient for the purpose shall not be found within said limit of ten miles upon each side of said railroad, then the said alternate sections of land are hereby granted, and may be selected within the limits of twenty miles on each side of the centre line of said railroad. And on the line of said railroad from the city of Mobile, in the State of Alabama, to the western boundary of the State of Louisiana, every alternate section of public land, designated by odd numbers, to the amount of ten alternate sections per mile for each mile of said railroad, from the said city of Mobile to the western boundary of the State of Louisiana, such alternate section of land to be selected within the limits of ten miles upon each side of the centre line of said railroad; and if public lands sufficient for the purpose shall not be found within said limits of ten miles upon each side of said railroad, then the said alternate sections of land are hereby granted, and may be selected within the limits of forty miles north of the centre line of said railroad, excepting, however, from this grant, all mineral lands, and lands sold by the United States, or lands in which a preëmption or homestead claim may have attached at the time the line of said railroad shall have been located and established.'

"What will the railroad docket of the Senate not contain by the time the Southerners have brought up their side of the railroad jobs to the present proud height of their Northern friends, and shall have added to the Washington lobby, its own army of blood-suckers, plausible gentlemen of unquestioned honor, and thieves?—for it takes all these, and more, to make a lobby. What a nice thing it will be for taxpayers!

"All this presents the railroad interest merely in outline. Every bill deserves a separate letter to show the means used to get it before the Senate, the persons engaged in pressing it, and the parties to be benefited by it; and in due time the principal ones at least will get that chapter.

"When the railroad jobs are disposed of, then the deck is only cleared for action against jobs in general. There are, aside from these, the Niagara Ship Canal with a coupon of twelve millions attached; the Commercial Navigation Company, with half as much on its coupon; the bills and schemes for getting damages paid to Southern men for property destroyed during the war, in all hundreds of millions; and then the lobby upon the more modest sum of five millions due from Southern railroads, and in which radical Republicans from Tennessee are deeply interested. The Osage Treaty is a nice plum; and one new feature is, that some Kansas men who showed a vast amount of righteous indignation over it, before their reëlection, are, now that their places are assured, helping the swindle on. Alta Veta is coming up again, and to crown all things, if it is possible, the change in the Indian Bureau is to be so managed as to place the present Indian ring on a firmer foundation than ever.

"The Republican party can now afford to rectify the irregularities which have crept into all portions of the Government while the great political battle with Rebellion was going on. If those here, as its Congress, will not free themselves from such things, the party need not die if it only throws them overboard. There are honest men enough who can take their places. Let the press watch jobs when the recess closes and the outside lobby, which is here in force, begins to work through its Senators and Representatives."

Commenting upon the same subject, the New York *Herald* said, editorially:

"The corruptions which have grown up in the national government, from the general demoralizations of our late civil war, are fearful to contemplate. One hundred millions a year lost to the Treasury from the spoliations of the whiskey rings 'beats out of sight' any thing in the line of whiskey frauds under any other government on the face of the globe; but on a corresponding scale with their field of operations, the Indian rings, the Post-Office and Interior Department rings, the tobacco rings, the frontier smuggling rings, and various other rings, insiders and outsiders, jobbers, contractors, Government officials and private speculators, are pretty well up to the percentage of the enormous stealings of the whiskey rings. The latest developments, however, show, that in the grandeur and number of their schemes of spoils and plunder, the Congressional rings of railroad jobbers throw into the shade all the other rings of the lengthy catalogue of confederate Treasury robbers.

"A Washington correspondent, who has been looking into the business, reports that one hundred and fifty-nine railroad bills and resolutions have been introduced in the Fortieth Congress (the term of which expired on the 4th of March, with that of President Johnson), and that twice as many more were in preparation in the lobby; that one thousand millions of acres of the public lands, and two hundred millions in United States bonds, would not supply the demands of these cormorants. In other words, their stupendous budget of railway jobs would require sops and subsidies in lands and bonds, which, reduced to a money valuation, swell up to the magnificent figure of half the national debt.

"Among the jobs of this schedule is the Atchison and Pike's Peak Railroad Company, or Union Pacific Central Branch, which, after having received Government sops to the extent of six millions, puts in for seven millions more. Next comes the Denver Pacific Railway and Telegraph Company, which, having feathered its nest to the figure of thirty-two millions, puts in for a little more; and this company is reported to be a mere gang of speculators, 'without any known legal organization whatever'—a lot of mythical John Does and Richard Roes, who cannot be found when called for. Next we have the Leavenworth, Pawnee and Western Railroad Company, now known as the Union Pacific, Eastern Division, chartered by the Kansas Territorial Legislature, in 1855, subsidized with

Delaware Indian reserve lands in 1861, and then in 1862, by a rider on the Pacific Railroad law, granted sixteen thousand dollars per mile in United States bonds, and every alternate section of land within certain limits, on each side of the road, and the privilege of a second mortgage. This is cutting it pretty fat. But it further appears that a clique of seceders from the old company illegally formed a new company, and, having by force of arms taken possession of the road, are pocketing the spoils which legally belong to the old company. All this, too, with the consent of the President, the Secretary of the Treasury, and Congress. Are they all birds of a feather, that they thus flock together?

"From another source we learn that some half dozen other Pacific branch or main stem railroads, Northern and Southern, are on the anvil, involving lands and bonds by tens and twenties and hundreds of millions; that of all these schemes fully three-fourths come from the Republicans in both Houses; that Senator Pomeroy, of Kansas, has seven of these jobs on the docket; Senator Ramsey, of Minnesota, four, Senator Conness, of California, five, and Senator Harlan, of Iowa, four. Senator Pomeroy, however, distances all competitors in the number and extent of his jobs; for, as it appears, they include a line from Kansas to Mexico, three bills for roads from Fort Scott to Santa Fé, in Texas, a South Carolina line through the Sea Island cotton section, two or three lines from the Mississippi River through to Texas, and 'a little private Atchison Pacific, one of the nicest and fattest speculations ever worked through.'

"Is not this a magnificent budget, and is not the

audacity of these railroad jobs and jobbers positively sublime? Some of these schemes are in successful operation, but many of them are still in the caterpillar, or chrysalis, state, and there is a prospect that very few of this class will come out as the full-blown butterfly."

Well might the eloquent Illinois Congressman, who now so ably represents the Republic in France, exclaim in indignant rebuke of these schemes of plunder, as he denounced them from his place in the House of Representatives:

"While the restless and unpausing energies of a patriotic and incorruptible people were devoted to the salvation of their Government, and were pouring out their blood and treasure in its defence, there was the vast army of the base, the venal and unpatriotic, who rushed in to take advantage of the misfortunes of the country and to plunder its treasury. The statute-books are loaded with legislation which will impose burdens on future generations. Public land enough to make empires has been voted to private railroad corporations; subsidies of untold millions of bonds, for the same purposes, have become a charge upon the people, while the fetters of vast monopolies have been fastened still closer and closer upon the public. It is time that the representatives of the people were admonished that they are the servants of the people and are paid by the people; that their constituents have confided to them the great trust of guarding their rights and protecting their interests; that their position and their power are to be used for the benefit of the people whom they represent, and not for their own benefit and the benefit of the lobbyists, the gamblers, and the speculators who have come to Washington to make a raid upon the Treasury."

CHAPTER III.

WATERED STOCKS.

Adroitness of Railroad Managers in securing Valuable Privileges from the Public—Recklessness of the People in granting the Demands of the Road—The only Restraints imposed—How the People made it possible for the Corporations to fleece them—How to Build a Road without Subscribing the necessary Funds—A False System—The Story of the Crédit Mobilier Swindle—How the Pacific Railroad bled the National Treasury—New System of Railroad Financiering—The Process of "Stock Watering"—Instances of successful Stock Watering—How a Bankrupt Road was made to pay Good Dividends—Successful Policy of the Pennsylvania Railroad Company—Vanderbilt's Master Stroke—Who pays for Watered Stock—A Lesson for the People.

We have seen the eagerness and success with which the railroad corporations of the present day seize upon the public lands, paying the public nothing in return for the immense wealth thus given them; and we have asserted that it is in reality the people who build the roads for the benefit of these corporations. The facts presented sustain our assertion.

It is the boast of the modern railway director that he is the friend of the public, and that his work is entirely for the good of the community. He modestly keeps himself in the background, and speaks of his road only, and of the immense advantages that it will bestow upon the regions through which it is to pass, and so plausible and well delivered are his words, that he succeeds in making the public believe that he is really working for the good of others, and investing his own capital for the benefit of the community, without thought for himself.

Thus influenced, the confiding people grant unhesitatingly all the privileges asked for. The right of way is given, lands are donated by Congress, money is subscribed by counties and cities, and the inhabitants of the region through which the road is to be built imagine themselves on the highway to wealth and prosperity. Trade, they are told, is to come pouring over the new route, a direct market is to be provided for the products of the region, and an era of general prosperity is to be inaugurated. Figures are not wanting to encourage these expectations. The most plausible calculations are made, the cost of the road is given, the annual expenses are estimated, and it is shown to the satisfaction of all that a system of moderate charges for the transportation of passengers and freight will secure to the road a revenue sufficient to meet its expenses, and, in time, to pay a fair dividend upon the capital invested in the undertaking.

A little reflection would cause the confiding public to be suspicious of the men who profess to have its interests so much at heart. Capitalists do not undertake railway enterprises from such benevolent motives. Like other men, they seek their individual profit, and the welfare of the public is with them a consideration only so far as it influences their undertaking. They look to receive ample dividends, and are careless of the thanks of a grateful country.

"In America, as in England, the private corporation owning the thoroughfare is the basis of the whole railroad system. In thus surrendering the control of this system out of its hands, the community as a rule made one and but one reservation in its own favor; it was almost universally stipulated that the rate of profit upon the capital invested in the work of construction should not exceed a certain annual per centage, varying, according to locality, from 10 to 20 per cent. Within this limit the corporations were free to earn and divide all they could."

But having set this limit, the community allowed the corporations to retain the exclusive management of their roads, and made no arrangement whereby the latter could be held up to the limit thus fixed. No provision was made for ascertaining the real, as contradistinguished from the nominal cost of the roads, nor was anything arranged in regard to arrears of unpaid dividends. A recent writer upon the subject well says: "It was absurd to suppose that even the most honest capitalist would accept the strict construction of a law which insured him a certain loss in each bad year or unprofitable enterprise, and limited him in case of success to a reasonable profit. Of course, therefore, the law was no sooner enacted than it was circumvented. . . . . The doubt raised was whether the stipulated per centage was to be paid upon what the property cost its holders, or upon what it was actually worth. Interpreting it in the way last specified, the capitalist proceeded to act accordingly. . . . . It therefore devolved upon the owners of the property to cast up the balance sheet themselves, and to decide all nice points undoubtedly in their own favor. Where a people so provides for its own interests it needs no prophet to foretell the consequences. No landlord deals in this way with a tenant." The community has voluntarily placed itself in the power of the railways, and it must pay the penalty of such absurd conduct.

In the early days of railroad financiering it was customary to make sure that funds would be forthcoming in sufficient quantities to finish the undertaking before embarking in it. But the times have changed since then. In the Western States the public guarantee the cost and furnish the means of paying for the road by the land grants we have been considering. In the Eastern States other expedients are resorted to. In both sections, however, many of the tactics employed are the same.

Funds must necessarily be procured before the work can be begun, and the manner in which these are generally obtained reveals at once a mastery of the science of railroad financiering. The road for which aid is sought promises well, but it does not yet exist. It is to be constructed. Yet in spite of this the directors of the scheme proceed to pledge the road for the cost of its construction; or in other words they mortgage a work which does not exist. The stock is worth nothing, but there is another means at hand. Bonds are created and put in the market at a certain stated price. They are usually placed in the hands of some leading banking house in the principal financial centres of the country to be sold. The bonds are sold, and the work of constructing the road goes on with the money obtained for them. "The stock itself then passes as a gratuity into the hands of those advancing money upon the bonds. The result is, that by this ingenious expedient the capitalist holds a mortgage, paying a secured and liberal interest, on his own property, which has been conveyed to him forever for nothing. The stock is at once nothing and everything. Given away, the donees own and manage the road, and, receiving a fixed and assured interest upon their bonds, enjoy a further right to exact an additional sum, and one as large as they are able to make it, from the developing business of the country, as dividends on the stock. Instances of this form of railroad financiering need not be specified, for it is now the common course of Western railroad construction."

Perhaps the best instance on record of the manner in which skilful directors of a railroad can procure the construction of their road at the cost of other parties, and secure the profits to themselves, is afforded by the history of the notorious Crédit Mobilier Company, which constructed the Union Pacific Railway; and though the story is now old and known to the reader, it will bear repeating here.

The early history of the Pacific Railroad is a story of constant struggles and disappointments. It seemed to the soundest capitalists a piece of mere fool-hardiness to undertake to build a railroad across the continent and over the Rocky Mountains, and, although Government aid was liberally pledged to the undertaking, it did not, for a long time, attract to it the capital it needed. At length, after many struggles, the doubt which had attended the enterprise was ended. Capital was found, and with it men ready to carry on the work. In September, 1864, a contract was entered into between the Union Pacific Company, and H. W. Hoxie for the building by said Hoxie of one hundred miles of the road, from Omaha west. Mr. Hoxie at once assigned this contract to a company, as had been the understanding from the first. This company, then comparatively unknown, but since very famous, was known as the Crédit Mobilier of America.

D. LANCELOT.

BUILDING THE CENTRAL PACIFIC RAILROAD IN THE SIERRA NEVADAS.

The company had bought up an old Charter that had been granted by the Legislature of Pennsylvania to another company in that State, but which had not been used by them.

"In 1865 or 1866, the late Oakes Ames, then a Member of Congress from the State of Massachusetts, and his brother Oliver Ames, became interested in the Union Pacific Company, and also in the Crédit Mobilier Company, as the agent for the construction of the road. The Messrs. Ames were men of very large capital, and of known character and integrity in business. By their example and credit and the personal efforts of Mr. Oakes Ames, many men of capital were induced to embark in the enterprise, and to take stock in the Union Pacific Company and also in the Crédit Mobilier Company. Among them were the firm of S. Hooper & Co. of Boston, the leading member of which (Mr. Samuel Hooper) was then and is now a member of the House; Mr. John B. Alley, then a member of the House from Massachusetts, and Mr. Grimes, then a Senator from the State of Iowa. Notwithstanding the vigorous efforts of Mr. Ames and others interested with him, great difficulty was experienced in securing the required capital.

"In the Spring of 1867, the Crédit Mobilier Company voted to add 50 per cent. to their capital stock, which was then $2,500,000, and to cause it to be readily taken, each subscriber to it was entitled to receive as a bonus an equal amount of first mortgage bonds of the Union Pacific Company. The old stockholders were entitled to take this increase, but even the favorable terms offered did not induce all the old stockholders to take it, and the stock of the Crédit

Mobilier Company was never considered worth its par value until after the execution of the Oakes Ames contract hereinafter mentioned. On the 16th day of August, 1867, a contract was executed between the Union Pacific Railroad and Oakes Ames, by which Mr. Ames contracted to build 667 miles of the Union Pacific road at prices ranging from $42,000 to $96,000 per mile, amounting in the aggregate to $47,000,000. Before the contract was entered into, it was understood that Mr. Ames was to transfer it to seven trustees who were to execute it, and the profits of the contract were to be divided among the stockholders in the Crédit Mobilier Company, who should comply with certain conditions set out in the instrument transferring the contract to the trustees. Subsequently, all the stockholders of the Crédit Mobilier Company complied with the conditions named in the transfer, and thus became entitled to share in any profits said trustees might make in executing the contract. All the large stockholders in the Union Pacific were also stockholders in the Crédit Mobilier, and the Ames contract and its transfer to trustees were ratified by the Union Pacific and received the assent of the great body of stockholders, but not of all. After the Ames contract had been executed, it was expected by those interested that, by reason of the enormous prices agreed to be paid for the work, very large profits would be derived from building the road, and very soon the stock of the Crédit Mobilier was understood to be worth much more than its par value. The stock was not in the market, and had no fixed market value, but the holders of it, in December, 1867, considered it worth at least double the par value, and in January or February, 1868, three or four times the par value; but it does not appear that these facts were generally or publicly known, or that the holders of the stock desired they should be."

As will be seen from the above statement, the stockholders of the Crédit Mobilier were also stockholders in the Union Pacific Company.

Like all great corporations of the present day, the Union Pacific road was largely dependent upon the aid furnished by the Government for its success. The managers of the company, being shrewd men, succeeded in placing all the burdens and risks of the enterprise upon the General Government, while they secured to themselves all the profits to be derived from the undertaking. "The Railroad Company was endowed by Act of Congress with 20 alternate sections of land per mile, and had Government loans of $16,000 per mile for about 200 miles; thence $32,000 per mile through the Alkali Desert, about 600 miles, and thence in the Rocky Mountains $48,000 per mile. The Railroad Company issued stock to the extent of about $10,000,000. This stock was received by stockholders on their payment of five per cent. of its face. When the Crédit Mobilier came on the scene, all the assets of the Union Pacific were turned over to the new company in consideration of full-paid shares of the new company's stock and its agreement to build the road. The Government, meanwhile, had allowed its claim for its loan of bonds to become a second instead of a first mortgage, and permitted the Union Pacific road to issue first mortgage bonds, which took precedence as a lien on the road. The Government lien thus became almost worthless, as the new mortgage which took precedence amounted to all the value of the road. The proceeds of this extraordinary transaction went to swell the profits of the Crédit Mobilier, which had nothing to pay out except for the mere cost of construction. This also explains why some of the dividends of the latter company were paid in Union Pacific bonds. As a result of these processes, the bonded debts of the railroad exceeded its cost by at least $40,000,000."

Mr. Ames was deeply interested in the scheme, being, indeed, one of its principal managers. Being a Member of Congress, he was peculiarly prepared to appreciate the value of Congressional assistance in behalf of the Crédit Mobilier. It would seem that the object of the Crédit Mobilier was to drain money from the Pacific road, and consequently from the Government, as long as possible. Any legislation on the part of Congress designed to protect the interests of the Government, would, as a matter of course, be unfavorable to the Crédit Mobilier, and it was the aim of that Corporation to prevent all such legislation. The price agreed upon for building the road was so exorbitant, and afforded such an iniquitous profit to the Crédit Mobilier, that it was very certain that some honest friend of the people would demand that Congress should protect the Treasury against such spoliation. It was accordingly determined to interest in the scheme enough Members of Congress to prevent any protection of the National Treasury at the expense of the unlawful gains of the Crédit Mobilier. Mr. Oakes Ames, being in Congress, undertook to secure the desired hold upon his associates. The plan was simply to secure them by bribing them, and for this purpose a certain portion of the Crédit Mobilier stock was placed in the hands of

Mr. Ames, as trustee, to be used by him as he thought best for the interests of the company.

Provided with this stock, Mr. Ames went to Washington in December, 1867, at the opening of the Session of Congress. The story of his exploits there is now familiar to every one.

Reduced to plain English, the story of the Crédit Mobilier is simply this: The men entrusted with the management of the Pacific road made a bargain with themselves to build the road for a sum equal to about twice its actual cost, and pocketed the profits, which have been estimated at about THIRTY MILLIONS OF DOLLARS—this immense sum coming out of the pockets of the tax payers of the United States. This contract was made in October, 1867.

"On June 17, 1868, the stockholders of the Crédit Mobilier received 60 per cent. in cash, and 40 per cent. in stock of the Union Pacific Railroad; on the 2d of July, 1868, 80 per cent. first mortgage bonds of the Union Pacific Railroad, and 100 per cent. stock; July 3, 1868, 75 per cent. stock, and 75 per cent. first mortgage bonds; September 3, 1868, 100 per cent. stock, and 75 per cent. first mortgage bonds; December 19, 1868, 200 per cent. stock; while, before this contract was made, the stockholders had received, on the 26th of April, 1866, a dividend of 100 per cent. in stock of the Union Pacific Railroad; on the 1st of April, 1867, 50 per cent. of first mortgage bonds were distributed; on the 1st of July, 1867, 100 per cent. in stock again."

After offering this statement, it is hardly necessary to add that the vast property of the Pacific road, which should have been used to meet its engagements, was soon swallowed up by the Crédit Mobilier.

The history of the Crédit Mobilier is instructive upon another point. It presents to us a skilful and successful instance of what is now a common practice with railroad companies—the fictitious increase or watering of the stock of the company.

Stock watering has become so common, and has been indulged in with such success, that many persons have come to regard it as a legitimate transaction. A competent writer has defined the practice as "the re-appraisal by its owners of a corporate property which has, or is alleged to have, increased in value on their hands, without any new outlay, and the issue to themselves of new evidences of value equal to such supposed increase." But the popular definition—and the true one—would be the increase of the stock of a corporation at the expense of the public, and for the purpose of earning dividends upon money never invested.

It will be well to consider how this operation of "watering" the stock of a road is performed. Such knowledge will be of value farther on.

"The history of the companies which have been consolidated into what is known as the Pittsburg, Fort Wayne and Chicago Railroad, furnishes a very fair illustration. Here the process of watering was early commenced as a simple and desperate expedient for raising money at an enormous discount for the purpose of completing an enterprise of doubtful success. In the earliest history of one of these companies we read: 'The stock subscriptions which were paid in cash into the treasury of the company were very small—amounting, perhaps, in all, to less than three per cent. on the final cost of building and equipping the road. The stock subscriptions were paid for mostly in uncultivated lands, farms, town lots and labor upon the road.' Of the whole road as it stands we are told that, of the $18,663,876, now representing the cost of the road and equipment, etc., the shareholders contributed in cash only about ten per cent., or less than $2,000,000; and their contributions in cash, bonds, notes, lands and personal property, labor, etc., amounted to something less than $4,000,000, or rather more than twenty per cent. of the present cost of the work. The difference between this sum and the capital stock, as now shown by the books of the company, is made up of dividends which were *paid in stock*, interest on stock *paid in stock*, premium on stock allowed to stockholders at the time of consolidation, which was *paid in stock*, and a balance of stock still held by the trustees.

"This, however, was in the early days of the enterprise, the days of doubtful success, when the stock was thought worthless, and was almost given away. But in 1866 a new era dawned upon the Fort Wayne road; it began to pay dividends. In 1870 the stock of the company, the history of a portion of which has just been given, stood at $11,500,000, while its indebtedness amounted to about $13,600,000 more, being in all some $1,150,000 above the cost of road and equipment as they stood upon the books of the company. In June of this year a lease was effected of the entire property by the Pennsylvania Railroad Company. The Fort Wayne stockholders had their option between annual dividends of 12 per cent. on the stock then in existence, or the more moderate rate of 7 per cent. on a proportionately increased amount. They wisely chose the latter, and forthwith the $11,500,000 of stock became $19,714,000, while the road, which was claimed to have cost only $24,000,000, was suddenly represented by $33,400,000 of securities, none of which bore a less interest than 7 per cent.

"The great masterpieces of Commodore Vanderbilt have, however, so eclipsed all other performances in this line that they may be said to constitute an epoch in the history of paper inflation—it might also be said of bubble-blowing. It is only necessary, therefore, in this connection, to recount the history of the chain of roads, now reduced in number to two, which connect New York and Chicago by way of Albany. The distance between these points is 982 miles. It is useless to begin the story further back than 1852, when the through line was completed, consisting of sixteen independent links, several of which were themselves made up of numerous smaller and once independent roads. That was a year of active and much-needed consolidation. The New York Central led off under a special act of legislature. Eleven roads went into the consolidation, with an aggregate capital of $23,235,600. The stock lowest in value of the eleven was settled upon as the par of the new concern, and the stocks of the other ten companies were received at a premium varying from 17 to 55 per cent. By this simple financial arrangement, $8,894,500 of securities, of which not one cent was ever represented by property, but which in reality constituted so much guaranteed stock, was made a charge, principal and interest, against future income. This was the price paid to get rid of the vested rights which had been allowed to settle down upon this thoroughfare. Between 1852 and 1868, the stock and indebtedness of the consolidated company had been increased, for one reason or another, until, when Mr

Vanderbilt became President in the latter year, they amounted in round numbers to $40,000,000, representing a road which its construction account showed had cost $36,600,000. Vanderbilt had for some years been President of the Hudson River road, and, as such, in 1867 had doubled its capital stock ($7,000,000), calling in 50 per cent. of the increased amount, and thus watering to the extent of $3,500,000. Extending his control over the Central, he now proceeded to better his previous instructions. A stock dividend of 80 per cent., not a dollar of which was called in, was suddenly declared. Over $23,000,000 of securities were thus created at once. Operations stood still at this point, but only for a moment. The next measure was a consolidation of the Central and the Hudson River Railroads. This was effected in the succeeding year upon a stock basis of $90,000,000—a further watering of 27 per cent. being allotted to the Central—while the turn of the Hudson River road now having come again, there was provided for it the munificent amount of 85 per cent. The result of these astounding feats of financial legerdemain was that a property which in 1866 appeared from its own books to have cost less than $50,000,000, and which was then represented by over $54,000,000 of stock and indebtedness, was suddenly shot up to over $103,000,000 in 1870, upon the whole of which interest and dividends were paid. At the same time the cost of the road stood upon the books of the company at less than $60,000,000, or about $70,000 per mile, while in evidences of property each mile was charged with no less than $122,000. The average cost of railroads throughout the world has been somewhat less than $100,000 per mile, while in

America it has stood at about half of that amount. According to the books of the company over $50,000 of absolute water has been poured out for each mile of road between New York and Buffalo.

"The next step towards Chicago was one of 88 miles to Erie. This was made up of a consolidation of two roads effected in 1867, which went in with $2,800,000 of capital and came out with $5,000,000. The total capital account of the company was then a trifle over $3,200,000. In 1869, the consolidation of the lines between Buffalo and Chicago was effected, and this road became a party to it with $6,000,000 of stock and $4,000,000 of indebtedness—at least 30 per cent. of water in excess of all cost of construction.

"The next step in the line is one of 96 miles to Cleveland; this was filled by the celebrated Cleveland, Painesville and Ashtabula road, which, in the six years between 1862–'67, divided 120 per cent. in stock, 33 per cent. in bonds, and 79 per cent. in cash. Having really cost less than $5,000,000 in money, it was consolidated at nearly $12,000,000.

"The next step was from Cleveland to Toledo, 148 miles. Here it was that Vanderbilt began his operations, for in 1866 he secured possession of this road, and signalized his administration of its affairs by the issuing of a scrip dividend of 25 per cent. upon its $5,000,000 of capital.

"The last two roads were consolidated into the Lake Shore road, 258 miles in length, in 1867; the stock and indebtedness of the new company was $22,000,000. In 1869 the work of consolidation was perfected from Buffalo to Chicago by the merging of all the connecting links into the Lake Shore and Michigan Southern

Railroad Company, operating nearly 1300 miles of road, represented by $57,000,000 of stock and indebtedness, which was increased to $62,000,000 in 1871, and which it had the further privilege of increasing to about $73,000,000. These figures throw a very curious light upon the real cost of railroad construction in America. They represent a nominal outlay of but $48,000 per mile, and yet it is not denied that in the amount was included $20,000,000 of fictitious capital. These roads, not improbably, may have cost those who constructed them in cash, actually paid in either directly in money or in dividends which had never been drawn out, the full amount of the consolidation capital. The profits had, it is true, been very unequally divided, but substantial justice was done in the end; what had been lost in one road was made good in another, but as a whole the community was, perhaps, paying for nothing which it had not received. No credit on this account is due to those managing the affairs of the company. They undoubtedly regarded the Vanderbilt operations as masterpieces of railroad management, and only regretted that the earnings of the company under their control could by no possibility justify any similar performances; and yet the contrast between the results hitherto arrived at upon this line, under a system of moderate, average watering, and those achieved further east by Vanderbilt, is singularly suggestive. It is probably safe to say that the Vanderbilt stock waterings between Buffalo and New York annually cost the American people not less than $3,000,000 in excess of all remuneration which ever, under any construction of right, belonged to the owners of the lines. Under these circumstances it would seem, judging by the example of the Lake Shore road, that comparatively legitimate and reasonable waterings should satisfy any one not inordinately rapacious.

". . . . . . The Pacific Railroad furnishes a fine example of all these ingenious devices. In speaking of this enterprise it is not pleasant to adopt a tone of criticism toward the able and daring men who with such splendid energy forced it through to completion. It was a work of great national import and of untold material value. Those who took its construction in hand incurred great risk, and at one time trembled on the verge of ruin. This enterprise was to them a lottery, in which they might well draw a blank, but, should they draw a prize, the greatness of the prize must justify the risk incurred. The community asked them to assume the risk, and was willing to reward their success. Success was thought to be well worth all it might cost. At the same time the process of construction afforded a curious example of the methods through which fictitious evidences of value can be piled upon each other. The length of the united road was 1919 miles, and the cost of construction was estimated at $60,000,000. To meet this outlay a stock capital was authorized of $100,000,000 for each of the two great divisions of the line; upon this, however, no dependence was placed as a means of raising money; it was only a debt to be imposed, if possible, on the future business of the country. A curious mystery hangs over this part of the financial arrangements of the concern. Probably not $20,000,000 ever has been, or ever will be, derived from this source. The rest is very clear. There was the Government subsidy of $30,000 a mile, and $30,000 a mile of mortgage indebtedness; there was a land grant of 12,800 acres a mile, and, where there were States, there were bonds, with interest guaranteed by the State and gifts of real estate from cities, where cities existed; and there were even millions of net earning applied to construction. The means to build the road were not grudgingly bestowed Meanwhile, of the real cost of construction but little is correctly known; absolutely nothing indeed of the western division, or Central Pacific. Managed by a small clique in California, the internal arrangements of this company were involved in absolute secrecy. The eastern division was built, however, by an organization known as the Crédit Mobilier, which received for so doing all the unissued stock, the proceeds of the bonds sold, the government bonds, and the earnings of the road—in fact, all its available assets. Its profits were reported to have been enormous, and they made the fortunes of many, and perhaps of most of those connected with it. Who, then, constituted the Crédit Mobilier? It was but another name for the Pacific Railroad ring. The members of it were in Congress; they were trustees for the bondholders, they were directors, they were stockholders, they were contractors; in Washington they voted the subsidies, in New York they received them, upon the Plains they expended them, and in the Crédit Mobilier they divided them. Ever-shifting characters, they were ubiquitous—now engineering a bill, and now a bridge—they received money into one hand as a corporation, and paid it into the other as a contractor. As stockholders they owned the road, as mortgagees they had a lien upon it, as directors they contracted for its construction, and as members of the Crédit Mobilier they built it. What is the community to pay for it?

"At the close of 1870, with $103,000,000 of their capital yet unsubscribed, and thus reserved for issue, should the earnings of the roads at any future period make watering practicable; with this amount of stock in reserve, the two companies operated 2083 miles of road, represented by stock and debt to the amount of $240,000,000. Thus the last results of Vanderbilt's genius have been surpassed at the very outset of this enterprise. The line from Chicago to New York represents now but $60,000 to the mile, as the result of many years of inflation, while the line between Omaha and Sacramento begins life with the cost of $115,000 per mile. It would be safe to say that the road cost in money considerably less than one half of this sum. The difference is the price paid for every vicious element of railroad construction and management; costly construction, entailing future taxation on trade; tens of millions of fictitious capital; a road built on the sale of its bonds, and with the aid of subsidies; every element of real outlay recklessly exaggerated, and the whole at some future day is to make itself felt as a burden on the trade which it is to create.

"Enough has been said to illustrate the bearing which stock-watering and extravagant construction have upon taxation. It would be useless to attempt to estimate the weight of the burden imposed through these means upon material development. The statistics which should enter into any reliable estimate are not accessible, and any approximation would be simply a matter of guess-work. A table was published, during the year 1869, in a leading financial organ, comparing the capital stocks of twenty-eight roads as they stood on July 1, 1867, and May 1, 1869. During those twenty-two months it was found that the total had increased from $287,036,000 to $400,684,000, or 40 per cent. Carrying the comparison on nine of these roads back two years further, it was found that, in less than four years, their capitals had increased from less than $84,000,000 to over $208,000,000, or 150 per cent. A portion of this, perhaps 25 per cent. of the whole, represents private capital actually paid in and expended; another portion, perhaps equally large, represents dividends the payment of which was foregone and the money applied to construction; the whole of the remainder may be set down as pure, unadulterated 'water,' which calls for an annual tax-levy of some three or four millions a year."

It will, of course, be clear to the reader that the motive influencing the men who thus increase their stock is simply to increase the amount drawn from the public as earnings upon the sums supposed to be invested in the road. A road operating with a capital of $3,000,000, and earning ten per cent. upon this, increases its capital, by the watering process to $6,000,000 and claims ten per cent. upon this valuation. In plain English, the road is extorting from the community the $300,000 represented by the earnings upon the $3,000,000 of watered stock. That sum is drawn from the people and transferred to the pockets of the stockholders.

Commenting upon this, the Chicago *Tribune* in a recent issue said:

"The railroads have the full protection of the law in the decisions of the United States courts, which hold their charters to be in the nature of a contract which the State cannot violate. They can set up the law in any case where a State Legislature or the people endeavor to deprive them of any of the privileges specifically conferred by their charters. But they will make a mistake if they presume that any law bars out the people from ascertaining whether or not they are complying with their obligations under the contract. The courts will extend to the people the same protection that they extend to the corporations; and, in the conflict between the railroads and the farmers, the principal thing to be determined by evidence is whether the rates charged by the railroads represent a profit on the actual investment, or a percentage on fictitious capital not authorized under the charters, but created in a variety of ways without the investment of money. If the former, the railroad rates will be sustained; if the latter, the rates will be changed, in one way or another, and the railroads will be forced to be content with earnings that will pay a fair interest on the actual investment.

"In the eyes of the law a corporation is a fictitious person, created for special purposes and strictly limited to the terms of its charter. It can take nothing by implication. It can form no copartnerships, enter into no business transactions, spread out into no field not explicitly defined in the law which originally brought it into being, or in amendments thereto. Now, we know of no railroad charter which authorizes the corporation to earn a percentage on fictitious capital, and the courts will not construe this to be an unexpressed or implied privilege of the railroads. On the contrary, the law expressly holds that railroads must make fair and reasonable rates—and rates can be neither fair nor reasonable which represent dividends on capital that has never been invested or profits on stocks fictitiously issued for the benefit of speculators. It is on this ground that farmers can make their best fight, and if they keep close to this line of battle they will be certain of victory.

"We have no means of knowing what proportion of the capital stock of the railroads of this country is fictitious. An estimate made some two years ago placed it at 33 per cent. of the aggregate railroad stocks. The proportion is certainly not less to-day, and probably is much larger. If this be true then the average railroad rates are 33 per cent. higher than they would be if the railroad stocks of the country represented the capital actually invested in constructing and operating them. It is the work of the people to ascertain the precise difference between the actual investments and the fictitious stocks, and when this shall have been done there will be a solid basis for determining what reasonable rates are.

"The way and means adopted for creating fictitious railroad stocks are at once numerous and ingenious. A popular method is to declare stock dividends. If the Rock Island road, for instance, is earning more money than it cares to have the people know of, it declares a stock dividend. The capital stock is thereby increased and the earnings appear to be less. The fact is, that a means has been provided whereby the earnings may be increased without arousing the suspicion of the public. The new stock represents no investments of capital whatever, but thenceforth it constitutes a basis on which the railroads claim the right of earning the current rate of interest. The rates of transportation are thus increased to pay interest on stock originally issued for the purpose of covering up excessive earnings. Another favorite way of issuing fictitious stock is by the leasing of other railroads. But the most common means of obtaining fictitious stock is by what is known as the Crédit Mobilier plan of building railroads. Starting with a land grant from Congress, or subsidies from State or municipal governments, the construction company issues sufficient bonds to cover the cost of building the road, outside of all shrinkage from depreciation, brokers' commissions, etc. These bonds are sold, and the road is built and equipped from the proceeds. The construction company then have the capital stock of the road intact. Whether it be $1,000,000 or $10,000,000, it has not cost them one dollar. They then commence operating the road, and claim that it should not only earn money to pay the interest on the bonds, but also enough to pay dividends on capital stock that does not represent a single dollar of actual investment. The cost of the road is entirely comprised in the bonds that have been issued, and the capital stock is altogether fictitious. How large a proportion of the 63,000 miles of railroads in the United States has been constructed in this way, it is not possible to say. But the time has come when the people will undertake to find out. The people are willing that the railroads shall earn a fair profit on actual cost, but they can no longer be forced to pay a royalty on fraudulent issues. That time has passed, and the sooner the railroads make up their minds to it the better it will be for them."

CHAPTER IV.

THE CONSOLIDATION PROCESS.

A Railroad of necessity a Monopoly—George Stephenson's Views—The Interests of the Roads naturally Hostile to those of the People—Foolish Prodigality of the People—Competition disastrous to the Roads—Consolidation of Railroads inaugurated to stop Competition—Success of the Efforts for Consolidation—The Four Enemies of Free Trade—Vanderbilt's Success with the New York Central—The Pennsylvania Company—Its History—The Reign of Monopoly successfully inaugurated.

A RAILROAD is of necessity a monopoly. It is built for the express purpose of monopolizing the trade of the region through which it passes, and its first necessity is to prevent or destroy competition. Competition means cheap freights, low fares, and is in the interest of the community. It deprives a corporation of its power to tax the public with excessive rates, and compels it to make only such charges as are fair and reasonable. The interests of the road demand that there shall be no interference with it from any quarter, that its directors shall have the sole power to fix and arrange the rates for the transportation of passengers and freight, and that nothing shall occur to interfere with the monopoly they seek to establish.

The true nature of the railway system was plainly understood and stated by its great founder and advocate in England, George Stephenson. "He saw that a line once built must impose a tax on the community, if only to keep itself in existence. He also saw that if a competing road was built to divide any given business which could by any possibility be done over a road already constructed, in the end that business must support two roads instead of one. A very slender knowledge of human nature would have enabled him to take the next step, and conclude that any number of competing roads would ultimately unite to exact money from the community, rather than continue a ruinous competition."

It may be plainly stated then, that from the very outset, the interests of the public and those of the railroad companies were antagonistic. There was and is an irrepressible conflict between them, and it will require a more than ordinary degree of forebearance and patriotism on the part of the railroads to bring about a compromise.

At the outset, the people of the various States, in their eagerness to obtain the roads, granted important privileges without demanding any equivalent. Few restrictions were placed upon the proposed schemes. Corporations were given the right of way, and other important privileges the granting of which often involved the sacrifice of valuable private interests, and the State or the people demanded and received practically nothing in return. The eagerness of the people to obtain the roads was so great that nearly every projected enterprise received the sanction of the State, and was put into operation. In this way many useless roads were built, and the present generation is called upon to suffer for this folly.

In one respect, however, the system, as originally inaugurated, was correct. It provided for and allowed the construction of competing roads, and thus gave the public an opportunity to pay a fair price for transportation.

This was not as the corporations wished. Their risk was increased by it and their dividends diminished. They began to cast about for ways and means of putting an end to this state of affairs, and at length hit upon an expedient which has well nigh realized their wildest dreams of power and wealth. They inaugurated a policy of consolidation of roads. The great corporations of the East set to work to lease or buy up the lines connecting with them, by which they must reach the Western States, or which acted as feeders to their routes. They succeeded in their object, and soon the railroad system of the country was narrowed down to a few great lines, the minor enterprises disappearing as independent roads and forming parts of the great consolidated companies.

The principal railroad enterprises of the United States were undertaken with one common object—to bring the produce of the West to the Atlantic markets, and to provide the Western States with the manufactures and wares of the Eastern States. This was the grand prize for which so many plans were laid, and so much skilful work performed.

By the process of consolidation, the communication between the seaboard and the West has been limited to four great lines—the New York Central, the Erie, the Pennsylvania, and the Baltimore & Ohio Railroads. At the first glance it would seem that these four lines are sufficient to furnish all the competition necessary to secure fair rates in the matter of transportation. But such is not the case. These four consolidated companies were formed for the express purpose of destroying competition, and they stand like four gigantic sentinels over the avenues of trade to enforce their will. They offer the only means of communication between the East and West, and shippers and travellers are compelled to choose between them. Having disposed of their rivals, they have now a common interest—to keep rates up to the highest point, and they have power and wealth enough to carry out their wishes. With power to prevent the construction of any rival lines, these four companies hold the transportation business of the country in their grasp; and, being subject to no practical restraint, they may make such regulations, and compel the public to pay such rates as they may see fit.

It will be well to glance at the manner in which the consolidation of the two most powerful corporations was effected. It reveals some curious facts in our railroad history. We tell the story in the language of a brilliant writer,* from whom we have quoted before:

"Twenty-one years ago, the New York Central road, which forms the nucleus of the Vanderbilt combination, was not in existence as a corporation. In 1853 it was chartered, and eleven distinct corporations were merged into it. Five of these corporations, the longest of which could boast but of 76 miles of track, divided among them the 300 miles which separate Albany from Buffalo. The corporation created out of these elements was again, in its turn, merged in 1869 into the larger New York Central & Hudson River Railroad Company, which controls within the State of New York but little less than a thousand miles of track, and is represented by rather more than $100,000,000 of capital. The con-

* Charles F. Adams, Jr.

solidation, so far, was perfect, and had taken place under a State law and within State limits. Growth, however, did not stop here; the combinations of capital simply adapted themselves to the forms of a political system. Beyond the limits of New York, the corporation held, in the eye of the law, no property; it did not control a mile of track. At Buffalo, however, the Central connected with another company, itself made up of four separate primal links which had once connected Buffalo with Chicago, and which had united in obedience to the same law of development which had built up the Central. West of Chicago came yet other links in the trans-continental chain. Three lines competed to fill the gap which lay between Chicago and the eastern terminus of the Pacific road,—the Northwestern, the Rock Island, and the Burlington & Missouri. In the autumn of 1869, the consolidation of the Central and the Hudson River took place. Immediately afterwards, at the annual election of the Lake Shore & Michigan Southern, the Vanderbilt interest took open possession of that corporation, controlling a majority of its stock. In May, 1870, it in like manner assumed control of the Rock Island and Chicago & Northwestern. The same parties in interest were now practically the owners of a connected line of road from New York to Omaha; there was no consolidation as yet, but, so far as the public and competing roads were concerned, the close of 1870 found the six parties, which but a short time before had been in possession of the trans-continental thoroughfare, reduced to three. Without taking into consideration the immense influence which their position necessarily gave to them over other and less powerful members of the railroad system, here was a single combination of capital representing the control of at least 4500 miles of road and not less than $250,000,000 of capital.

"This, however, is but the result of a loose alliance between men notorious for their feuds and their selfishness; the combination is temporary, depending perhaps upon the continued life of one who lacks little of being an octogenarian. The men who control it not infrequently evince talents of a very high order, and their course is made continually interesting by episodes of dramatic surprise. They lack, however, the greatest and most indispensable element of permanent success,—some underlying, indissoluble bond of union. In this respect they differ entirely from the great combination which has gradually taken shape in the neighboring State of Pennsylvania. What is commonly known as the Pennsylvania Central Railroad Company is probably to-day the most powerful corporation in the world, as, indeed, it owns and operates one of the oldest of railroads. Its organization, as compared with that of its great rival, the New York Central, bears the relation of a republic to an empire. Cæsarism is the principle of the Vanderbilt group; the corporation is the essence of the Pennsylvania system. The marked degree in which the character of the people have given an insensible direction to the management of their corporations in these two States is well deserving of notice. In New York politics the individual leader has ever been the centre; in Pennsylvania, always the party. The people of this last State are not marked by intelligence; they are, in fact, dull, uninteresting, very slow and very persevering. These are qualities, however, which they hold in common with the ancient Romans, and they possess, also, in a marked degree, one other characteristic of that classic race, the power of organization, and through it of command. They have always decided our Presidential elections; they have always, in their dull, heavy fashion, regulated our economical policy. Not open to argument, not receptive of ideas, not given to flashes of brilliant execution, this State none the less knows well what it wants, and knows equally well how to organize to secure it. Its great railroad affords a striking illustration in point. It is probably the most thoroughly organized corporation, that in which each individual is most entirely absorbed in the corporate whole, now in existence. With its president and its four vice-presidents, each of whom devotes his whole soul to his peculiar province, whether it be to fight a rival line, to develop an inchoate traffic, to manipulate the Legislature, or to operate the road,—with this perfect machinery and subordination, there is no reason why the corporation should not assume absolute control of all the railroads of Pennsylvania.

"Such is this great corporation, high in credit in the money-markets of the world, careful withal of its outward repute, apparently unbounded in its resources. Organized so long ago as 1831, it had thirty miles of road ready for operation in the succeeding year. Not until 1854, however, was the Pennsylvania Railroad proper completed. It then controlled the line from Harrisburg to Pittsburg, 210 miles, which had cost a little less than $17,000,000, and was represented by about $12,000,000 of stock and $7,000,000 of indebtedness. This might be considered the starting-point; $3,500,000 of annual gross earnings on a capital a little less than $20,000,000. For many years its growth was confined to Pennsylvania. In 1869, however, its policy in this respect underwent a change, and it burst through State limits, extending its field of operations over the vast region lying between the great lakes and the Ohio upon the north and south, and the Missouri on the west. This sudden development was, as usual, the immediate result of competition, and was almost forced upon the corporation in spite of itself, as a measure of defence. The secret history of the railway intrigues and legislative manipulations of 1869 would make a very singular narrative could the whole of it be disclosed. That year was, in fact, a turning-point in our railway progress. The Erie management had then fallen into confessed discredit, and was beginning its remarkable attempt under Messrs. Gould and Fisk to carry on a great commercial enterprise in absolute disregard of every principle of good faith, commonly supposed to be at the basis of civilized transactions. Those managing this thoroughfare were desperately thrusting out in every direction, contracting, buying, and leasing all adjoining roads with a rashness only surpassed by their easy disregard of the obligations thus contracted. Early in 1869 they sought to cut off the connections of the Pennsylvania road, and to shut it up within the limits of that State. For a brief time the battle seemed to go in their favor, but suddenly the tide turned. The result showed that they were no match for the powerful antagonist they had provoked; —their overthrow was so effectual as to have in it some elements of the ludicrous. Bills in the interest of the Pennsylvania company, which it was doubtful if it were in the power of any legislature to pass, were pushed through their various stages, and received executive approval, with a speed unprecedented; contracts, arranged with the Erie managers by boards of directors, were unexpectedly rejected in meetings of stockholders; and for a time this irresistible power even threatened to wrest from the Erie road its own peculiar and long-established connections. The result of these operations was that the Pennsylvania Central soon controlled by perpetual lease a whole system of roads radiating to all points in the West and Southwest. By one it reached Chicago, by another St. Louis, and by a third Cincinnati. At Indianapolis it had absorbed a network of routes; at Chicago and St. Louis it had formed close connections looking directly towards the Pacific. Here for a time it rested, declaring that its policy did not look to any expansion beyond the Mississippi. The corporation rested, perhaps, but not the ambitious men who controlled it; their individual operations now commenced. They obtained the control of roads endowed with vast land grants in Michigan and in Minnesota; they were the directors of the Northern Pacific; and when the men who had constructed the Union Pacific broke down under the multiplicity of their engagements, the first vice-president of the Pennsylvania road appeared as the new president of that road also. The very land grants belonging to the companies these men now controlled amounted to 80,000 square miles, or an area equivalent to the aggregate possessions of four of the existing kingdoms of Europe.

"Meanwhile the Pennsylvania Railroad Company, distinct from its individual directors, now owned or held by lease 400 miles of road in Pennsylvania, and directly controlled 450 miles more, almost entirely within the same State; beyond its limits it leased and operated nearly two thousand miles in addition, holding the stock and bonds of railroads, canals, towns, and cities, like some vast Crédit Mobilier; it had, indeed, no less than $20,000,000 standing on its books as represented by these investments. In the sixteen years its own capital and indebtedness had swollen from $20,000,000 to $65,000,000, with a liberty secured to increase them to nearly $100,000,000; at the same time the system of roads which it held in its hands returned a yearly income of hardly less than $40,000,000, of which about $10,000,000 was claimed as net profit.

"If, however, as its direction had officially declared, the corporation had no distinct interests to push west of the Mississippi, the same could not be said of the region east of the Susquehanna. In the closing days of 1870 New York was suddenly startled by the announcement that the Pennsylvania Railroad had effected a perpetual lease of the whole famous railroad monopoly known as the United Companies of New Jersey. The rumor proved true, and some 450 miles of additional track, besides 65 miles of canals and some 30 steamers, in all some $35,000,000 of property, was by this transaction added to the vast consolidation, and brought it to the shores of New York harbor.

"It is unnecessary to consider how much further this combination will carry its operations, or in what they will result. The Pennsylvania road now controls directly and as itself owner or proprietor, and wholly distinct from its directors, more than 3000 miles of track, claiming to represent $175,000,000 of securities, and returning a gross income of at least $40,000,000 per annum. It is far from impossible that this combination may, from its very magnitude, lead to its own downfall."

The Erie Railway properly extends from Jersey City to Dunkirk, New York, a distance of 451 miles, but with its various branches it now operates a total length of 1032 miles. Until July, 1871, it was the lessee of the Atlantic and Great Western road, which, connecting with the track of the Erie at Salamanca, New York, carried the line to Cincinnati, a distance of 447 miles. Though the lease has been surrendered, the two roads are practically one as regards the question of transportation. The Erie Company own property to the amount of $118,295,979, and, in 1872, the gross earnings were $18,371,887.

The Baltimore and Ohio Railroad extends properly from Baltimore to the City of Wheeling, on the Ohio River, a distance of 379 miles, but, with its branches and leased roads, it controls and operates a total of 1067 miles of road. It already touches Lake Erie at Sandusky, and has now in construction a branch road extending from a point 90 miles north of Newark, Ohio, on the Lake Erie Division, to Chicago. The Company own property to the amount of $56,014,481, and, in 1872, the gross earnings were $13,626,677.

Here we have four corporations representing a total ownership of nearly $600,000,000, and an aggregate annual income of over $100,000,000.

It would be impossible to mention in detail all the various attempts at consolidation, successful and unsuccessful, that have been made in this country. What we have given will sufficiently illustrate this part of our subject. All such efforts have a common object, and that object is the extortion from an already overtaxed community of the highest rates of transportation that can be obtained.

We shall again refer to this portion of our subject, to point out some of the evils arising from the monopoly we have been considering.

CHAPTER V.

THE TRANSPORTATION TAX SWINDLE.

Sources of Railroad Earnings—The Freight Business—Enormous Tribute paid by the People to the Roads—The Railroads irresponsible to the Public—The Necessity of the Roads to the Country—Anomalous Position of the Railroads—What are Legitimate and what are Fictitious Earnings—Carelessness of the People respecting their Rights—Their Punishment—Arbitrary Course of the Roads in levying Freights—How the Railroads tax the People—The Community made to pay the Losses of the Roads—Instructive Lessons—How Competition is killed—Efforts of the State of Illinois to protect its Citizens—The Railroads refuse to obey the Law—The Railroad Yoke fastened upon the People.

THE object for which railways are constructed is the earning of interest on the amount of capital invested in them. The only means by which such enterprises can earn money, are by the transportation of freight and passengers. All roads are built with a view to the ultimate freight business that will come to them, the passenger traffic being with most corporations a secondary consideration.

As the wealth and productiveness of the country increase, the transportation increases also. In 1840, when there were less than 3000 miles of railroad in operation in the United States, the transportation business of the country amounted to about $8,000,000, or about fifty cents to each inhabitant of the Union. In 1860, it had increased to about $150,000,000, or about $5 to each inhabitant. In 1871, it had grown to the enormous sum of $450,000,000, or nearly $12 to each inhabitant.

This enormous sum of $450,000,000 may be taken as a fair annual average of the value of our internal commerce. It comes directly from the earnings of the whole people of the United States, and is gathered into the treasuries of the various railway corporations in the form of sums paid for the transportation of passengers and of the products and manufactures of the country.

In chartering the railways of the Union, the people have given to the corporations conducting these enterprises, the sole right to regulate the freight charges of their roads. In some cases there has been a stipulation that the earnings of the road should not exceed a certain percentage upon the capital invested, but, as we have shown, it has been left to the road not only to regulate its charges, but to make such returns of its earnings as it may see fit. The people have surrendered the right to scrutinize the proceedings of the corporation, and the corporation charges whatever rates it pleases, and as much as it thinks the public will pay.

Men may travel or not, as they are inclined, but the farmer must send his products to market, and the merchant and manufacturer must transport their wares to the point where there is the greatest demand for them. So the road is sure of its freight traffic. Men are compelled to use it, for it is the only means of transportation open to them. They are fully aware of this, and the corporation is equally aware of it.

The railroads then occupy the position of a body within the State, and almost, if not quite, independent of it, levying a tax upon its citizens. Not one man in the community can escape the necessity of in some way contributing to the earnings of the road. The aggregate amount annually contributed is, as we have seen, enormous. How much of this represents a legitimate profit upon the capital invested in railways? and how much is a wanton robbery of the public? These are questions of the deepest interest to the public, and yet they have attracted so little attention that none of the State Governments have made an effort to obtain the requisite data from which to answer them. We only know what the roads choose to allow us to learn, and they are very careful to keep us from knowing too much; while staggering under this enormous tax, and vaguely comprehending that it is excessive and unjust, no one has undertaken to introduce measures which will lay before the people the full extent of the evil from which they are suffering. We only know that "certain private individuals, responsible to no authority and subject to no supervision, but looking solely to their own interests, or to those of their immediate constituency, yearly levy upon the internal movement of the American people a tax, as a suitable remuneration for the use of their private capital, equal to about one-half of the expenses of the United States Government, —army, navy, civil list, and interest on the national debt included."

The power to levy such charges as they think proper on the transportation of freight being entrusted tc the railway corporations, they are not slow to use it. At certain periods of the year the movement of freights is very brisk, as when the year's harvest is finding its way to market, or when merchants and dealers are sending home the stock they have purchased in the great centres of commerce. Then, when men are compelled to use the roads, the corporations advance the rates. Complaint is useless. The directors know that goods and grain must be carried over the road, and they fix the rates at an extravagant figure, and the shipper is forced to submit to the extortion.

The "through rates," as they are called, are high enough, but they do not affect the majority of shippers as much as the local rates. It has become a maxim with railroad men that if in the war of competition a loss is incurred in the "through rates," it must be made up in the amount received for local freights. It is always possible to ship a case of goods or a sack of grain from New York to Chicago at a proportionately cheaper rate than is charged for the same article from New York to Syracuse. Often times the amount charged for local freight is double that charged for through freight. The reason is that in the through freight transportation, the competition of a few great lines keeps the rate down to a comparatively lower figure; while a given road, enjoying a monopoly of the local business, can charge what it pleases. It is utterly irresponsible, and the shipper is at its mercy.

This irresponsibility leads to continual change, especially in the through freight business, and introduces an element of chance into mercantile transactions which sound business men find it hard to contend against. Merchants find it difficult to regulate their purchases, and producers are sometimes utterly at sea in their efforts to calculate their profits, when the tariff may be changed in a day, and all their calculations destroyed. "Just this fluctuation took place in

September, 1870, when it at one time cost far more to send goods from Boston to Chicago than from New York, and shortly after the New York firms had to ship their goods to Boston as the cheapest way of getting them to the West." During the year 1869, freights between New York and Chicago fluctuated between $5 and $37.60 per ton; and between New York and St. Louis, between $7 and $46 per ton. At one time during that year, the Erie Railroad carried freights from New York to Chicago for $2 per ton, and soon after advanced the rate to $37 per ton.

The year 1870 was remarkable for its fluctuations of this kind, and it led to a singular warfare between the rival lines connecting New York with the West. Each met with considerable losses, but each undoubtedly made these good at the public expense by some operation entirely within its control.

"During that year competition was bitter in the extreme; the rates made East and West were simply ruinous. On certain descriptions of freight they literally were reduced to nothing, and cattle were carried over the Erie road at a cent a head, as against one dollar a car, the rate charged on the Central. On other articles the reduction was not so great, but, both on passengers and goods, rates were purely nominal, and hardly averaged a third of the usual amounts. Of course this could not last. Early in September, 1870, representatives of the competing lines met in New York, and proceeded to put a stop to competition in the one way possible among monopolists,—by combination. The parties in interest were the Central, the Erie, and the Pennsylvania railroads. The competition was mainly from Illinois to New York. In both

Illinois and New York, laws forbidding the consolidation of competing lines were in force, and all the roads were carrying on operations in one or both of those States. At the meeting in question, it was decided to 'pool' the earnings of the colored lines to all competing points; in other words, all receipts from that business which was supposed to receive a peculiar benefit from competition, were to be paid into a common fund, competition was immediately to cease, fixed rates were to be charged, and thus, at last, all the great trunk lines were to be practically consolidated, in so far as the business community was concerned. This arrangement was agreed to, but broke down for the moment because of quarrels among certain of the individual contracting potentates. The irreconcilables were Messrs. Gould and Vanderbilt, two New York men, who represented two New York roads; and yet the New York statute-book contained a recently enacted law intended to prevent and render impracticable any combination like the one agreed upon. Not being able to effect the desired arrangement there, certain of the same parties went to Chicago, in a State where a similar provision to that in force in New York had been made a part of the Constitution, and there they actually did enter into an agreement, under which all the roads between Chicago and Omaha 'pooled' their receipts between those points, and this contract went into effect. . . .

"The failure of the New York negotiators was, however, only temporary; and, moreover, it is by no means clear that its failure was not a disaster to the community. In this combination would at least have been found some degree of certainty and of responsibility.

Rates would no longer have varied with every season and to every city; points destitute of competition would not have been plundered, as they now habitually are, that competing points might be supplied for nothing. During the summer of 1870, accordingly, many towns in New England were charged upon Western freights heavily in advance of the sums charged for carrying the same freights on the same roads a hundred or two miles farther on. All because the farther point was served at a loss to the carrier, and, therefore, the nearer had to pay the road profits for both, besides replacing the loss. The agents of the roads do not seek to deny this; they acknowledge and defend it. They say, and say truly: 'We must live. If our through business is done at a loss (and they show that it was done for nothing), then our local business must pay for all.' This was the case in New England. The cities of central New York fared no better. During a war of rates, almost any manufactured article will be carried from the seaboard to the West for perhaps one half of the amount charged for carrying the article there from a semi-interior point. So also as regards Eastern freights. Syracuse, Rochester, and the like class of cities can neither compete on equal terms with Boston in the markets of the West, nor with Chicago in those of the East. The discrimination against them is said to amount in certain cases to ten per cent. of the whole value of the article transported. Neither, under the competing system, is there any remedy for this evil, and a consciousness of this fact, of the risk to which they are continually exposed, has caused the breaking up of many manufacturing establishments at interior points."

The State of Illinois has undertaken to investigate the management of its railroads, and to impose upon the railroad companies a series of regulations for the protection of the public.

The new railroad law passed by the Legislature of Illinois in May, 1873, directs the Railroad and Warehouse Commissioners to prepare for each railroad in the State a schedule of reasonable maximum rates; prohibits as extortion more than a fair and reasonable rate to be charged for transportation of passengers or freight, or for the use of track; prohibits as unjust discrimination any difference in the prices charged for equal services of these three kinds rendered at different points or to different persons, the penalty being fines (recoverable in an action of debt, in the name of the People) of $1000 to $5000 for the first offence, $5000 to $10,000 for the second offence, $10,000 to $20,000 for the third offence, and for every subsequent offence $25,000, either party having the right of trial by jury. Moreover, the overcharged person may recover in any form of action thrice the damages sustained, with costs and attorney's fee.

As may be supposed, this law gave great offence to the railroad interest of the State, and every obstacle has been thrown in the way of its execution. Indeed the roads have steadily disregarded it. Governor Palmer, of Illinois, in a speech delivered at Springfield, on the 4th of July, 1873, said:

"The last Legislature enacted a law for the government of railroads in this State, which is a monument of the patience and reasonableness of the people. It merely declares that the railroads shall not charge for their services more than a fair and reasonable rate; that they shall make no unjust discriminations in their charges for any kind of service; that to charge a greater sum for services rendered to one person than is charged to another for greater services, shall be presumptive evidence of extortion; and the whole law merely assumes that the relation of the railroads and their customers shall hereafter exist and continue upon the footing of equality and justice, and that like services shall be presumed to be worthy a like compensation. I regret to be compelled to say that the railroad managers have as yet shown no disposition to accept this law in the just spirit in which it was enacted. On the contrary, they have found in its passage a new pretext for extortion. They assume, in the first place, that they must have ten per cent. net profit on the nominal capital invested in their roads, and the large sums furnished to them by the people is a part of the aggregate upon which the same people are required to pay them the interest. Such a claim is most unreasonable. Their capital was invested in railroads, subject to the fluctuations and casualties of business, and that is all that will be conceded to them. They must also submit the cost and methods of their management to the scrutiny of the juries of the State, and must account for all unnecessary expenses incurred in efforts to counteract rivals, or to force business into unwilling channels. In their pretended obedience to law, it is manifest they are merely acting a part, intended to test the firmness of the people. They no longer discriminate, they say. They now apply the knife to the root of every branch of industry. I have seen the proposed tariffs of many of the roads, and they are avowals of a distinct purpose to crush out every interest with the utmost impartiality. They intend to compel the next Legislature to repeal the railroad law of last winter; they mean to make war upon every effort to curb them, and to use the people as the agents of their own undoing."

CHAPTER VI.

RAILROAD TYRANNY.

Dangers arising from the Railroad Monopoly—Irresponsibility of the Roads—Their Disregard of Individual Rights—A Man's Fight with a Railroad—A Corporation's Idea of a Contract—What a Railroad Ticket is Worth—Brutal Assault on Mr. Coleman—A Struggle for Justice—The Policy of Railroad Corporations Announced—The Public to be tied Hand and Foot—Railroad Testimony—How to Manufacture Evidence—What a Negro got by Losing his Ticket—A Specimen Railroad Murder—A Life for a Lost Ticket—A new Penalty for Drunkenness—Startling Details—The Avenue of Death—Railroad Killing not considered Murder—Unjust Treatment of Passengers—The Palace Car Swindle—Baggage Smashers—The War on the Merchants—How a Railroad endeavored to ruin a Business Firm—The Power of the Corporations.

We have seen the gradual growth of the railroad system of the country; how many of the roads have been built at the public expense by means of the immense land grants they have obtained; how fictitious capital has been created by the issuing of watered stock for the purpose of concealing the impositions of the road upon the public; how that which is a monopoly in itself has been made a more odious monopoly by the process of consolidation; and how these corporations have committed to them the right to tax the whole community, without being responsible to any one. We come now to consider some of the evils springing from this immense system of monopolies.

Conceding all the good results that have been brought about by the successful growth of our railways; admit-

THE MEN WHO BUILD THE RAILROADS ON THE PACIFIC COAST.

ting all that they have done towards furnishing a rapid and convenient method of communication between distant points, and all that they have accomplished in developing new sections of country, we are sure it will be admitted by the majority of the thinking men of the country that the railroads of the present day are as much of a danger as a convenience to the country, and that unless they are soon subjected to some system of regulation by which they can be compelled to respect the rights of the people to whom they owe their existence, they will become not only sources of danger, but the most annoying tyrannies that have ever cursed a land. That there is danger from this source we hope to show.

Practically the railroads of the United States are subject to no restraint. Nominally they are acting under the law, but in reality they make themselves superior to it, and when occasion suits them, they do not hesitate to violate and defy it. They claim the right to manage their road for their own benefit only, and are utterly regardless of the rights of others. The sole object of the directors is to wring money from those who are forced to use the line, and the public, for whose convenience the road is supposed to have been built, are denied the simplest privileges. Scarcely a day passes that some individual's rights are not violated by these companies, and if the injured party is bold enough to carry the matter before the courts, he has a hard task before him to obtain the simplest justice. He has to encounter the immense power of the road, backed by its wealth, and the chances are ten to one against him. He will either be beaten by the money of the corporation, or he will be forced to drag his case along, at ruinous expense, until he abandons it in despair.

A fair specimen of the disregard of the railroads for individual rights was afforded a few years ago in the case of Mr. John A. Coleman, of Providence, Rhode Island. This gentleman was shamefully maltreated and thrown from a train on the New York and New Haven Railroad, and thereby injured for life, merely for demanding to ride over the road with a ticket for which he had already paid, instead of buying a new one. The case is so characteristic that we shall let Mr. Coleman tell the story in his own words:

"About four years ago" (the matter occurred in 1868), says Mr. Coleman, "I purchased a ticket from Providence to New York *via* Hartford and New Haven. At New Haven my business detained me until too late in the evening to resume my journey by rail. I therefore took the eleven o'clock boat, in order to pass a comfortable night and to be able to meet my engagements the next day. That left the railway coupon ticket from New Haven to New York on my hands. I afterwards had no opportunity to use the ticket in the direction in which it was marked—always happening thereafter to travel with through tickets from Boston to New York. In returning to Boston from New York, June 11, 1868, I applied at the office of the New York and New Haven Railroad, in Twenty-seventh street, New York, for a ticket to Boston *via* Springfield; the ticket master refused to sell me one unless I would wait three hours for the train, which left at three o'clock, P. M., going through to Boston. He said he would sell me a local ticket to Springfield, and I could buy another from there to Boston. This would cost me more than seven dollars to Boston, instead of six dollars, the regular through fare, which of course I did not want to pay. I told him expressly that I wished to stop over at a way station one train to do some telegraphing, but without avail; he would not sell the ticket. As I could not wait three hours, I thought it would be a good time to use my old coupon, as I was accustomed to do upon other roads under similar circumstances. Accordingly I presented the coupon to the guard stationed at the entrance to the cars. He rudely and imperiously refused me admittance, stating that the ticket was 'good for nothing.' Some warm words passed between us, and he finally called the conductor, who stood near. The conductor was, if possible, more imperious than the guard. He said the ticket was 'good for nothing,' and peremptorily ordered me not to go on board the cars. I told him I thought the ticket was good, and that I was accustomed to use coupons in that way upon all other roads over which I travelled. He replied that 'it was no such thing; he travelled more than I did and knew all about it;' and concluded by saying that if I 'attempted to get upon the cars' he 'would put me off.' Severe remarks were made by several gentlemen standing near to the conductor during this time, to the effect that this was another manifestation of the general spirit of insolence and meanness towards passengers for which that road was noted. I then purchased a ticket to Providence *via* New Haven and Hartford, and got on board the train. I felt irritated at the treatment I had received, and having a constitutional objection to being browbeaten, I determined to ascertain why the practice with regard to tickets on this road was so unlike that upon other roads. Having had time to recover my equanimity somewhat after the cars had started, and supposing the conductor might be still angry and unreasonable, I determined to put the case to him, as one gentleman would to another, and to exercise self-control, that my manner should be quiet and give him no cause for offence. Accordingly, as he approached me in taking up his tickets, I said, 'Mr. Conductor, there is no use for you and me to quarrel about this ticket. This is a plain business matter, an affair of dollars and cents only. The case stands like this: I am travelling nearly all the time; and being frequently compelled to diverge from the route that I intended to take in starting, I am left with unused coupons. These coupons all cost me money; and by the end of the year they would accumulate to such an extent that they would represent too large a sum for me to lose.'

"The conductor replied, 'That coupon is good from New Haven to New York, but it is not good from New York to New Haven. My directors ordered me, three years ago, not to take such tickets, and I shall not do it.' I then said, 'My position is this; I have paid this road a certain amount of money for a certain amount of service, and I think I am entitled to that amount of service, whether my face is turned east or west. You say this ticket is good from New Haven to New York, which is seventy-four miles; I think it is good from New York to New Haven, which is also seventy-four miles; and I cannot understand the distinction which you make.' A gentleman who sat before me remarked at this moment, 'If there is any meanness which has ever been discovered upon a railroad, it is sure to be found upon this one, for it is the meanest railroad ever laid out of doors.' I replied, 'If this is so, I hope they will make an exception in my case, as all I require are the common courtesies of the road and an equivalent for my money.' The conductor said, 'I see you are all linked together to make me trouble.' And he went along.

"The gentleman who had spoken to me requested to see my coupon, and remarked that he had never heard the question raised before, and certainly had never heard the case put in that way. He further remarked that, 'Whether it was law or not, it was common sense.' A part of the Board of Trade delegation of Boston was in the car, returning from the Philadelphia Convention. Among these were Mr. Curtis Guild, Mr. Eugene H. Sampson, and a prominent railroad director of Boston, Mr. B. B. Knight, a cotton manufacturer of Providence, and other gentlemen from both cities. Several of these gentlemen, who had become interested in the discussion, requested to see the coupon, and they took the same view of the matter that I did.

"As we were approaching Stamford, the conductor again came to me, and said in a very abrupt manner, 'Well, sir! how shall we settle this matter?' I said, 'Just as before; there is the ticket, and I wish to go to New Haven; the circumstances have not altered in the least.' I had determined to take the matter quietly; the conductor saw that it was useless to attempt to frighten me by his imperious manner, and then began to remonstrate, saying, 'You have no business to make me disobey my directors, and lose my place upon the road; I have to get my living in this way, and it is mean for you to do so.' This was a new aspect of the case, and I replied, 'That is the only embarrassing question which has arisen in this discussion. I have no quarrel with you, and I would not do you a personal injury upon any consideration; but you and I both have travelled long enough to know that this matter is wholly within your discretion. You can take this coupon and turn it in at New York where you turn in your other tickets, and no one will know whether it is taken going east or going west, and no one will care.' My meaning was, that, as no injury was done, no injury could be known. He took the remark the other way; and said, in a sneering tone, evidently for the benefit of the other passengers, 'You might just as well ask me to steal ten dollars from the company, because they would not know it.' I replied, 'Theoretically, that may be true; but, practically, it is nonsense; you very well know that I have no intention to defraud this road; but in order to relieve you of all embarrassment about your position, I will make you a proposition: Here is my address, and these gentlemen know that I am responsible; you take the ticket and turn it in, and if you are even reprimanded for it by your directors, write to me, and I will send you the money for the ticket, upon your promise as a gentleman that you will send the ticket to me again; for I shall want the ticket.'

"The passengers said, 'That was very fair and would avoid all trouble.' The conductor said, 'It is very fair, but I sha'n't do it, that's all; I want another ticket out of you, sir.' I said, 'I shall not give you one.' He said, 'Then I shall request you to get off this train at Stamford.' I replied, 'I shall just as politely decline to do so.' He said, 'Then I will put you off.' I replied in general terms, and with some natural heat, that I did not believe he was able to do it. He said, 'I guess I can put you off if I get help enough.' I told him that was undoubtedly true, but warned him that I would pursue the matter further, if he brought his roughs into the car and laid hands upon me.

"At this moment the elderly gentleman who sat in front of me rose and said, 'Mr. Conductor, I am a "railroad man" and in my judgment this gentleman's position is correct. If he brings it to an issue, I think he will beat you; but if you think he is not correct, but trying to evade his fare, the proper way is to telegraph to New Haven, and have a policeman come aboard and quietly arrest him; that is business-like; but don't you take the law into your own hands and throw him off the train, for that is not done nowadays upon any respectable railroad.' I said, 'Certainly, I will submit to a policeman, but I will not be thrown off by him.' The conductor sneeringly replied, 'We don't do business in that style on this road.' I said, 'I have been aware of that for ten years past; and I propose to see if you cannot be compelled to do business in that style upon this road.' He said we were all against him, and he would leave it to the superintendent.

"The train had stopped in the mean time at Stamford. I paid no further attention to the conductor, but commenced reading. Very soon some one shouted, 'They are coming for you.' The conductor came in at the head of five or six rough brakemen and baggagemen, and said, pointing to me, 'This is the man; pull him out, and put him out on the platform.' They seized my coat and tried to roll me out of the seat. My coat tore, and they did not move me. This seemed to enrage them, and they sprang upon me like so many tigers. Two of them seized me by the legs, and as many as could got in back of the seat and seized me by the shoulders and commenced violently wrenching me from the seat. I instinctively grasped the arms of the seat, and they took the cushion and frame up with me. When they got me into the aisle, and had me completely at their mercy, three heavy blows with the clinched fist were struck upon the back of my head. Every individual in the car jumped to his feet the instant the blows were struck. The ladies screamed, and some of the gentlemen rushed to stop the conductor and his roughs from striking me. Fearing for my life, I struck one of the ruffians under the chin, and planted a blow square in the face of another. We had a hard struggle until they overpowered me. They carried me horizontally until they reached the car door, when they dropped my feet a little to pass through singly. I struck another away from me, and he went over between the cars. They fiercely grasped me again and threw me broadside from the platform of the car down upon the platform of the depot. I struck heavily on my side, my whole length. In this struggle they tore the flesh upon my arm and legs, and they ruptured me for life. The passengers swarmed out of the cars, and gave me their addresses. The superintendent came up, and I told him I would give him a dose of common law, and see if I could not teach him something. He said he would give me all the law I wanted, if I wished to test the case. I then ran and jumped on the train as it was in motion. The superintendent and his son and another man ran after and seized me around the body, stripped me off the car, and held me by main strength until the train was clear of the depot. As soon as they released me, I drew my through ticket from my pocket, and asked them why they held me. The superintendent started as though I had struck him, and said, 'Why didn't you show that ticket before, sir?' I said, 'Because it is not customary to show tickets in getting on at the way-stations, and you did not give me a chance.' He said, 'If you had been a gentleman, you would have shown that ticket.' I replied, 'I do not ask your opinion as to who is a gentleman, for you are no judge.' He said, 'You tried to steal your ride to New Haven and sell your ticket; and now we will give you all the law you want; and we'll show you that the laws in Connecticut are different from where you came from.'

"I took that for granted, and returned to New York. When I reached Boston again, I attached the New York and Boston express-train, partly owned by the New Haven road, in the Boston and Albany depot, and brought suit against them in the Superior Court of Massachusetts for ten thousand dollars damages. The first trial of the case occurred in April, 1869. The judge charged directly against passengers upon every point. He ruled that the ticket was a contract. That the road had a right to make any rule it pleased for its own government; and if a passenger broke a rule, he was a trespasser; and, being a trespasser, the road had the same right to eject him from its cars that one of the jurymen had to eject a man from his private house if he did not want him there. The only question for the jury to consider was, whether an excess of violence had been used by the road in the maintenance of a right. The jury, after being out only one hour, awarded me thirty-three hundred dollars damages. The judge, at the request of the road, after several weeks' delay, set the verdict aside on the exclusive ground that the amount was excessive.

"The second trial occurred in the same court in January, 1870, and resulted in a disagreement of the jury. They stood eleven to one for me, and it was afterwards understood that the man who disagreed had been connected in some capacity with the road. The third trial took place in May, 1870, and resulted in an award of thirty-four hundred and fifty dollars damages. Again the road demanded a new trial, which the judge refused to grant. The road then appealed to the Supreme Court upon points of *law*. The judge in charging the jury had happened to say, that if the resistance of the plaintiff to ejectment from the car consisted in simply refusing to walk out when he was told to go by the conductor, of course blows on the head, such as had been testified to, were unnecessary; and if the jury were satisfied that such blows had been given, a verdict should be rendered accordingly. This bit of common sense gave a new opportunity for the exhibition of that wonderful subtlety called 'law.' The Supreme Court, after the usual tedious delay of several months, in which plaintiff and witnesses had abundant time to die, gave the New York and New Haven Railway Corporation another opportunity to fulfil their threat of making it 'terrible for the public to fight it, right or wrong.' It decreed that the judge had no right to give an opinion as above, but should have left the question for the jury. Accordingly a new trial was granted, which took place in June, 1871. Up to this time three people connected with the suit had died, and one witness for the plaintiff had moved to Kansas; while young girls who were on the train when the outrage was committed had passed from girlhood through long courtships and were already matrons. However, with the impetuosity of a youthful temperament and the knowledge of a just cause, I made another onslaught upon the corporation after only thirteen months' delay since the last trial, and eventually obtained a verdict of thirty-five hundred dollars damages, after one hour's deliberation by the jury.

"For the fifth time the road demanded another trial, which being refused by the judge, they again appealed from his ruling to the Supreme Court. They asked the judge to charge the jury, that if the plaintiff had a tendency to hernia, or any physical disability that was liable to be increased by violence, the plaintiff ought to have so informed the employés of the road; and failing in that, he, and not the road, was responsible for the consequences. According to the railroad theory, therefore, if a gentleman is attacked by a scoundrel, unless the victim gives a complete diagnosis of his condition to the ruffian, he, and not the villian who struck him, is responsible for consequences when his skull is broken. To obtain the opinion of the Supreme Court of Massachusetts upon this important point has taken twelve months more, but I am happy to be able to state, that at last one point is established by the Massachusetts courts in favor of the rights of railroad passengers, namely, that it is *not* necessary for a man to inform a ruffianly aggressor what his grandmother died of, nor to describe his hereditary symptoms, even though it is the employé of a railroad corporation who comes to strike him.

"The case was of simple brutal assault in a public railroad car. The witnesses for the plaintiff were well-known merchants of Boston, who were members of the Board of Trade, railroad directors, and steamboat men, as well as others, including ladies. Their testimony was clear and consistent throughout every trial. Pitted against their testimony was that of the brakemen, the baggage-men and the conductor, every one of whom was in the employment of the road and a party to the assault. Not a passenger who saw the outrage committed in the car was brought forward by the road. The testimony of the employés was so absurd upon the first trial, that the court was repeatedly interrupted by laughter. No testimony of theirs upon any after trial has been like that of the first, but was manufactured to suit the theory of the railroad. It has been privately admitted by the road that 'the facts were with me, but the law,' meaning, I suppose, the judge's rulings, 'was with them.' So simple a case would

THE GREAT AMERICAN DESERT—THE COUNTRY THE PACIFIC RAILROADS PROPOSE TO **IMPROVE!**

have been disposed of at a single hearing in a minor court, had it occurred between two poor men. But I have been compelled to pass through four weary trials, lasting four years, gaining quick verdicts from juries, and being defeated only by the first judge, who granted a new trial to this railroad corporation, because thirty-three hundred dollars were excessive damages for the beating and rupturing of a man by their servants. Being the chief justice, his rulings, of course, were taken as the law by the associate judges who presided at the subsequent trials, and from whom I received great courtesy and fairness.

"But the contest is finished after the exhaustion of every legal advice, and there is something to be said about it in the interest of the public. I have been repeatedly told by parties interested in the road that the company had too much money to be beaten by me, and they would spend enough to defeat me. The paragraph at the head of this article is quoted from a statement made to me by an influential person connected with the corporation. These threats were of no consequence as applied to me, for their object was intimidation. They did not succeed. The corporation is beaten. I have received the money for damages which they said they would never pay, and my personal contest is ended. But these threats were not directed against myself alone, but against the public. If a limb is crushed by the negligence of the railroad men, *fight* instead of *pay* the victim, is their theory of dealing with the public; and they will remove all opposition by the power of wealth, influence with courts, and sheer terrorism. 'They may make any rules they please' for the public, and may carry out their arbitrary designs against the people, in spite of decency or common sense." *

During the progress of the trials of this case, one of the officials of the road, in a conversation with Mr. Coleman, arrogantly announced the policy of his corporation in such matters, and in doing so revealed the policy of the entire system of which his road forms a part. "*The Road*," he said, "*has no personal animosity against you, Mr. Coleman, but you represent the public; and the Road is determined to make it so terrible to the public to fight it, right or wrong, that they will stop it. We are not going to be attacked in this way.*"

Let it be remembered. This is the policy of the numerous roads that traverse our country. Each corporation represents a large amount of wealth and power. It claims the right to do as it pleases, to violate the rights of the public whenever they come in conflict with its own selfish ends, and when the public undertakes to assert its rights in the courts, the road, using its wealth and power for this purpose, "will make it so terrible for the public to fight it, right or wrong, that they will stop it." In plain English, the road assumes to be the master instead of the servant of the public, and it is rapidly making good this assumption.

"Every year the power of the railroad corporations to trample upon the rights of the public is becoming greater, notwithstanding its proportions are already frightful. The corporations are centralizing power, making themselves a unit against the public. They

* The reader will find the whole of Mr. Coleman's able and interesting article in *The Atlantic Monthly* for December, 1872.

overawe and control the entire business of the country. This is no mere figure of speech. Two men equal in intelligence and means own mills situated upon roads converging at a certain point and equidistant from that point. Their conditions may be precisely alike, and both compete for the same market. By the ruling of the judge (in the Coleman case) the 'railroad has a right to make any rule it pleases for its own government,' and one of the roads makes a rule that its freight tariff shall be double the rates upon the other road. The profit is swept from the manufacturer, and the field given to his competitor upon that other road, his business is ruined, his mill is idle, and becomes worthless; he is shut up by the railroad. The freights may afford the road an exorbitant profit, but the 'road has a right to make any rule it pleases.' Does the public charter railroad corporations as pecuniary speculations against itself? Does the public take away private property and give it to a company of private individuals called a railway corporation, so that it may make any rule it pleases, and though it can carry the public at a handsome profit at two cents a mile, it may charge three, five, or ten cents per mile, at its pleasure? Does the public intend to furnish a set of men a weapon to cut its own throat? Does it intend deliberately to tax itself through them for a common service, so that a few favored individuals may become inordinately arrogant and rich?"

Mr. Coleman was very fortunate in securing substantial justice at the end of his long fight with the railroad. The company fought him persistently, resorting to every artifice, and it would seem that it did not hesitate to introduce manufactured testimony for the purpose of defeating him. In the *Atlantic*, for May, 1873, in describing some of the incidents of the trial, Mr. Coleman says:

"The first witness was as prompt as a well-drilled recruit. He described the incidents of my ejection: the conductor called upon him and some of the other 'boys' to take a man out of the car; they attempted to carry out his order quietly, but the man refused to go; therefore they laid gentle hands on him, whereupon the man kicked and struck and bit, and he (the witness) had to take hold of the man's hands to restrain his violence. He swore positively that it took six men to move the man. In answer to an inviting question, he eagerly testified that he saw Mr. Coleman bite one of the boys on the arm,—right through the woollen garment that the man wore. The story was clear, concise, and told with an air of confidence that was quite impressive. 'Mr. Witness,' said my lawyer, beginning the cross-examination, 'you said just now that you saw Mr. Coleman bite one of the men?' 'Yes, sir; on the arm.' 'Which arm?' The witness hesitated; he was well prepared in generalities, but not in details. Presently he answered, 'The left arm.' 'How many men had hold of Mr. Coleman at this time?' 'One man was on his left side and another on his right, others had him by his legs, and I was in front.' 'These men were *abreast* of Mr. Coleman, taking him out squarely through the car, were they?' 'Yes, sir.' 'Will you swear to that positively?' '*Yes*, sir,' said the witness, resolutely. 'Careful, now; are you *sure* of that?' 'Yes, *sir;* I am *sure* of it.' 'On which side of Mr. Coleman was the man who was bitten?' Again the witness hesitated, and his face, hitherto calm, grew flushed and anxious. But he answered at last, 'The left side, sir.' 'Will you swear positively to that also?' . '*Yes*, sir; I swear positively to it.' 'Now, sir,' resumed the lawyer, 'do you not know that a man of Mr. Coleman's breadth in that narrow car-aisle would completely fill it, so that neither two men nor one could stand at his side, as you swear they did?' Flustered, but not daunted, the witness explained, 'The men were a little *back* of Mr. Coleman;' and witness quitted the stand, leaving the court to meditate on the strange spectacle of a man curving his giraffe-like neck, and fastening his teeth in the *left* arm of a man who stood on his *left* side, and a 'little *back* of him!'

"Several other honest witnesses gave similar testimony as to the biting, and as to the violent behavior of the plaintiff, and the gentle but firm deportment of the railroad men; these latter struck no blows, but several were delivered by the plaintiff. The harmony of the witnesses was beautiful. They seemed to have beheld the scenes which they described with a single eye: as to the biting, the arm bitten, and the position of the biter, their agreement was perfect. At this stage of the proceedings a recess was taken. On the reassembling of the court, other witnesses for the railroad were examined; but, strange to say, not one of them could give any particular information as to the biting; they swore that Mr. Coleman *did* bite, but though they had enjoyed the same opportunities for observation with their predecessors on the stand, they 'couldn't exactly remember the details.' Such is the effect of lunch.

"The conductor told a plausible story, modelled carefully on my own statement, but differing in certain points that could be turned against me. It will be remembered that he told me in the cars that the directors had made a 'rule' forbidding him to take tickets backward. On cross-examination, my counsel asked him where he was accustomed to turn in his tickets to the company. He attempted to evade the question again and again, but finally answered, with painful reluctance, 'in New York.' It was further extorted from him that the tickets were turned in at New York whether taken in going to or from that city; *that it made no difference which way my coupon was used;* and, finally, that the directors of the road had never given him (as he asserted to me) a rule against taking coupons 'backward,' but that the superintendent had verbally ordered him not to take them, about three years before! This superintendent, who, with his son, wrenched me from the train at Stamford when I attempted to re-enter it after my ejection, was obliged to swear that it was the exclusive right of the directors to make 'rules,' and, further, that they never had made a 'rule' touching the ticket question; he himself having verbally instructed the conductors not to take tickets 'backward,' which he had no shadow of authority to do. Thus it seems that the 'rule' for the violation of which I had been mildly rebuked by the servants of the railroad—a violation which was the soul of the defence, its single excuse and answer to my allegations—*was not a 'rule' at all, but a mere verbal order given by an unauthorized person.* Yet, in the face of the declaration, by one of the highest officers of the road, that there was no 'rule,' the judge charged the jury that a 'rule' had been broken, that I was a trespasser, and that the railroad company had a right to eject me from the train, employing the necessary force and no more! Such a charge concerns every person in the community; for it seems that any of us, *for disobedience to a non-existent rule*, may be brutally dragged from a railway car, and, seeking redress, shall be informed by the court that the railway company is responsible only for 'excess of violence.'

"The examination of the superintendent having been concluded, the counsel for the railroad stated to the court that the victim of Mr Coleman's carnivorous ferocity had been discharged from the road immediately after his mifortune; that diligent search had been made for him, but in vain. By one of those dramatic felicities, so frequent in fiction and so rare in real life, just at this juncture a telegram was brought in announcing that the bitten man had been found, and would arrive on a train due in ten minutes. The judge granted the delay asked for, and the spectators brightened up in anticipation of new and measurably tragic revelations. The delay was brief. In a few minutes the door of the court room was thrust open, and in rushed the witness, breathless with haste. A brisk, bronzed person he was, self-contained and self-satisfied, with locomotive gait, and a habit of gesture suggestive of brake-rods. He mounted the witness stand, was sworn, and delivered his direct testimony with easy indifference, coupling his sentences as he would couple cars, with a jerk. This is his story in brief: 'The conductor c'm out the car 'n' said, "'S man in there want ye t' take out." Went in the car, and he said, "That's th' man: put 'im out!" I jes' took 'im up and carried him out through the car out on t' th' platform th' depôt, an' took 'n' set 'im down, an' never hurt him a mite.' 'Did Mr. Coleman bite you?' inquired the counsel for the railroad. 'Yes, sir.' 'Did he bite you on the arm?' 'Yes, sir.' The lawyer asked him no more questions, evidently satisfied with the effect of his evidence thus far, and possibly remembering that, unlike the other witnesses for the road, he had not enjoyed the benefit of lunch. Remitted to my counsel for cross-examination, the witness, well pleased with his success, and confident in his own powers, met the inquisitorial onset with calm dignity.

"'Mr. Witness,' said the lawyer, 'you were in the car on the day when Mr. Coleman was taken out, were you?' 'Yes, sir; I took him out myself.' 'Ah! you assisted the men to take him out, did you?' 'No, *sir;* didn't have no men; took him out myself.' 'Oh! you took him out alone, then?' 'Yes, sir; took him out alone.' 'You swear to that?' 'Yes, sir; swear to it.' 'Nobody helped you?' 'No, sir; took him out myself.' 'Well, sir,' pursued the lawyer, 'you must be a stout fellow to handle a man like that. Won't you please describe just how you took him out?' 'Well, I jes' went up to th' man, reached one arm 'round his neck, so fashion, had his head right up here on my arm, 'n' I jes' took 'im right through the car out on t' the platform th' depôt, an' set 'im down and never hurt 'im a mite.'

"Every face was intent upon the witness, and not a sound was heard save his voice, though there were premonitory symptoms of laughter. With a suavity delightful to see, the lawyer said, while he scanned the compact frame of the witness, 'Why, you must be a powerful fellow!' 'Yes, *sir;* I'm big enough for him.' 'Well, now, will you be kind enough to tell the jury, did Mr. Coleman strike anybody?' 'No, sir; I didn't give 'im no chance; I had 'im.' 'You swear to that positively?' 'Yes, sir.' A look of dismay and disgust settled upon the faces of the earlier witnesses for the road, who had graphically and minutely described my violent resistance, my kicks and blows. The spectators giggled, and even the judge relaxed the solemnity of his visage. 'Did anybody strike Mr. Coleman?' continued the lawyer. '*No*, sir; I had 'im and didn't give 'em no chance.' 'You swear to that, too?' 'Yes, sir.' 'Well, Mr. Witness, when you had Mr. Coleman's head upon your arm, as you described, I suppose you had his face turned a little toward your breast?' The witness, eagerly following this description of the situation and the gestures which illustrated it, his face now flushed and beaded with perspiration (for the work was harder than he had thought it), nodded assent. 'Mr. Coleman's mouth, then, would come about there?' inquired the lawyer, pointing to the inside of the arm, next to the body. 'Yes, sir; that's just the place where he bit me.' 'You swear to that positively?' 'Yes, sir, positively.' All the witnesses for the road, except the conductor, who did not commit himself as to the biting, swore emphatically that the bite was on the outside of the left arm, some of them placing the bitten man upon the left of the biter; and now comes a third untutored witness, who claimed to be the sufferer, and who of course ought to know the place of the bite, testifying with equal positiveness that the bite was on the inside of his arm. Even the counsel for the road could not refuse to join in the universal merriment which ensued.

"On subsequent trials all this testimony as to the biting was rearranged. The victim of my ferocity was obliged to share the honor of taking me out with five auxiliaries, and the bite was transferred to his *right* arm. Being a draughtsman, I had measured the car, and was ready with a drawing to show that the new theories of the defence as to the method of taking me out left just three inches for the movement of each stalwart brakeman as he walked at my side.

"I suppose that I need give no extended report of the argument of the road's counsel. He took the highest ground—the ground that the public had no right to question the management of the road; that the company owned it, and had the right to manage it as any other property is managed by a private corporation; that is, he denied the fact that *the public is virtually a partner* in railroad companies, which it creates and lifts into power by grants of franchises and land. Indeed, this distinction between public and private corporations has been carefully ignored by the judiciary of the country; and to this the present alarming domination of railroad corporations is mainly traceable.

"I may say, for the encouragement of those who look to the courts for deliverance from a railroad tyranny, whose bonds the judiciary seems willing enough to rivet, that, in every trial, my counsel carried the jury with him, one single juror of the forty-eight excepted. This juror was said to have been formerly an employé of the New York and New Haven Railroad. The action of the several juries, so far as the public is concerned in it, is satisfactory and cheering; for it indicates unmistakably that the spring of railroad power in our courts is not in the deliberate judgment of intelligent men; but the judges' charges were in effect restatements of the arguments of the counsel for the railroad touching the general question of the rights and powers of railroads. The juries were instructed that the public has no voice in the affairs of railroads; that contracts with passengers were to be made on conditions fixed by one party, the railroad; that if a passenger violated its regulations, an assault upon him by the agents of the corporation was justifiable, though these latter must be careful to avoid excess of violence. The juries were also instructed that if they found that, in this case, the defendants had employed an excess of violence, they must not allow punitive damages, but only such as would compensate the plaintiff for his injuries. Despite these instructions the four juries promptly brought in verdicts in my favor, each one giving heavier damages than its immediate predecessor. On the second trial the jury disagreed, owing to one of its members; I am informed that many of his associates desired to award me $15,000. The first jury agreed upon a verdict of $10,000; but one of their number, versed in the ways of courts, suggested that it would probably be set aside, and that I would consequently be subjected to great trouble and expense; so they reduced the figures to $3300, which was increased to $3500 on the last trial."

All persons are not so lucky as Mr. Coleman. Very few of those who have the courage to seek redress for injuries sustained, succeed in obtaining justice. They must be possessed of either the patience of Job or the wealth of Crœsus to maintain their cases against the roads. Instances will occur to every reader of these pages of acts of railroad tyranny. He may himself have been the victim of some outrage of this kind.

A few years previous to the war, the writer chanced to be travelling on the Washington Branch of the

Baltimore & Ohio Railroad, coming from the Capital to Baltimore. Among the passengers was a negro man, who had no ticket. When the conductor demanded of him his ticket, he was unable to produce it, and the official at once and very properly told him he must pay his fare. The poor fellow was terribly confused, and began a stammering explanation of his position. The conductor lost patience, and pulled the bell cord to stop the train. Then, summoning a brakesman to his aid, he seized the poor negro, who made no resistance, and pushed him out upon the platform. There the unfortunate wretch was seized by the two "officials," and, before the train had fairly stopped, was literally thrown from the platform to the ground beyond the track. He fell heavily, and was doubtless injured, but it was impossible to tell, for the train shot forward again, and the unfortunate victim of railroad brutality was left behind.

Now, suppose this man had been killed by the fall, when thrown from the train, does any one suppose the conductor would have suffered for his crime? The whole power of this very powerful road would have been exerted to shield him. The victim was but a negro, and in those days a poor African had no right—not even the right to his life. In case he had been killed, his master might have demanded his value in money from the road; it would have been refused, and a suit would have been necessary to recover it. Even then the chances would have been in favor of the road.

Some three years ago—perhaps not so long—a train on the New Jersey Railroad was crossing the Hackensack Meadows, which lie between Newark and

Jersey City. It was night, and a very dark night at that. One of the passengers was found to have lost his ticket. The conductor refused to accept his explanation. He must pay his fare a second time or leave the train. He refused to submit to the outrage. He was hustled to the platform; but no effort was made to stop the train, which at this moment swept on to the open bridge which crosses the Hackensack river on the outskirts of Jersey City. Another push from the ruffians in charge of the train, and the man was thrown to the floor of the bridge. The momentum of the train made it impossible for him to secure a foothold upon the bridge. He rolled helplessly over the side and into the river, where he was drowned. There was no one to blame, in the opinion of the officials of the road, and every effort was made to prevent an investigation and "keep the matter quiet." No one was punished. The murdered man had dared to refuse to pay twice for his ride, and his life was forfeit to the company.

Another instance is that of a man who embarked upon a train in a neighboring State, and was too drunk either to pay his fare or to answer the questions of the conductor. The train was stopped, and he was thrust from it, at a considerable distance from any station. In his helpless condition he staggered on to the track and fell upon it in a drunken stupor. An hour later, a train, following that from which he had been ejected, ran over him as he lay on the rails, and killed him.

In the State of Vermont, not long since, an old lady and her daughter, believing that railroad tickets are "good until used," took passage on one of the night trains on a certain road, and, securing berths in the

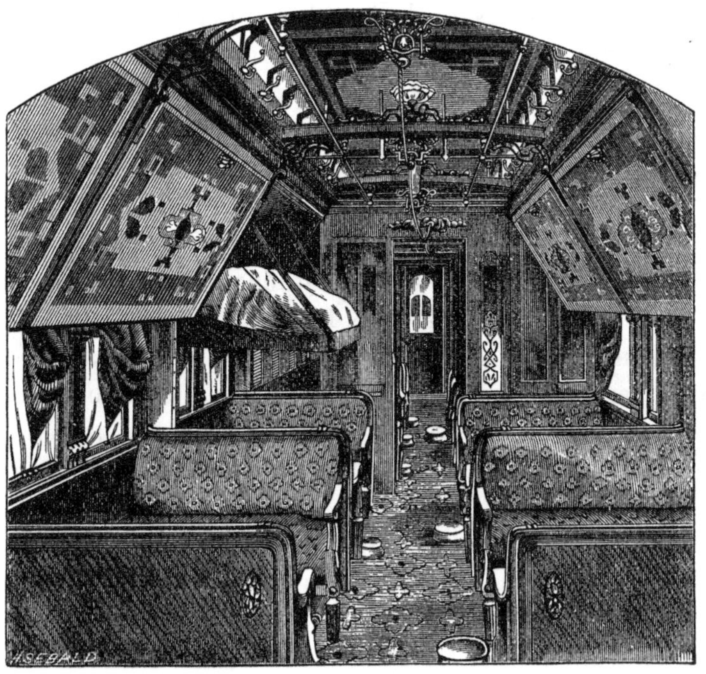

INTERIOR OF PALACE CAR.

sleeping car, proceeded to make themselves comfortable for the night. About midnight they were aroused by the conductor, who had discovered that their tickets were two or three days old, or, in other words, had been purchased two or three days before. He pronounced the tickets worthless, and demanded that they should pay their fares in money. This they declined to do. In spite of the lateness of the hour, and the fact that a heavy rain was falling at the time, the conductor compelled them to leave the train at a little wayside station, where they could procure only shelter from the storm in the cold and dirty waiting-room of the depôt.

In the city of New York, the Fourth avenue, from Forty-second street to the Harlem river, is used by the trains of the New York Central & Hudson River, the New York, Harlem & Albany, and the New York, New Haven & Hartford railroads. Along this thoroughfare, which is intersected by some ninety odd of the "cross streets" of the city, about eighty trains pass up and down every day. Although in the heart of the city, a dangerously high rate of speed is maintained. Within a single month, as many as three persons have been cut down and killed by the railroad trains, and others have been more or less injured. No one has been punished. The roads using the track hold their officials blameless, and exert their power—the power of the Vanderbilt Monopoly—to prevent investigation and screen the offenders from punishment.

The truth is, that railroads, having learned to disregard individual rights, have come to disregard human life. They kill and maim with utter recklessness, and insolently deny the right of the victims to seek redress for their injuries.

Descending to minor points, we find the roads intent upon making money and careless of the comfort of their passengers. A traveller purchasing a ticket is promised by the road a seat in the car in which he is to travel. Frequently the train is crowded, and there are no seats left to the late-comers, who must stand during the entire journey. Should the injured party in this case complain to the company of a breach of contract on their part, he would simply be laughed at.

Very many lines are now using Palace and Drawing-

THE PARLOR CAR—EXTRA CHARGE FOR ITS USE.

room cars, for admission to which passengers are required to pay a sum in excess of the regular fare. These cars are luxurious and comfortable, and few, who are able to afford it, neglect to use them. There is no reason why they should not be attached to every train. The roads using them, however, do not leave their occupancy to the discretion of the passenger. They compel him to use them by providing an insufficient number of ordinary cars, or cars which are so filthy and uncomfortable that men gladly pay the extra charge to escape from them.

Who is there that has travelled but could tell an eloquent tale of loss in the matter of baggage? It is safe to say, that thousands of dollars are lost annually by the travelling public in the way of trunks and portmanteaus, broken or injured by the carelessness of railroad employés.

But travellers are not the only victims of the "Company's" disregard of private rights. The vast army of shippers of freight throughout the Union represent another class of sufferers; and each man of this class could tell his tale of individual wrong. We could multiply instances; but to do so, would simply fatigue the reader. One characteristic case shall serve to illustrate this feature of railroad tyranny. It is told in the circular of a Boston firm addressed to its Western correspondents, and it shows in a vigorous light the utter helplessness of the individual in his struggle with the great corporations:

"BOSTON, *October* 31, 1872.

"GENTLEMEN: On account of the unusual and unwarranted action of the Boston & Albany Railroad

Co., in sending broadcast through the West public notice that no property consigned us would be received by them at Albany for transportation to us, unless freight and charges on such were prepaid, we are forced to take this course to set us right with our friends and shippers throughout the West. During the past two years we have received considerable grain over the Red, White, and Blue Transit Lines, such coming to this city over the B. & A. R. R., one of the co-partners to such lines. This grain has been largely short in weight, the losses in transit on cars being many times large and often excessive. We have repeatedly called attention of the R. R. Co. to such shortages, but they have invariably, and usually in an arrogant and arbitrary way (a way peculiar to this corporation, as our merchants all can testify to), refused to pay any attention to our demands. We have submitted to this species of robbery as long as we feel inclined to, and now, having been thus forced to it, take the stand, that, as common carriers, the railroads are liable, and should be held responsible, for failure to deliver property intrusted to them, in like good order and quantity as received by them; that, when we can prove a certain quantity shipped in a car at the West, we are entitled to a like quantity delivered us here, or payment for the shortage. We therefore declined paying the B. & A. R. R. Co. a lot of their freight bills unless they would allow our shortages, which we were desirous of having them look into, to satisfy themselves as to the justice of. They, however, most positively refused to notice our claims against them, but said we must pay their bills as presented, *right* or *wrong*, and, if wrong, trust to their refunding them when they see fit; and as we have not submitted to their arbitrary demands, but have decided to hold out, and let our courts settle the question, they have taken the course—as it seems to us out of sheer malice, to injure us—of notifying all their Western connections to refuse all property consigned us unless freight was prepaid. This is not through fear that they shall lose by us on freight their due, as they have commenced suit against us for amount of their bills, and we have given them a bond to cover the same, so they are secure on that score; but it is done simply so to annoy us as to make us *surrender unconditionally* to them. We propose to see, however, if we have any rights at all in the matter, or whether the railroad corporations are the supreme law in themselves, and everything must yield to them. The B. & A. R. R. Co. have even gone so far as to refuse to receive at Albany grain for which we hold through bills of lading, contracting to deliver such at East Boston; and through their influence flour and bran in transit to us, and for which we also hold through bills of lading, contracting to deliver such at East Boston, have been stopped at Toledo and Cleveland. We are also daily in receipt of advices from our friends, that cars for shipments intended for us are being refused by them at all points throughout the West.

"SCUDDER, BARTLETT & CO."

The instances given are enough to sustain our position—that the railroads recognize no such thing as individual rights. Neither do they recognize nor respect the rights of the public as represented by the State. They are humble enough, plausible enough until the charter is granted and the road built. Then they do not hesitate to defy even the power of the State. They have little to fear from the courts; and they understand the art of managing State Legislatures. Nay, even the National Congress is regarded as subject to them.

CHAPTER VII.

THE CAPTURE OF THE COURTS.

Sources of Redress for the People against Railroad Tyranny—Failure of the Courts to afford Protection—Efforts of the Railroads to debauch the Courts of Justice—The Free Pass System—Judicial Stockholders—Designs of the Railroads upon the Law—A Case in Point—How the Erie Road managed the Courts—A New System of Railroad Jurisprudence—Curious Details—How Boss Tweed became a Director of Erie—Efforts of Fisk & Co. to lock up Money—Daniel Drew beaten—The Government intervenes—The War in the Courts—The Value of an Injunction—How the Law was made to aid Sharp Practice—Mr. Jas. Fisk's little Journey—The Country Judge *vs.* The City Judge—The Railroad makes War on the Press—Arrest of Mr. Samuel Bowles—Justice turned against the People.

When the individual, or the community, is aggrieved by the railroad, redress may be sought from two sources—the legislature and the courts. These august bodies are popularly supposed to be the very centres and fountains of justice; *but are they?*

People are beginning to lose their faith in the courts —in judges and juries. Recent developments have shown that men who should be spotless are not fit to sit in judgment upon a case involving such issues as are presented in a matter between an individnal and a great corporation. Judges, attorneys, and jurors are often directly or indirectly interested in the cause of the corporation, because they are holders of stock or bonds of some similar enterprise. Or, if the judge be not a stock or bond holder, he has no doubt received at various times courtesies from the road, or some road, in the shape of free passes and the like, which incline him toward the side of the corporation. Railroad men are keen judges of human nature. They understand the use and effect of a free pass. They never give the use of them and its facilities from mere friendship or admiration for a man, be he never so famous. They grant the favor with the distinct expectation of some day asking and receiving an equivalent. It has been charged in the public press that a judge of one of the Western Supreme Courts permits a railroad corporation, which is a party to several suits pending before him, to transport, free of charge, building material for his new house, thereby saving him from five hundred to one thousand dollars in freight money. Railroad companies are always glad to number judges of the State courts, and Members of Congress among their stockholders, and it is common to make very great concessions to these gentlemen in their purchases. As the late Mr. Oakes Ames expressed it, they are "let in on the ground floor."

Appreciating the fact that their interests and those of the public are antagonistic, the railroads of this country have deliberately undertaken to beat the people, and to cheat them out of their rights. In order to accomplish this they have set to work to corrupt and capture both the legislative and the judiciary powers of the States.

The history of the Erie Railroad is very instructive of the daring of railroad corporations, and the lengths to which they are willing to go in their outrages upon the public. It is also suggestive as showing just what can be done in the courts when "properly managed." The following account of one of the "Erie fights," will show how easily the courts can be manipulated by the great corporations.

WILLIAM M. TWEED, FORMERLY A DIRECTOR IN THE ERIE RAILROAD COMPANY.

The Erie road had, at the time of the occurrences related below, settled its first quarrel with Mr. Vanderbilt, and had passed into the hands of Messrs. James Fisk, Jr., and Jay Gould, Mr. Drew having retired from the treasurership of the corporation. The first step of Messrs. Gould and Fisk, upon obtaining possession of the road, had been to dismiss the old Board of Auditors, and to concentrate all the power in their own hands as president, treasurer, and controller. Fortunately for them, it being summer, the receipts of the road were very heavy at that time, and the stock had suddenly come into great favor in the English market, and was selling rapidly in London. A new feature was now introduced into the road, and Peter B. Sweeney and William M. Tweed were admitted to the Board of Directors. Erie had formed an alliance with Tammany. The infamous Ring of New York, then in the height of its power, had bound itself to sustain the road in any of its outrages. The annual election of the Board of Directors was at hand, and the only fear felt by Messrs. Fisk, Gould & Co. was, that their powerful rival, Commodore Vanderbilt, who was supposed to cherish still his designs upon the road, might obtain possession of a sufficient amount of stock to give him control of the election. In order to prevent this, the transfer books of the company were closed about thirty days ahead of the usual time previous to an election. The device was successful; the election passed off quietly, with no opposition. Fisk and Gould succeeded in reëlecting themselves and their friends, and Tweed and Sweeney were included in the board, and the alliance with Tammany formed as above stated.

The month of October, 1868, witnessed the formation of this new combination. The same month witnessed the beginning of one of their most determined efforts to rob the community and enrich themselves. Their plan was to take advantage of the stringency existing in the money market, in consequence of the demand for ready money from the rural districts for the purpose of moving the year's harvest, and by suddenly throwing a new issue of stock into Wall street, produce such a depression in the stock of their road as would enable them, subsequently, to buy up the stock of the road at their own figures, and by producing a greater stringency, compel the dealers to pay them a usurious rate of interest for the use of money and the carrying of stocks.

"Towards the end of October it had become perfectly notorious in Wall street that large new issues of Erie had been made, and that these new issues were intimately connected with the sharp stringency then existing in the money market." On the 27th of October, the Stock Exchange appointed a committee to wait upon the officers of the road for the purpose of procuring such information respecting these new issues as they might be willing to afford. The committee waited upon Mr. Gould, but received only vague assurances. "Mr. Gould informed them that Erie convertible bonds for ten millions of dollars had been issued, half of which had already been, and the rest of which would be, converted intc stock; that the money had been devoted to the purchase of Boston, Hartford & Erie bonds for five millions, and also—of course—to payments for steel rails." The committee endeavored to ascertain if any further issue of stock was contemplated, but were told by Mr. Gould that no new issue was contemplated at that time, except "in certain contingencies;" which mysterious phrase the acute financier declined to explain. The committee went back to the Exchange with the conviction that Erie meant mischief and was not to be depended on. Meanwhile, vague rumors concerning the new issue began to pervade the street, and to alarm the brokers. "It was not until months afterwards that a sworn statement of the secretary of the Erie Railway revealed the fact that the stock of the corporation had been increased from $34,265,300 on the 1st of July, 1868, the

JAY GOULD.

date when Drew and his associates had left it, to $57,-766,300 on the 24th of October of the same year, or by two hundred and thirty-five thousand shares in four months. This, too, had been done without consultation with the Board of Directors, and with no other authority than that conferred by the ambiguous resolution of February 19th. Under that resolution the stock of the company had now been increased one hundred and thirty-eight per cent. in eight months."

The suspicions of the committee of the Stock Exchange were soon verified, for the Erie managers at once threw off all reserve, and by forcing new issues of stock upon the market, gradually forced the price of Erie down to 35. The banks, taking the alarm, and knowing what a terrible disaster to them a general panic in the stock market foreboded, held on to their greenbacks, until the enormous sum of twelve millions of dollars was locked up and withdrawn from circulation. The effect upon the money market was terrible, and the business of the whole country suffered in sympathy with it. Prices of all kinds declined, and trade in every branch began to drop off. The movement of the crops of the year was brought to a sudden stop, it was almost impossible to negotiate a loan; and as much as one and a half per cent. a day was paid for carrying stocks. Wall street and its gamblers were lost sight of in the general distress of the country, and it was evident that unless some relief was speedily found, the reckless men who had brought about the trouble would drive the entire mercantile community into one of the most terrible convulsions it had ever experienced. When matters had reached this alarming point, the General Government intervened in the interests of legitimate business, and the Erie managers were informed that fifty millions of additional currency would be issued if necessary to relieve the community. This threat—and only this—brought the conspirators to a halt.

They had calculated their movements well, however, and they now wheeled about and began to run up the stock, and instantly sent it from 40 up to 50.

"At this point Mr. Daniel Drew once more made his appearance on the stage. . . . At first he had combined with his old friends, the present directors, in their 'locking-up' conspiracy. He had agreed to assist them to the extent of four millions. The vacillating, timid nature of the man, however, could not keep pace with his more daring and determined associates, and after embarking a million, becoming alarmed at the success of the joint operations and the remonstrances of those who were threatened with ruin, he withdrew his funds from the operators' control, and himself from their councils. But though he did not care to run the risk or to incur the odium, he had no sort of objection to sharing the spoils. Knowing, therefore, or supposing that he knew, the plan of campaign, and that plan jumping with his own bearish inclinations, he continued, on his own account, operations looking for a fall. One may easily conceive the wrath of the Erie operators at such a treacherous policy; and it is not difficult to imagine their vows of vengeance. Meanwhile all went well with Daniel Drew. Erie looked worse and worse, and the golden harvest seemed drawing near. By the middle of November he had contracted for the delivery of some seventy thousand shares at current prices, averaging, perhaps, 38, and probably was counting his gains. He did not appreciate the full power and resources of

MR. DREW CALLS ON MR. FISK.

his old associates. On the 14th of November their tactics changed, and he found himself involved in terrible entanglements,—hopelessly cornered. His position disclosed itself on Saturday. Naturally the first impulse was to have recourse to the courts. An injunction—a dozen injunctions—could be had for the asking, but, unfortunately, could be had by both parties. Drew's own recent experience, and his intimate acquaintance with the characters of Fisk and Gould, were not calculated to inspire him with much confidence in the efficacy of the law. But nothing else remained, and, after hurried consultations among the victims, the lawyers were applied to, the affidavits were prepared, and it was decided to repair on the following Monday to the so-called courts of justice.

"Nature, however, had not bestowed on Daniel Drew the steady nerve and sturdy gambler's pride of either Vanderbilt or of his old companions at Jersey City. His mind wavered and hesitated between different courses of action. His only care was for himself, his only thought was of his own position. He was willing to betray one party or the other, as the case might be. He had given his affidavit to those who were to bring the suit on the Monday, but he stood perfectly ready to employ Sunday in betraying their counsels to the defendants in the suit. A position more contemptible, a state of mind more pitiable, can hardly be conceived. After passing the night in this abject condition, on the morning of Sunday he sought out Mr. Fisk for purposes of self-humiliation and treachery.* He then partially revealed the difficulties of his situation, only to have his confidant prove to him how entirely he was caught, by completing to him the revelation. He betrayed the secrets of his new allies, and bemoaned his own hard fate; he was thereupon comforted by Mr. Fisk with the cheery remark that "he (Drew) was the last man who ought to whine over any position in which he placed himself in regard to Erie." The poor man begged to see Mr. Gould, and would take no denial. Finally Mr. Gould was brought in, and the scene was repeated for his edification. The two must have been satiated

---

* It ought perhaps to be stated that this portion of the narrative has no stronger foundation than an affidavit of Mr. Fisk, which has not, however, been publicly contradicted.

with revenge. At last they sent him away, promising to see him again that evening. At the hour named he again appeared, and, after waiting their convenience,—for they spared him no humiliation,—he again appealed to them, offering them great sums if they would issue new stock or lend him of their stock. He implored, he argued, he threatened. At the end of two hours of humiliation, persuaded that it was all in vain, that he was wholly in the power of antagonists without mercy, he took his hat, said, 'I will bid you good night,' and went his way.

"With the lords of Erie forewarned was forearmed. They knew something of the method of procedure in New York courts of law. At this particular juncture Mr. Justice Sutherland, a magistrate of such pure character and unsullied reputation that it is inexplicable how he ever came to be elevated to the bench on which he sits, was holding chambers, according to assignment, for the four weeks between the first Monday in November and the first Monday in December. By a rule of the court, all applications for orders during that time were to be made before him, and he only, according to the courtesy of the Bench, took cognizance of such proceedings. Some general arrangement of this nature is manifestly necessary to avoid continual conflicts of jurisdiction. The details of the assault on the Erie directors having been settled, counsel appeared before Judge Sutherland on Monday morning, and petitioned for an injunction restraining the Erie directors from any new issue of stock or the removal of the funds of the company beyond the jurisdiction of the court, and also asking that the road be placed in the hands of a receiver. The suit was brought in the name of Mr.

August Belmont, who was supposed to represent large foreign holders. The petition set forth at length the alleged facts in the case, and was supported by the affidavits of Mr. Drew and others. Mr. Drew apparently did not inform the counsel of the manner in which he had passed his leisure hours on the previous day; had he done so, Mr. Belmont's counsel probably would have expedited their movements. The injunction was, however, duly signed, and, doubtless, immediately served.

"Meanwhile Messrs. Gould and Fisk had not been idle. Applications for injunctions and receiverships were a game which two could play at; and long experience had taught these close observers the very great value of the initiative in law. Accordingly, some two hours before the Belmont application was made, they had sought no less a person than Mr. Justice Barnard, caught him, as it were, either in his bed or at his breakfast, whereupon he had held a *lit de justice*, and made divers astonishing orders. A petition was presented in the name of one McIntosh, a salaried officer of the Erie road, who claimed also to be a shareholder. It set forth the danger of injunctions and of the appointment of a receiver, the great injury likely to result therefrom, etc. After due consideration on the part of Judge Barnard, an injunction was issued, staying and restraining all suits, and actually appointing Jay Gould receiver, to hold and disburse the funds of the company in accordance with the resolutions of the Board of Directors and the Executive Committee. This certainly was a very brilliant flank movement, and testified not less emphatically to Gould's genius than to Barnard's law; but most of all did it testify to the efficacy of the new combination between Tammany Hall and the Erie

Railway. Since the passage of the bill 'to legalize counterfeit money,' in April, and the present November, new light had burst upon the judicial mind; and as the news of one injunction and a vague rumor of the other crept through Wall street that day, it was no wonder that operators stood aghast and that Erie fluctuated wildly from 50 to 61 and back to 48.

"The Erie directors, however, did not rest satisfied with the position which they had won through Judge Barnard's order. That simply placed them, as it were, in a strong defensive attitude. They were not the men to stop there: they aspired to nothing less than a vigorous offensive. With a superb audacity, which excites admiration, the new trustee immediately filed a supplementary petition. Therein it was duly set forth that doubts had been raised as to the legality of the recent issue of some two hundred thousand shares of stock, and that only about this amount was to be had in America; the trustee therefore petitioned for authority to use the funds of the corporation to purchase and cancel the whole of this amount at any price less than the par value, without regard to the rate at which it had been issued. The desired authority was conferred by Mr. Justice Barnard as soon as asked. Human assurance could go no further. The petitioners had issued these shares in the bear interest at 40, and had run down the value of Erie to 35; they had then turned around, and were now empowered to buy back that very stock in the bull interest, and in the name and with the funds of the corporation, at par. A law of the State distinctly forbade corporations from operating in their own stock; but this law was disregarded as if it had been only an injunction. An injunction forbade the treasurer from making any disposition of the funds of the company, and this injunction was respected no more than the law. These trustees had sold the property of their wards at 40; they were now prepared to use the money of their wards to buy back the same property at 80, and a judge had been found to confer on them the power to do so."

The result of the fight in the stock market was that Drew was beaten. He made good his contracts at 57, and lost, as was generally supposed at the time, a million and a half of dollars.

From the Stock Board the battle was shifted to the Courts.

"On Monday, November 23d, Judge Sutherland vacated Judge Barnard's order appointing Jay Gould receiver, and, after seven hours' argument and some exhibitions of vulgarity and indecency on the part of counsel, which vied with those of the previous April, he appointed Mr. Davies, an ex-chief justice of the Court of Appeals, receiver of the road and its franchise, leaving the special terms of the order to be settled at a future day. The seven hours' struggle had not been without an object; that day Judge Barnard had been peculiarly active. The morning hours he had beguiled by the delivery to the grand jury of one of the most astounding charges ever recorded; and now, as the shades of evening were falling, he closed the labors of the day by issuing a stay of the proceedings then pending before his associate. Tuesday had been named by Judge Sutherland, at the time he appointed his receiver, as the day upon which he would settle the details of the order. His first proceeding upon that day, on finding his action stayed by Judge Barnard, was to grant a motion to show cause, on the next day, why Barnard's order should not be vacated. This style of warfare, however, savored too much of the tame defensive to meet successfully the bold strategy of Messrs. Gould and Fisk. They carried the war into Africa. In the twenty-four hours during which Judge Sutherland's order to show cause was pending, three new actions were commenced by them. In the first place, they sued the suers. Alleging the immense injury likely to result to the Erie road from actions commenced, as they alleged, solely with a view of extorting money in settlement, Mr. Belmont was sued for a million of dollars in damages. Their second suit was against Messrs. Work, Schell, and others, concerned in the litigations of the previous spring, to recover the $429,250 then paid them, as was alleged, in a fraudulent settlement. These actions were, however, commonplace, and might have been brought by ordinary men. Messrs. Gould and Fisk were always displaying the invention of genius. The same day they carried their quarrels into the United States courts. The whole press, both of New York and of the country, disgusted with the parody of justice enacted in the State courts, had cried aloud to have the whole matter transferred to the United States tribunals, the decisions of which might have some weight, and where, at least, no partisans upon the bench would shower each other with stays, injunctions, vacatings of orders, and other such pellets of the law. The Erie ring, as usual, took time by the forelock. While their slower antagonists were deliberating, they acted. On this Monday, the 23d, one Henry B. Whelpley, who had been a clerk of Gould's, and who claimed to be a stockholder in the Erie and a citizen of New Jersey, instituted a suit against the Erie Railway before Judge Blatchford, of the United States District Court. Alleging the doubts which hung over the validity of the recently issued stock, he petitioned that a receiver might be appointed, and the company directed to transfer into his hands enough property to secure from loss the plaintiff as well as all other holders of the new issues. The Erie counsel were on the ground, and, as soon as the petition was read, waived all further notice as to the matters contained in it; whereupon the court at once appointed Jay Gould receiver, and directed the Erie Company to place eight millions of dollars in his hands to protect the rights represented by the plaintiff. Of course the receiver was required to give bonds with sufficient sureties. Among the sureties was James Fisk, Jr. The brilliancy of this move was only surpassed by its success. It fell like a bombshell in the enemy's camp, and scattered dismay among those who still preserved a lingering faith in the virtue of law as administered by any known courts. The interference of the court was in this case asked for on the ground of fraud. If any fraud had been committed, the officers of the company alone could be the delinquents. To guard against the consequences of that fraud, a receivership was prayed for, and the court appointed as receiver the very officer in whom the alleged frauds, on which its action was based, must have originated. It is true, as was afterwards observed by Judge Nelson in setting it aside, that a *prima facie* case, for the appointment of a receiver, 'was supposed to have been made out,' that no objection to the person suggested was made, and that the right was expressly reserved to other parties to come into court, with any allegations they saw fit against Receiver Gould. The collusion in the case was, nevertheless, so evident, the facts were so notorious and so apparent from the very papers before the court, and the character of Judge Blatchford is so far above suspicion, that it is hard to believe that this order was not procured from him by surprise, or through the agency of some counsel in whom he reposed a misplaced confidence. The Erie ring, at least, had no occasion to be dissatisfied with this day's proceedings.

"The next day Judge Sutherland made short work of his brother Barnard's stay of proceedings in regard to the Davies receivership. He vacated it at once, and incontinently proceeded, wholly ignoring the action of Judge Blatchford on the day before, to settle the terms of the order, which, covering as it did the whole of the Erie property and franchise, excepting only the operating of the road, bade fair to lead to a conflict of jurisdiction between the State and Federal courts.

"And now a new judicial combatant appears in the arena. It is difficult to say why Judge Barnard, at this time, disappears from the narrative. Perhaps the notorious judicial violence of the man, which must have made his eagerness as dangerous to the cause he espoused as the eagerness of a too swift witness, had alarmed the Erie counsel. Perhaps the fact that Judge Sutherland's term in chambers would expire in a few days had made them wish to intrust their cause to the magistrate who was to succeed him. At any rate, the new order staying proceedings under Judge Sutherland's order was obtained from Judge Cardozo,—it is said, somewhat before the terms of the receivership had been finally settled. The change spoke well for the discrimination of those who made it, for Judge Cardozo is a very different man from Judge Barnard. Courteous but inflexible, subtle, clear-headed, and unscrupulous, this magistrate conceals the iron hand beneath the silken glove. Equally versed in the laws of New York and in the mysteries of Tammany, he had earned his place by a partisan decision on the excise law, and was nominated for the bench by Mr. Fernando Wood, in a few remarks concluding as follows: 'Judges were often called on to decide on political questions, and he was sorry to say the majority of them decided according to their political bias. It was therefore absolutely necessary to look to their candidate's political principles. He would nominate, as a fit man for the office of Judge of the Supreme Court, Albert Cardozo.' Nominated as a partisan, a partisan Cardozo has always been, when the occasion demanded. Such was the new and far more formidable champion who now confronted Sutherland, in place of the vulgar Barnard. His first order in the matter—to show cause why the order of his brother judge should not be set aside—was not returnable until the 30th, and in the intervening five days many events were to happen.

"Immediately after the settlement by Judge Sutherland of the order appointing Judge Davies receiver, that gentleman had proceeded to take possession of his trust. Upon arriving at the Erie building, he found it converted into a fortress, with a sentry patrolling behind the bolts and bars, to whom was confided the duty of scrutinizing all comers, and of admitting none but the faithful allies of the garrison. It so happened that Mr. Davies, himself unknown to the custodian, was accompanied by Mr. Eaton, the former attorney of the

Erie corporation. This gentleman was recognized by the sentry, and forthwith the gates flew open for himself and his companion. In a few moments more the new receiver astonished Messrs. Gould and Fisk, and certain legal gentlemen with whom they happened to be in conference, by suddenly appearing in the midst of them. The apparition was not agreeable. Mr. Fisk, however, with a fair appearance of cordiality, welcomed the strangers, and shortly after left the room. Speedily returning, his manner underwent a change, and he requested the new-comers to go the way they came. As they did not comply at once, he opened the door, and directed their attention to some dozen men of forbidding aspect who stood outside, and who, he intimated, were prepared to eject them forcibly if they sought to prolong their unwelcome stay. As an indication of the lengths to which Mr. Fisk was prepared to go, this was sufficiently significant. The movement, however, was a little too rapid for his companions; the lawyers protested, Mr. Gould apologized, Mr. Fisk cooled down, and his familiars retired. The receiver then proceeded to give written notice of his appointment, and the fact that he had taken possession; disregarding, in so doing, an order of Judge Cardozo, staying proceedings under Judge Sutherland's order, which one of the opposing counsel drew from his pocket, but which Mr. Davies not inaptly characterized as a 'very singular order,' seeing that it was signed before the terms of the order it sought to affect were finally settled. At length, however, at the earnest request of some of the subordinate officials, and satisfied with the formal possession he had taken, the new receiver delayed further action until Friday. He little knew the resources of his opponents, if he vainly supposed that a formal possession signified anything. The succeeding Friday found the directors again fortified within, and himself a much enjoined wanderer without. The vigilant guards were now no longer to be beguiled. Within the building, constant discussions and consultations were taking place; without, relays of detectives incessantly watched the premises. No rumor was too wild for public credence. It was confidently stated that the directors were about to fly the State and the country—that the treasury had already been conveyed to Canada. At last, late on Sunday night, Mr. Fisk with certain of his associates left the building, and made for the Jersey Ferry; but on the way he was stopped by a vigilant lawyer, and many papers were served upon him. His plans were then changed. He returned to the office of the company, and presently the detectives saw a carriage leave the Erie portals, and heard a loud voice order it to be driven to the Fifth Avenue Hotel. Instead of going there, however, it drove to the ferry, and presently an engine, with an empty directors' car attached, dashed out of the Erie station in Jersey City, and disappeared in the darkness. The detectives met and consulted; the carriage and the empty car were put together, and the inference, announced in every New York paper the succeeding day, was that Messrs. Fisk and Gould had absconded with millions of money to Canada.

"That such a ridiculous story should have been published, much less believed, simply shows how utterly demoralized the public mind had become, and how prepared for any act of high-handed fraud or outrage. The libel did not long remain uncontradicted. The next day a card from Mr. Fisk was telegraphed to the newspapers, denying the calumny in indignant terms. The eternal steel rails were again made to do duty, and the midnight flitting became a harmless visit to Binghamton on business connected with a rolling-mill. Judge Balcom, however, of injunction memory in the earlier records of the Erie suits, resides at Binghamton, and a leading New York paper not inaptly made the timid inquiry of Mr. Fisk, 'If he really thought that Judge Balcom was running a rolling-mill of the Erie Company, what did he think of Judge Barnard?' Mr. Fisk, however, as became him in his character of the Mæcenas of the bar, instituted suits claiming damages in fabulous sums, for defamation of character, against some half-dozen of the leading papers, and nothing further was heard of the matter, nor, indeed, of the suits either. Not so of the trip to Binghamton. On Tuesday, the 1st of December, while one set of lawyers were arguing an appeal in the Whelpley case before Judge Nelson in the Federal courts, and another set were procuring orders from Judge Cardozo staying proceedings authorized by Judge Sutherland, a third set were aiding Judge Balcom in certain new proceedings instituted in the name of the Attorney-General against the Erie road. The result arrived at was, of course, that Judge Balcom declared his to be the only shop where a regular, reliable article in the way of law was retailed, and then proceeded forthwith to restrain and shut up the opposition establishments. The action was brought to terminate the existence of the defendant as a corporation, and, by way of preliminary, application was made for an injunction and the appointment of a receiver. His Honor held that, as only three receivers had as yet been appointed, he was certainly entitled to appoint another. It was perfectly clear to him that it was his duty to enjoin the defendant corporation from delivering the possession of its road, or of any of its assets, to either of the receivers already appointed; it was equally clear that the corporation would be obliged to deliver them to any receiver he might appoint. He was not prepared to name a receiver just then, however, though he intimated that he should not hesitate to do so if necessary. So he contented himself with the appointment of a referee to look into matters, and, generally, enjoined the directors from omitting to operate the road themselves, or from delivering the possession of it to 'any person claiming to be a receiver.'

"This raiding upon the agricultural judges was not peculiar to the Erie party. On the contrary, in this proceeding it rather followed than set an example; for a day or two previous to Mr. Fisk's hurried journey, Judge Peckham, of Albany, had, upon papers identical with those in the Belmont suit, issued divers orders, similar to those of Judge Balcom, but on the other side, tying up the Erie directors in a most astonishing manner, and clearly hinting at the expediency of an additional receiver to be appointed at Albany. The amazing part of these Peckham and Balcom proceedings is, that they seem to have been initiated with perfect gravity, and neither to have been looked upon as jests, nor intended by their originators to bring the courts and the laws of New York into ridicule and contempt. Of course the several orders in these cases were of no more importance than so much waste paper, unless, indeed, some very cautious counsel may have considered an extra injunction or two very convenient things to have in his house; and yet, curiously enough, from a legal point of view, those in Judge Balcom's court seem to have been almost the only properly and regularly initiated proceedings in the whole case.

"These little rural episodes in no way interfered with a renewal of vigorous hostilities in New York. While Judge Balcom was appointing his referee, Judge Cardozo granted an order for a reargument in the Belmont suit,—which brought up again the appointment of Judge Davies as receiver,—and assigned the hearing for the 6th of December. This step on his part bore a curious resemblance to certain of his performances in the notorious case of the Wood leases, and made the plan of operations perfectly clear. The period during which Judge Sutherland was to sit in chambers was to expire on the 4th of December, and Cardozo himself was to succeed him; he now, therefore, proposed to signalize his associate's departure from chambers by reviewing his orders. No sooner had he granted the motion, than the opposing counsel applied to Judge Sutherland, who forthwith issued an order to show cause why the reargument ordered by Judge Cardozo should not take place at once. Upon which the counsel of the Erie road instantly ran over to Judge Cardozo, who vacated Judge Sutherland's order out of hand. The lawyers then left him and ran back to Judge Sutherland with a motion to vacate this last order. The contest was now becoming altogether too ludicrous. Somebody must yield, and when it was reduced to that, the honest Sutherland was pretty sure to give way to the subtle Cardozo. Accordingly the hearing on this last motion was postponed until next morning, when Judge Sutherland made a not undignified statement as to his position, and closed by remitting the whole subject to the succeeding Monday, at which time Judge Cardozo was to succeed him in chambers. Cardozo, therefore, was now in undisputed possession of the field. In his closing explanation Judge Sutherland did not quote, as he might have done, the following excellent passage from the opinion of the court, of which both he and Cardozo were justices, delivered in the Schell case as recently as the last day of the previous June: 'The idea that a cause, by such manœuvres as have been resorted to here, can be withdrawn from one judge of this court and taken possession of by another; that thus one judge of the same and no other powers can practically prevent his associate from exercising his judicial functions; that thus a case may be taken from judge to judge whenever one of the parties fears that an unfavorable decision is about to be rendered by the judge who, up to that time, had sat in the cause, and that thus a decision of a suit may be constantly indefinitely postponed at the will of one of the litigants, only deserves to be noticed as being a curiosity in legal tactics,—a remarkable exhibition of inventive genius and fertility of expedient to embarrass a suit which this extraordinarily conducted litigation has developed. . . . . Such a practice as that disclosed by this litigation, sanctioning the attempt to counteract the orders of each other in the progress of the suit, I confess is new and shocking to me, ... and I trust that we have seen the last in this high tribunal of such practices as this case has exhibited. No apprehension, real or fancied, that any judge is about, either wilfully or innocently, to do a wrong, can palliate, much less justify it.'* Neither did Judge Sutherland state, as he might have stated, that this admirable expression of the sentiments of the full bench was written and delivered by Judge Albert Cardozo. Probably also Judge Cardozo and all his brother judges, rural and urban, as they used these bow-strings of the law, right and left, —as their reckless orders and injunctions struck deep into business circles far beyond the limits of their State,—as they degraded themselves in degrading their order, and made the ermine of supreme justice scarcely more imposing than the motley of the clown,—these magistrates may have thought that they had developed at least a novel, if not a respectable, mode of conducting litigation. They had not done even this. They had simply, so far as in them lay, turned back the wheels of progress and reduced the America of the nineteenth century to the level of the France of the sixteenth. 'The advocates and judges of our times find bias enough in all causes to accommodate them to what they themselves think fit. . . . . What one court has determined one way another determines quite contrary, and itself contrary to that at another time; of which we see very frequent examples, owing to that practice admitted among us, and which is a marvellous blemish to the ceremonious authority and lustre of our justice, of not abiding by one sentence, but running from judge to judge, and court to court, to decide one and the same cause.'†

"It was now very clear that Receiver Davies might

* *Schell* v. *Erie Railway Co.*, 51 Barbour's S. C. 373, 374.
† Montaigne's Works, vol. ii. p. 316.

abandon all hope of operating the Erie Railway, and that Messrs. Gould and Fisk were borne upon the swelling tide of victory. The prosperous aspect of their affairs encouraged these last-named gentlemen to yet more vigorous offensive operations. The next attack was upon Vanderbilt in person. On Saturday, the 5th of December, only two days after Judge Sutherland and Receiver Davies were disposed of, the indefatigable Fisk waited on Commodore Vanderbilt, and, in the name of the Erie Company, tendered him fifty thousand shares of Erie common stock at 70.* As the stock was then selling in Wall street at 40, the Commodore naturally declined to avail himself of this liberal offer. He even went further, and, disregarding

---

*Throughout these proceedings glimpses are from time to time obtained of the more prominent characters in their undress, as it were, which have in them a good many elements both of nature and humor. The following description of the visit in which this tender was made was subsequently given by Fisk on the witness stand: "I went to his (Vanderbilt's) house; it was a bad, stormy day, and I had the shares in a carpet-bag; I told the Commodore I had come to tender 50,000 shares of Erie, and wanted back the money which we had paid for them and the bonds, and I made a separate demand for the $1,000,000 which had been paid to cover his losses; he said he had nothing to do with the Erie now, and must consult his counsel; . . . . Mr. Shearman was with me; the date I don't know; it was about eleven o'clock in the morning; don't know the day, don't know the month, don't know the year; I rode up with Shearman, holding the carpet-bag tight between my legs; I told him he was a small man and not much protection; this was dangerous property, you see, and might blow up; . . . . besides Mr. Shearman the driver went in with the witnesses, and besides the Commodore I spoke with the servant girl; the Commodore was sitting on the bed with one shoe off and one shoe on; . . . . don't remember what more was said; I remember the Commodore put on his other shoe; I remember those shoes on account of the buckles; you see there were four buckles on that shoe, and I know it passed through my mind that if such men wore that kind of shoe I must get me a pair; this passed through my mind, but I did not speak of it to the Commodore; I was very civil to him."

his usual wise policy of silence, wrote to the New York *Times* a short communication, in which he referred to the alleged terms of settlement of the previous July, so far as they concerned himself, and denied them in the following explicit language: 'I have had no dealings with the Erie Railway Company, nor have I ever sold that company any stock or received from them any *bonus*. As to the suits instituted by Mr. Schell and others, I had nothing to do with them, nor was I in any way concerned in their settlement.' This was certainly an announcement calculated to confuse the public; but the confusion became confounded, when, upon the 10th, Mr. Fisk followed him in a card, in which he reiterated the alleged terms of settlement, and reproduced two checks of the Erie Company, of July 11, 1868, made payable to the Treasurer and by him endorsed to C. Vanderbilt, upon whose order they had been paid. These two checks were for the sum of a million of dollars. He further said that the company had a paper in Mr. Vanderbilt's own handwriting, stating that he had placed fifty thousand shares of Erie stock in the hands of certain persons, to be delivered on payment of $3,500,000, which sum he declared had been paid. Undoubtedly these apparent discrepancies of statement admitted of an explanation; and some thin veil of equivocation, such as the transaction of the business through third parties, justified Vanderbilt's statements to his own conscience. Comment, however, is wholly superfluous, except to call attention to the amount of weight which is to be given to the statements and denials, apparently the most general and explicit, which from time to time were made by the parties to these proceedings. This short controversy merely added a little more discredit to what was already not deficient in that respect. On the 10th of December the Erie Company sued Commodore Vanderbilt for $3,500,000, specially alleging in their complaint the particulars of that settlement, all knowledge of or connection with which the defendant had so emphatically denied.

"None of the multifarious suits which had been brought as yet were aimed at Mr. Drew. The quondam Treasurer had apparently wholly disappeared from the scene on the 19th of November. Mr. Fisk took advantage, however, of a leisure day, to remedy this oversight, and a suit was commenced against Drew, on the ground of certain transactions between him, as Treasurer, and the railway company, in relation to some steamboats concerned in the trade of Lake Erie. The usual allegations of fraud, breach of trust, and other trifling, and, technically, not State prison offences, were made, and damages were set at a million of dollars.

"Upon the 8th the argument in Belmont's case had been reopened before Judge Cardozo in New York, and upon the same day, in Oneida county, Judge Boardman, another justice of the Supreme Court, had proceeded to contribute his share to the existing complications. Counsel in behalf of Receiver Davies had appeared before him, and, upon their application, the Cardozo injunction, which restrained the receiver from taking possession of the Erie Railway, had been dissolved. Why this application was made, or why it was granted, surpasses comprehension. However, the next day, Judge Boardman's order having been read in court before Judge Cardozo, that magistrate suddenly revived to a full appreciation of the views expressed by him in June in regard to judicial interference with judicial action, and at once stigmatized Judge Boardman's action as 'extremely indecorous.' Neglecting, however, the happy opportunity to express an opinion as to his own conduct during the previous week, he simply stayed all proceedings under this new order, and applied himself to the task of hearing the case before him re-argued.

"This hearing lasted many days, was insufferably long and inexpressibly dull. While it was going on, upon the 15th, Judge Nelson, in the United States Court, delivered his opinion in the Whelpley suit, reversing, on certain technical grounds, the action of Judge Blatchford, and declaring that no case for the appointment of a receiver had been made out; accordingly he set aside that of Gould, and, in conclusion, sent the matter back to the State court, or, in other words, to Judge Cardozo, for decision. Thus the gentlemen of the ring, having been most fortunate in getting their case into the Federal court before Judge Blatchford, were now even more fortunate in getting it out of that court when it had come before Judge Nelson. After this, room for doubt no longer existed. Brilliant success at every point had crowned the strategy of the Erie directors. For once Vanderbilt was effectually routed and driven from the field. That he shrunk from continuing the contest against such opponents is much to his credit. It showed that he, at least, was not prepared to see how near he could come to the doors of a State prison and yet not enter them; that he did not care to take in advance the opinion of leading counsel as to whether what he meant to do might place him in the felons' dock. Thus Erie was wholly given over to the control of the ring. No one seemed any longer to dispute their right and power to issue as much new stock as might seem to them expedient. Injunctions had failed to check them; receivers had no terrors for them. Secure in their power, they now extended their operations over sea and land, leasing railroads, buying steamboats, ferries, theatres and rolling-mills, building connecting links of road, laying down additional rails, and, generally, proving themselves a power wherever corporations were to be influenced or legislatures were to be bought.

"Christmas, the period of peace and good-will, was now approaching. The dreary arguments before Judge Cardozo had terminated on December 18, long after the press and the public had ceased to pay any attention to them, and already rumors of a settlement were rife. Yet it was not meet that the settlement should be effected without some final striking catastrophe, some characteristic concluding tableau. Among the many actions which had incidentally sprung from these proceedings was one against Mr. Samuel Bowles, the editor of the Springfield *Republican*, brought by Mr. Fisk in consequence of an article which had appeared in that paper, reflecting most severely on Fisk's proceedings and private character—his past, his present, and his probable future. On the 22d of December, Mr. Bowles happened to be in New York, and, as he was standing in the office of his hotel, talking with a friend, was suddenly arrested on the warrant of Judge McCunn, hurried into a carriage, and driven to Ludlow Street Jail, where he was locked up for the night. This excellent jest afforded intense amusement, and was the cause of much wit that evening at an entertainment given by the Tammany ring to the newly-elected Mayor of New York, at which entertainment Mr. James Fisk, Jr., was an honored guest. The next morning the whole press was in a state of high indignation, and Mr. Bowles had suddenly become the best-advertised editor in the country. At an early hour he was, of course, released on bail, and with this outrage the second Erie contest was brought to a close. It seemed right and proper that proceedings which, throughout, had set public opinion at defiance, and in which the Stock Exchange, the courts, and the Legislature had come in for equal measures of opprobrium for their disregard of private rights, should be terminated by an exhibition of petty spite, in which bench and bar, judge, sheriff and jailer, lent themselves with base subserviency to a violation of the liberty of the citizen.

"It was not until the 10th of February that Judge Cardozo published his decision setting aside the Sutherland receivership, and establishing on a basis of authority the right to over-issue stock at pleasure. The subject was then as obsolete and forgotten as though it had never absorbed the public attention. And another 'settlement' had already been effected. The details of this arrangement have not been dragged to light through the exposures of subsequent litigation. But it is not difficult to see where and how a combination of overpowering influence may have been effected, and a guess might even be hazarded as to its objects and its victims. The fact that a settlement had been arrived at was intimated in the papers of the 26th of December. On the 19th of the same month a stock dividend of eighty per cent. in the New York Central had been suddenly declared by Vanderbilt. Presently the Legislature met. While the Erie ring seemed to have good reasons for apprehending hostile legislation, Vanderbilt, on his part, might have feared for the success of a bill which was to legalize his new stock. But hardly a voice was raised against the Erie men, and the bill of the Central was safely carried through. This curious absence of opposition did not stop here, and soon the two parties were seen united in an active alliance. Vanderbilt wanted to consolidate his roads; the Erie directors wanted to avoid the formality of annual elections. Thereupon two other bills went hastily through this honest and patriotic Legislature, the one authorizing the Erie Board—which had been elected for one year—to classify itself so that one-fifth only of its members should vacate office during each succeeding year, the other consolidating the Vanderbilt roads into one colossal monopoly. Public interests and private rights seem equally to have been the victims. It is impossible to say that the beautiful unity of interests which led to such results was the fulfilment of the December settlement; but it is a curious fact that the same paper which announced in one column that Vanderbilt's two measures, known as the consolidation and Central scrip bills, had gone to the Governor for signature, should, in another, have reported the discontinuance of the Belmont and Whelpley suits by the consent of all interested.* It may be that public and private interests were not thus balanced and traded away in a servile Legislature, but the strong probabilities are that the settlement of December made white even that of July. Meanwhile the conquerors—the men whose names had been made notorious through the whole land in all these

* See the New York *Tribune* of May 10, 1869.

infamous proceedings—were at last undisputed masters of the situation, and no man questioned the firmness of their grasp on the Erie Railway. They walked erect and proud of their infamy through the streets of our great cities; they voluntarily subjected themselves to that to which other depredators are compelled to submit, and, by exposing their portraits in public conveyances, converted noble steamers into branch galleries of a police office; nay, more, they bedizened their persons with gold lace, and assumed honored titles, until those who witnessed in silent contempt their strange antics were disposed to exclaim, in the language of poor Doll Tearsheet: 'An Admiral! God's light, these villains will make the word as odious as the word "occupy," which was an excellent good word before it was ill sorted; therefore, Admirals had need look to't'"*

Such was the success of a desperate and reckless corporation in managing the courts of a great State. So thoroughly had they performed their work that they were popularly declared to "run the courts." Justice was turned against the people and used as an instrument of oppression.

There are other corporations just as reckless and other courts as complaisant. What has been done in New York has been repeated on a smaller scale elsewhere.

The railroads, in their determined effort to fasten their yoke upon the people, are seeking to poison the very fountain of justice and equity. Let the people look to it that they do not succeed.

---

* For the complete account of these extraordinary transactions the reader is referred to "A Chapter on Erie," published by J. R. Osgood & Co., of Boston. The book will repay a careful perusal.

CHAPTER VIII.

RAILROAD LEGISLATION.

Success of the Railroads in managing Legislatures—Efforts to corrupt Congress—The Railroad Lobby at Washington—How the State Legislatures are managed—A Case in Point—The Camden & Amboy Monopoly and the New Jersey Legislature—Erie Legislation—Exploits of the Erie Ring at Albany—The Story of a Check Book—A Disappointed Legislature.

WE have shown the efforts of the Railroads to corrupt the courts of justice. We propose now to glance at some of their exploits in the legislative bodies of the country.

Railroad companies are constantly asking new favors of and fresh privileges from the representatives of the people. They do not content themselves with resting their claims upon their merits. They "work them through" these bodies by unfair means. They keep a corps of regularly employed secret agents at each State capital, and at Washington, whose express duty is to corrupt the representatives of the people and influence their votes by unlawful means.

The late developments in the Crédit Mobilier investigation are familiar to all. They show the persistent and systematic manner in which the Pacific Railway endeavored to corrupt the highest legislative body in the land. The President of that corporation testified under oath before a Congressional committee at the same session, that he had contributed $10,000 toward securing the election of a United States Senator from

Iowa. Mr. Burbridge testified that he had contributed $5000 toward the election of a Senator from Nebraska. Members of Congress are the constant recipients of courtesies from the various railroads. Free passes are given them, special and luxuriously appointed cars are placed at their disposal, and the result is a demoralization on the part of our national legislators that is constantly developing itself in scandalous affairs which bring the blush of shame to the cheek of every true lover of his country.

It is known that shrewd men and women are annually sent to Washington for the purpose of engineering some corporation scheme through Congress. It is known that these persons are entrusted by their principals with large sums of money, and the whole country is satisfied that these sums constitute a gigantic corruption fund, to be used for the purpose of debauching the Congress of the United States.

As for the State Legislatures, the people have long since come to regard them as hopelessly corrupt. It is common to denounce the Legislature of the State of New York as the chief of sinners in this respect, but the sad truth is that the Legislature of New York is but a representative body in this respect. Your experienced railroad manager knows that there is not a legislative body in the country in which his compatriots have not attempted bribery with more or less success. He can tell you the exact market value of each legislature in the Union, and when he enters upon the conquest of one of these bodies, he can estimate very near the exact sum it will be necessary to expend upon it. If we choose our illustrations from the New York Legislature, it is merely for convenience.

In the winter of 1872–73, the Third Avenue Street Railroad Company of New York sought of the legislature a charter for the construction of an elevated road along the line of that thoroughfare. The road was and is badly needed by the Metropolis, and the company, with the economy for which it is noted, "refused to pay a cent for the passage of their bill." The scheme had received the endorsement of the public, the entire metropolitan press demanded its adoption, but the legislature rejected the bill. The Third Avenue Company had refused to buy the votes of members, and it must be punished for its insolence. The indignant members rejected the bill. They had been so carefully trained by the railway corporations seeking their aid, to regard their votes as merchantable property, that they turned upon the first corporation refusing to buy, and crushed it.

It is popularly believed that the Legislature of New Jersey was for many years in the pay of the famous Camden & Amboy monopoly and its successor, the Pennsylvania Company. It was not until the outraged and indignant people of the State rose against these monopolies, and threatened to put an end to the official existence of the legislators, that the General Assembly of New Jersey saw fit to take steps for the discontinuance of the wrongs from which the State had suffered for more than a generation.

The conflicts of the Erie Railroad under the management of Messrs. Gould and Fisk, afforded many instances of the art of manipulating a legislature, and it is popularly believed that many a dollar of Erie funds found its way into the pockets of the honorable gentlemen who assemble at Albany to make laws for the

Empire State. During the session of 1868, it was believed that Erie spent its money lavishly at Albany. It was feared that Commodore Vanderbilt would succeed in his attempt to secure the Erie road, and that the New York Legislature would so influence the investigation of the affairs of Erie, which had been begun by the Senate, as to oust the parties into whose hands the road had fallen, and pave the way to his long desired victory. Vanderbilt's power at Albany was well known, and the Erie managers found it necessary to defeat him at all hazards. The friends of Erie introduced into the legislature sundry measures for the promotion of their interests, among others a bill virtually prohibiting the consolidation of the Erie and the Central in the hands of Vanderbilt. This bill was referred to the Committee on Railroads on the 13th of March. On the 20th a public hearing was begun. On the 27th the report of the committee, adverse to the bill, was adopted in the Assembly by a vote of 83 to 32. It was generally understood that this vote was a broad hint that the Assembly would do nothing for Erie without being paid for it.

The Erie Directors were at this time sojourning in Jersey City, whither they had fled to escape arrest for contempt of court, Judge Barnard having issued warrants for their apprehension. They desired to return to New York, but before doing so it was advisable that a recent issue of convertible bonds, which had been the source of their trouble with the courts, should be legalized; and this could be done only by the Legislature of New York. Accordingly, Mr. Gould, though liable to arrest upon Judge Barnard's process, was sent to Albany to procure the passage of the desired law. Mr. Gould reached Albany on the 30th of January. The next day he was arrested for contempt of court, upon Judge Barnard's process. Bail was fixed at half a million of dollars, and it was immediately given, he being ordered to appear at New York on the following Saturday. He spent the interval in attending to the business on which he had come, and the next Saturday presented himself before Judge Barnard in New York, and was placed in charge of the sheriff to answer certain interrogatories. Judge Barrett, of the Court of Common Pleas, was resorted to, and he granted a *habeas corpus*, by virtue of which Mr. Gould was once more brought into court. The hearing of the case was postponed. Judge Barrett then consigned Mr. Gould to the care of an officer, with the positive injunction not to lose sight of him for a moment. Mr. Gould's presence was necessary at Albany, and he at once returned there, accompanied by the officer. Upon reaching Albany he pleaded illness, and declared himself unable to go back to New York, although it is certain that he was well enough to go to the capitol in a heavy snow storm. The upshot of the matter was that, Gould declaring that he was too ill to travel, the officer returned to New York and reported to Judge Barrett that his prisoner had run away. The judge was very indignant, but the matter was hushed up, and Mr. Gould, though at Albany, was theoretically returned to the custody of the sheriff, but was allowed to remain at Albany until the restoration of his health, bail being given for his appearance. He employed his singular period of illness in prosecuting his work with the legislature.

"The full and true history of this legisiative campaign will never be known. If the official reports of investigating committees are to be believed, Mr. Gould at about this time underwent a curious psychological metamorphosis, and suddenly became the veriest simpleton in money matters that ever fell into the hands of happy sharpers. Cunning lobby members had but to pretend to an influence over legislative minds, which every one knew they did not possess, to draw unlimited amounts from this verdant *habitué* of Wall street. It seemed strange that he could have lived so long and learned so little. He dealt in large sums. He gave to one man, in whom he said he 'did not take much stock,' the sum of $5000, 'just to smooth him over.' This man had just before received $5000 of Erie money from another agent of the company. It would therefore be interesting to know what sums Mr. Gould paid to those individuals in whom he did 'take much stock.' Another individual is reported to have received $100,000 from one side 'to influence legislation,' and to have subsequently received $70,000 from the other side to disappear with the money; which he accordingly did, and thereafter became a gentleman of elegant leisure. One Senator was openly charged in the columns of the press with receiving a bribe of $20,000 from one side, and a second bribe of $15,000 from the other; but Mr. Gould's foggy mental condition only enabled him to be 'perfectly astounded' at the action of this Senator, though he knew nothing of any such transactions. Other Senators were blessed with a sudden accession of wealth, but in no case was there any jot or tittle of proof of bribery. Mr. Gould's rooms at the Develin House overflowed with a joyous company, and his checks were numerous and heavy; but why he signed them, or what became of them, he seemed to know less than any man in Albany. This strange and expensive hallucination lasted until about the middle of April, when Mr. Gould was happily restored to his normal condition of a shrewd, acute, energetic man of business; nor is it known that he has since experienced any relapse into financial idiocy.

"About the period of Mr. Gould's arrival in Albany the tide turned, and soon began to flow strongly in favor of Erie and against Vanderbilt. How much of this was due to the skilful manipulations of Gould, and how much to the rising popular feeling against the practical consolidation of competing lines, cannot be decided. The popular protests did indeed pour in by scores, but then again the Erie secret-service money poured out like water. Yet Mr. Gould's task was sufficiently difficult. After the adverse report of the Senate committee, and the decisive defeat of the bill introduced into the Assembly, any favorable legislation seemed almost hopeless. Both Houses were committed. Vanderbilt had but to prevent action,—to keep things where they were, and the return of his opponents to New York was impracticable, unless with his consent; he appeared, in fact, to be absolute master of the situation. It seemed almost impossible to introduce a bill in the face of his great influence, and to navigate it through the many stages of legislative action and executive approval, without somewhere giving him an opportunity to defeat it. This was the task Gould had before him, and he accomplished it. On the 13th of April a bill, which met the approval of the Erie party, and which Judge Barnard subsequently compared not inaptly to a bill legalizing counterfeit money, was taken up in the Senate; for some days it was warmly debated, and on the 18th was passed by the decisive vote of seventeen to twelve. Senator Mattoon had not listened to the debate in vain. Perhaps his reason was convinced, or perhaps he had sold out new 'points' and was again cheating himself or somebody else; at any rate, that thrifty Senator was found voting with the majority. The bill practically legalized the recent issues of bonds, but made it a felony to use the proceeds of the sale of these bonds except for completing, furthering, and operating the road. The guaranty of the bonds of connecting roads was authorized, all contracts for consolidation or division of receipts between the Erie and the Vanderbilt roads were forbidden, and a clumsy provision was enacted that no stockholder, director, or officer in one of the Vanderbilt roads should be an officer or director in the Erie, and *vice versa.* The bill was, in fact, an amended copy of the one voted down so decisively in the Assembly a few days before, and it was in this body that the tug of war was expected to come.

"The lobby was now full of animation; fabulous stories were told of the amounts which the contending parties were willing to expend; never before had the market quotations of votes and influence stood so high. The wealth of Vanderbilt seemed pitted against the Erie treasury, and the vultures flocked to Albany from every part of the State. Suddenly, at the very last moment, and even while special trains were bringing up fresh contestants to take part in the fray, a rumor ran through Albany as of some great public disaster, spreading panic and terror through hotel and corridor. The observer was reminded of the dark days of the war, when tidings came of some great defeat, as that on the Chickahominy or at Fredericksburg. In a moment the lobby was smitten with despair, and the cheeks of the legislators were blanched, for it was reported that Vanderbilt had withdrawn his opposition to the bill. The report was true. Either the Commodore had counted the cost and judged it excessive, or he despaired of the result. At any rate, he had yielded in advance. In a few moments the long struggle was over, and that bill which, in an unamended form, had but a few days before been thrown out of the Assembly by a vote of eighty-three to thirty-two, now passed it by a vote of one hundred and one to six, and was sent to the Governor for his signature. Then the wrath of the disappointed members turned on Vanderbilt. Decency was forgotten in a frenzied sense of disappointed avarice. That same night the *pro rata* freight bill, and a bill compelling the sale of through tickets by competing lines, were hurriedly passed, simply because they were thought hurtful to Vanderbilt; and the docket was ransacked in search of other measures, calculated to injure or annoy him. An adjournment, however, brought reflection, and subsequently, on this subject, the legislature stultified itself no more."

CHAPTER IX.

RAILROAD STOCK GAMBLING.

Who own the Railroads?—The Old-fashioned Method of building a Road—The Present Style—A Contrast—The Honest Policy not suited to the Present Ideas of Railroad Men—The Art of building Railroads with other People's Money brought to Perfection—The Era of Mortgages—The Land Grab System—Demoralization in Railroad Finances—The Gamblers in Power—The Real Owners of the Railroads robbed by the Directors—A Rotten System and its Consequences—The Banks Involved—The Railroads demoralizing the whole Country—The New York Herald's Picture of the United States Senate—Food for Patriotic Reflection—Railroad Senators.

THE question is often asked, "Who are the real owners of a railroad?" At the outset of our railroad enterprises an answer would not have been difficult. Now, however, so entirely has the whole system been changed that no one can tell who is the actual owner of any road in the country.

In former times, men proposing to build a railroad began their enterprise by subscribing certain sums of money for the purpose of constructing and equipping the road. With the funds thus subscribed the road was actually begun, and certificates of indebtedness were issued and delivered to the subscribers. These certificates were called "stock," and represented the capital invested in the undertaking. The holders of the stock were really the owners of the road, and very properly elected its officers and managed its affairs, as it was their own property they were dealing with. If their subscriptions were insufficient to complete the undertaking, they mortgaged the road and issued bonds representing the indebtedness they thus assumed. These various steps represented legitimate transactions, and were fair and proper.

As the country grew in wealth and population, the railroad system grew with it, and, unfortunately, sundry elements of demoralization crept into the system. Men became wiser in their generation, and railroad managers were quick to improve upon the progress of the age. A new system of constructing roads was introduced. Railroad bonds had become so popular with the public that the corporations came to the conclusion that they could be put to a use never dreamed of by the originators of the earlier railroads. The new plan was to mortgage the road before it was built, and before its incorporators had subscribed a dollar towards its construction. With the proceeds of the mortgage bonds the road could be built, the public paying for it, and the incorporators being put to no expense. The holders of the bonds thus became the real owners of the road, but the actual possession was with the incorporators. The bonds being sold, and the road mortgaged, the stock was issued and divided among the incorporators. When it became valuable, that is after the road had been built at the expense of the bondholders, the incorporators could sell their shares, the entire proceeds of which were so much gain to them. This was financiering extraordinary, and it became so popular and so profitable that it entirely superseded the old-fashioned method by which the stockholders built their roads with their own money.

The next step in advance was to secure the land

grants, which have been wrung from the nation through the connivance of delinquent Congresses. These lands had a certain value, and were so much additional property to mortgage. They made it all the more unnecessary for the incorporators to provide money of their own, the new advance constituted a double imposition upon the public, an additional gain to the stockholders.

The receipts from the sales of bonds and of lands having built the road, and the stock having acquired a definite value, the shareholders began to see that their interest lay in selling it and realizing a profit upon that which cost them nothing. Indeed, the only interest they now have in the stock of their road is to sell it at an advance, and they thus inaugurate a series of speculations which lead to corners, a watering of the stock, and a steady depreciation of the value of the property. Boards of directors in sympathy with such operations are chosen, and the processes of watering and stock gambling are carried on until the mountain of debt is piled so high—for the road, not for the stockholders—that the corporation is unable to meet the interest on the original mortgage.

The condition of affairs is singular. The road is in the hands of the stockholders and their directors, who have paid nothing for it. They have exclusive control of it. They constitute a solid, compact body, with a fixed and definite purpose, and are usually under the leadership of some shrewd and able mind. The real owners of the road, the bondholders, with whose money it was built, constitute a vast multitude of small capitalists, farmers, men unaccustomed to deal with financial matters, widows, orphans, and others. Sometimes large quantities of the bonds are owned in foreign countries. These bondholders are at the mercy of the stockholders, who are steadily rendering the property worthless. The only redress of the bondholders is to foreclose the mortgage upon the road, but it is almost impossible to obtain concert of action among them.

The stockholders meanwhile use the stock to enrich themselves. It is gambled for, flung up and pulled down, and tossed about the Exchange until no one knows what its value will be twenty-four hours ahead. It is good only for purposes of gambling, and the road is left to take care of itself, and the bondholders to get what return they can for their money. So the dividends are paid on the stock of the road, and that commodity thus kept on a respectable footing in the stock market, the speculators care very little whether the bondholders receive their interest or not.

This system of constructing and managing railroads being false and unsound from the first, it is not surprising that it should lead to very grave complications in the monetary affairs of the country. Sound business men, capitalists whose large and honorable experience entitles them to a respectful hearing, have often warned the people of the Republic that this rotten system could not last, and that the efforts of the railroad managers to wring money from the people by their shameful speculations in the stock of their roads must at some time, sooner or later, result in a terrible financial disaster. Such a climax has been reached, and while these pages are passing through the press, the whole country is reeling from the effects of one of the severest convulsions of the century, brought about by the recklessness of a combination of managers of railroads and speculators in railroad stocks.

In 1868, the Comptroller of the Currency, in his annual report, thus referred to the danger which even then threatened the country from this cause:

"It is scarcely possible to avoid the inference that nearly one half of the available resources of the national banks in the city of New York are used in the operations of the stock and gold exchange; that they are loaned upon the security of stocks which are bought and sold largely on speculation, and which are manipulated by cliques and combinations, according as the bulls or bears are for the moment in the ascendency. . . . . Taking advantage of an active demand for money to move the crops West and South, shrewd operators form their combination to depress the market by 'locking up' money,—withdrawing all they can control or borrow from the common fund; money becomes scarce, the rate of interest advances, and stocks decline. The legitimate demand for money continues; and, fearful of trenching on their reserve, the banks are strained for means. They dare not call in their demand loans, for that would compel their customers to sell securities on a falling market, which would make matters worse. Habitually lending their means to the utmost limit of prudence, and their credit much beyond that limit, to brokers and speculators, they are powerless to afford relief;—their customers by the force of circumstances become their masters. The banks cannot hold back or withdraw from the dilemma in which their mode of doing business has placed them. They must carry the load to save their margins. A panic which should greatly reduce the price of securities would occasion serious, if not fatal, results to the banks most extensively engaged in such operations, and would produce a feeling of insecurity which would be very dangerous to the entire banking interest of the country."

The warning thus plainly uttered was not heeded. The same unsafe manner of doing business was continued, the banks and trust companies involving themselves deeper than ever in the operations of the railroad gamblers.

Numerous roads were planned, as the time passed on. Large quantities of the public lands were filched from the people with the connivance of Congress. The press of the country again and again uttered its protest against these misappropriations of the national property, but Congress, with characteristic contempt of the popular will, continued its land grants. It has been publicly declared that the corporations have so thoroughly bought up the National Legislature that the people have no chance of protection in their property when the masters of the Honorable Members demand its appropriation to their uses. The New York *Herald*, of September 22d, 1873, thus states the view of this question held by a very large and respectable portion of the American people, regardless of party feeling:

"Instead of checking the extravagances of a popular assembly, the Senate has taken the lead of every phase of extravagant legislation. It has become the fountain of jobbery and corruption—the source of land grants and dishonest reserve proposals. The country would be amazed to know how many Senators are the paid attorneys of railway and other corporations—attorneys paid to 'practise in the Supreme Court.' The country does know that the most notorious men in our public life are in the Senate, that Senators who went to Washington poor have become rich; and they have seen its members do spiteful and mean things, like the rejection of Mr. Hoar, one of the conspicuous men of the time, simply because he had been ill-tempered with the politicians when they came to bother him as Attorney General. After rejecting Mr. Hoar we can readily understand why it removed Mr. Sumner, the best informed man in public life on foreign affairs, from the committee, and gave the place to a gentleman who probably does not know whether the Danubian Principalities are in Europe or Asia Minor.

"The Senate is no longer a compact representative body. It does not represent even the States. In one State a Senator is chosen by the money of a railroad; in another by his own money. One Senator is known to be the agent of this interest; another as the agent of a second interest. No shrewd railroad manager will be without his Senator. We should not like to guess at the number on the books of Thomas A. Scott or T. C. Durant or Dick Franchot. We know who represents the Bank of California; we should like to know all who were owned by Jay Cooke and the Northern Pacific. The glory of the old Senate has departed, and we have some greedy, selfish cliques. There is a small but mainly a feeble class of respectable men, like Frelinghuysen and Edmunds and Anthony. Then comes the muscular, aggressive class, with Carpenter, Morton, Chandler; the moneyed class, like Cameron, Hamilton, Sprague, and Jones, and the drift of adventurers from the Southern States, from Florida and Alabama and South Carolina, who presume to sit in the seats and vote themselves back pay as the successors of John C. Calhoun, Felix Grundy, J. P. Benjamin, Robert Hunter, and John C. Breckinridge. We shall not needlessly write names, but the country knows well to whom we refer. It knows that there is no feature of this deplorable time more marked than the lowering of the Senate. When we consider the manner of men in the Senate, their overruling motives, their greed for money and patronage, their enmity to any measure that will limit their power, we cannot marvel that even Grant has surrendered. He could do nothing without the Senate, could not even remove an officer of his Cabinet. Of course he surrendered. He might have fought the Senate; but he saw how Johnson failed. It required more civic courage and foresight than Grant possesses to see that while Johnson wounded and assailed the country he had the country with him.

"The Senate fought Johnson and ended in dividing the patronage with him. Then it fought no longer. It menaced Grant until he threw Hoar and Cox into its shambles. Then it became acquiescent. As long as Grant strove to give tone and majesty to his administration and to elevate the public service, the Senate stood in his path, like the ominous giant who threatened the pilgrim Christian on his way to the land of Beulah and the gates of the house called Beautiful. To-day it is an independent power, composed largely of audacious men, representing the lowest strata in our political life, owing allegiance to railroads and tariff combinations and monopolies—caring nothing for the people whom it does not represent, and to whom it only owes a remote and contingent responsibility—warring upon the Executive until it was appeased with patronage, and then sinking into complete obedience to his will. The anomaly of a body of men making laws, confirming and vetoing appointments, taking a direct part in every measure of peace and war, and holding no responsibility to the people, is a scandal to free government and the prolific source of many of the evils which now distress and wound the Republic."

The Senate is not involved alone. In the popular estimation the lower house is equally guilty.

But be this as it may, the evil has been working until it has at length brought forth its legitimate consequences. One result, and that which is now attracting most attention, it will be interesting to notice.

CHAPTER X.

THE GREAT RAILROAD PANIC.

A Railroad Gamblers' Plot—The New York Gold Clique make War on the Farmers—The attempt to lock up Money—Trouble in the New York Stock Market—A Railroad the first to succumb—The Money Market on the 17th of September—Scene in the Stock Exchange—The Panic begins—Failure of Jay Cooke & Co.—Effect of the Failure—The Stock Market demoralized—Run on the Union Trust Company—More Suspensions—Worthless Railroad Bonds the Cause of the Trouble—Spread of the Panic throughout the Country—The United States Government offers Aid—Suspension of the Union Trust Company—A Railroad the Cause of the Trouble—The Stock Exchange closed—An Anxious Sunday—The Railroad Gamblers demand that the United States Treasury be opened to them—Firmness of the Government—The Panic subsides—Its Lessons—A Warning to the Country.

TOWARDS the last of August, or early in September, 1873, a combination of speculators in railroad stocks led, as is popularly believed, by the shrewd and unscrupulous capitalists who had so often and successfully manipulated the stock of the Erie road, undertook to inaugurate a series of movements in the stock market of New York, for the purpose of depressing certain stocks in which they proposed to operate, and of enabling them to extort from less successful operators a heavy interest for the use of money by creating an artificial stringency in the money market. Their time was, from their point of view, well chosen. In the months of September and October there is always a real scarcity of money in New York, owing to the demands of the country banks for funds to bring the year's harvest to market. By taking advantage of this season of real stringency, and increasing the want by the artificial means at their command, the gamblers expected to reap a rich return. It mattered nothing to them that the country at large would be called upon to suffer, and that values of all kinds would be seriously endangered by the success of their infamous scheme. They looked only to their prospective gains, and cared nothing for the community. The movement was one of unusual magnitude, and the scarcity of funds in New York proved to be greater than even the conspirators had believed. Wall street early took the alarm, and for ten or twelve days a feeling of general uneasiness pervaded the community.

The first sign of the coming storm manifested itself on Wednesday, the 17th of September. On that day the paper of the New York & Oswego Midland Railroad was protested. The news of this misfortune threw the stock market into a fever of excitement. The *Tribune*, in its money article, thus describes the state of the market on the 17th:

"There was a dreadful sweeping away of stock margins to-day, the depreciation covering the entire list, and showing a decline from the closing quotations of last night to the lowest points reached to-day, of from ½ to 7 per cent., and averaging about 2½ per cent. . . . . There will no doubt be a general overhauling of brokers' ledgers to-night, and the mails will go out freighted with letters calling for more margins. Of course, many may be unable or unwilling to respond, in which case their stocks will be forced upon the market and sold for what they will fetch, the tendency of which will be to still further depress prices. As was foreshadowed in this column a month ago, numerous causes have been operating in favor of the bears. Every failure of a moneyed institution, every defalcation, every protested note of magnitude gives additional impetus to the downward course of prices. Conservative men stand aloof, while so much danger surrounds the centres of business. The banks are holding their money fast for their own protection, and next for that of legitimate trade, that the whole business of the country may not be demoralized for the sake of a few wildcat railroads and wildcat bankers who lend their name or their cash by the million to companies that have no immediate resources. As we have before stated, it will require the utmost caution on the part of our leading capitalists and heavy security owners to avert a panic, and perhaps a crash like that of 1857. The minds of

NEW YORK STOCK EXCHANGE.

capitalists and operators are surcharged with distrust, and the air of Wall street with rumors, started generally in the bear interest, of failures, defalcation, and of disasters dire, which renders it all the more necessary that the cool and clear heads should come to the front and take the direction of affairs. The street, already in a tremor of excitement from bygone troubles, found a new source of anxiety to-day when it became known that the New York & Oswego Midland Railway Company had come to grief, its paper having been dishonored, a more particular account of which will be found in another column. This information caused a wild and eager desire to get rid of stocks, and holders rushed their shares on to the market, bound to realize at whatever hazard of loss; contented, apparently, so long as they could see that their securities represented something in the way of value; and the result was such a tumbling in prices as we have already described. Although a better feeling and higher prices were established toward the close, as some of the shorts began to cover, yet the general market left off feverish and demoralized."

The fears of the moneyed men were not without foundation. Thursday, the 18th, brought only fresh trouble. The market opened in a state of great excitement. The floor of the large hall of the Stock Exchange was filled to overflowing by those possessing its privileges, and the gallery was so crowded that misgivings were entertained for its safety. The stairs and lobbies of the building were full of people, and in Broad street, in front of the Exchange, a large crowd had collected, eagerly and anxiously awaiting the events of the day.

The business of the Exchange began at ten o'clock, with a rapidly-falling market. The bears had it all their own way. Men had begun to lose confidence, and stocks fell with great rapidity. The Exchange was in a whirl of excitement. Towards noon the noise and confusion, which had reached a bewildering state, were suddenly brought to a pause by the sharp raps of the gavel of the presiding officer. He advanced to the front of his platform, and instantly the hall was as still as death. His face indicated that he had evil tidings to communicate, and the most daring operator present awaited in anxious suspense the announcement he had to make. In a few short, crisp words, every one of which struck upon the ears of the listeners like the blows of a sledge-hammer, the President stated that Jay Cooke & Co. had suspended payment, and a little later it was announced that the Philadelphia and Washington houses of this great firm, and the First National Bank of Washington, which was intimately connected with them, had also suspended.

These announcements completed the demoralization of the market. Stocks fell even lower, and were sacrificed with remorseless fury. The Exchange seemed like a bedlam, and outside a panic was rapidly extending through the street.

Messrs. Jay Cooke & Co. were known throughout the entire country as one of the first, if not the first, banking firms in the Union. They had been identified with the great loans of the General Government during the civil war, and by their judicious and vigorous management of them had rendered the country a genuine service, and had earned what may be termed a national reputation. They were popularly regarded as among

THE PRESIDENT OF THE NEW YORKSTOCK EXCHANGE ANNOUNCING THE SUSPENSION OF JAY COOKE & CO.

the wealthiest bankers in the Union. Upon the inauguration of the Northern Pacific Railroad they had undertaken to dispose of its bonds, and it may be said that all the confidence entertained in that scheme by the public was due to Messrs. Cooke & Co.'s endorsement of it.

The news of the suspension of this great house struck men with terror. They began to ask who was safe if the pressure was so great as to break down the foremost house in the country. The explanation of the firm that their suspension was due to their having made heavy advances to the Northern Pacific Road for bonds which they had not been able to dispose of, only increased the panic. Other houses were known to be deeply interested in the bonds of new railroad enterprises, and it was by no means sure that Jay Cooke & Co. would be the only sufferers.

A little later in the day, the failure of another house was announced, and still later it was stated officially that Richard Schell, a well-known capitalist, had failed. Mr. Schell was a director in the Union Trust Company, a leading banking corporation of the city, and suspicion at once attached itself to the company. The Union Trust Company was also known to be involved to a considerable extent in the affairs of the Lake Shore & Michigan Southern Railroad. The stock of this road had suffered heavily during the panic, and it was feared that the credit of the Trust Company would be impaired by these losses. The result was an immediate run upon the Trust Company, which lasted until the close of the business hours of the afternoon. Large sums were paid out to the alarmed depositors, but many were left unpaid when the hour for closing the doors arrived.

The morning of the 19th found matters in a most unhappy state. No remedy had yet been found for the trouble. The excitement was intense. The run on the Union Trust Company continued, and during the day immense sums were paid out to the depositors, who had now become alarmed in good earnest. But during the day all demands were met, and the directors of the company entertained the hope of successfully passing through their troubles.

In Wall street the same feverish anxiety prevailed. There had been such a terrible sacrifice of stocks on the previous day—some of the best of these securities having fallen as much as ten per cent. in value—that it was evident that the weaker houses must yield to the pressure upon them and close their doors.

Soon after the opening of the business of the day, the street reeled and staggered once more under as heavy a blow as had yet been struck it. The President of the Stock Exchange announced to the excited throng before him that Messrs. Fisk & Hatch had suspended payment. Now Fisk & Hatch were one of the most trusted and respected firms in the city, and their failure was second only to that of Jay Cooke & Co. It had been brought about by a similar cause—heavy advances to a railroad (the Chesapeake & Ohio road in this case) upon bonds which they had been unable to negotiate in time to meet their other obligations. The failures of eighteen other firms of greater or less prominence followed in quick succession, and when the closing hour arrived the confidence of the boldest operator had entirely departed. Men were bewildered. They knew not whom to trust; scarcely what to do.

Until this day the panic had been confined to New

York, but it now began to involve other cities in its effects. With the exception of the houses of Jay Cooke, no firm outside of New York had suffered; but on this day the effects of the crisis began to be felt elsewhere. In Philadelphia eleven suspensions were announced.

The only hopeful sign visible on Friday night was the offer of the United States Government to buy $10,000,000 of its bonds; but it was feared that this would not prove an effectual remedy.

Saturday, the 20th, witnessed no improvement in affairs as the day opened. When the hour of ten o'clock A. M. struck, the doors of the Union and National Trust Companies remained closed. Both had suspended payment.

The Union Trust Company had borne up bravely against the heavy pressure upon it on the previous days, but on Friday afternoon it had become evident to the directors that unless more funds could be had at once, the institution could not resume business on Saturday; and at the same time it was discovered that the Secretary of the company had disappeared with securities of the company in his possession amounting, it was said, to half a million of dollars. This was a terrible blow to the company, but it would not have caused its suspension in ordinary times. The chief cause of the trouble was an advance of $1,750,000 to the Lake Shore Railroad Company, to enable that corporation to pay a dividend of four per cent. in August, 1873. The total sum voted in this instance as a dividend was $2,000,000. It was stated at the time of the suspension that this dividend was unearned, and that the road had but $250,000 of the entire amount in hand, and was forced

PARK BANK, NEW YORK—THE FINEST BANK BUILDING IN THE UNITED STATES.

to borrow the remaining $1,750,000, which it obtained of the Trust Company. Horace F. Clark, now dead, was then the President of the Lake Shore road, and also of the Union Trust Company. Hence the loan from this company. The company neglected its legitimate business to aid the railroad in the matter of its dividends. When its troubles came upon it, it called upon the road to make good its loan; but the request was unheeded, and the company was forced to close its doors.

During the day eleven other firms suspended payment; and the National Bank of the Commonwealth was also compelled to close its doors.

Up to this time, with the exception of the Trust Companies and the Commonwealth Bank, the panic had been confined to the stock market; but the suspension of the Union and National Trust Companies and the Bank of the Commonwealth produced a serious fear that the banks of the city might become generally involved. This danger was averted, however, by the refusal of the general business community to be frightened by the panic; and with the exception of a slight run upon the Fourth National Bank, none of the banks of the city were subjected to any unusual demands. On Saturday, however, with a view to preparing for the crisis, if it should extend to them, there was a meeting of the Bank Presidents, at which it was resolved to pool the assets of their respective institutions, and to assist each other to the utmost by the issuing of temporary loan certificates, which could be used in case of need. This determined action increased the confidence of the public, and did much to avert the danger which at one time seemed imminent.

At noon on Saturday, so utterly unmanageable had the market become, and so general was the prospect of ruin among the dealers in stocks, that the Governing Committee of the Stock Exchange decided to close the doors of that institution until calmness could be restored to the street and some measures of relief devised. This action put an end to all transactions in stocks, and by thus protecting the dealers from greater losses aided in arresting the panic.

During the day the Union Banking Company of Philadelphia and other houses in various parts of the country announced their suspension.

Sunday, the 21st, was passed in feverish excitement. The President and the Secretary of the Treasury of the United States, having been summoned to New York, arrived in that city on Saturday night. On Sunday, they were waited upon by many of the capitalists of the city, and various measures for relief were urged upon them. The brokers united in a demand that the President should lend the whole or the greater part of the Treasury reserve of $44,000,000 of greenbacks to the banks, to furnish Wall street with funds for the resumption of its business and the settlement of its losses. The President promptly and properly declined to take so grave a step, as he had no warrant of law for such action; and, thanks to his firmness, the credit of the United States was not placed at the mercy of the railroad gamblers. On Sunday night it was officially announced that the Treasury would purchase any amount of five-twenty bonds that might be offered, and would also buy the issue of six per cent. bonds commonly known as '81's. This included the currency sixes, the ten-forties, and the new fives. The price was to be par in gold and accrued interest. In case the Stock and Gold Exchanges were not open on Monday, the ruling price on the street was to be taken as the market price.

By Monday morning, the 22d, the panic had begun

to subside. The banks were unshaken; the Government stood ready to prevent the sacrifice of United States bonds in the hands of those who desired to sell; and the Stock Exchange remained closed during the day. By Monday night the panic was substantially at an end. The trouble was not over, and the remainder of the week witnessed the suspension of several good houses, among them the firms of Howes & Macey and Henry Clews & Co. But the worst was passed. People began to take courage, and it was evident that the wild storm that had swept the money market so ruthlessly had spent its force.

It was clearly understood from the beginning that the affair was purely and simply a railroad panic, and it was hoped that it would be confined to the dealers in the bonds and stocks of railroads; but this hope was not destined to be realized. The reckless and unsound management of the railroad enterprises of the country had paved the way for the panic, and had made it possible; and the greed of the railroad gamblers had precipitated it; but they were not to be the only sufferers. The whole country was to be involved in it, and business of all kinds was to suffer from the effects of the great scare.

It was not alone the great scarcity of money, at a time when the free circulation of the currency of the country was of vital importance, that did harm; but the public confidence in financial enterprises of all kinds was shattered, and men fortunate enough to possess ready money held on to it, refusing to part with it. The movement of the crops was almost entirely stopped. The farmers found it well nigh impossible to procure money for their produce, and in spite of the enormous

OMAHA—EASTERN TERMINUS OF UNION PACIFIC RAILROAD.

European demand for American breadstuffs, it was with the greatest difficulty that shipments of grain could be made, so hard was it to negotiate exchange in the uncertainty which prevailed in the money market. The keenest financial distress prevailed for weeks after the panic had subsided, and all classes were severe sufferers. Nor has it subsided yet, and while these pages are passing through the press a general feeling of uneasiness prevails throughout the country, and men are fearing that something worse than has yet happened may be in store for us.

The panic revealed clearly the extent of the power of a few determined and reckless gamblers in railroad securities to mar the business of the entire country. At a period of unexampled prosperity, with a prosperous trade, a bountiful harvest, and an unequalled demand from abroad for our products, the country was thrown into a fever of alarm, and business of all kinds dealt a severe blow by the machinations of a few gamblers in the stock market of the principal city of the United States.

The panic also showed the extent to which railroad gambling had demoralized the business and people of the country. It showed that some of the strongest and most trusted houses of the country had lent themselves to the task of inducing people to invest their means in the securities of railroads, the success of which was doubtful, to say the least. It showed that the banks, the depositories of the people's money, had to an alarming extent crippled themselves by neglecting their legitimate business and making advances on securities which proved very uncertain, if not worthless, in the hour of trial. The money needed for the legitimate business of the country had been placed at the mercy of the railroad gamblers, and had been used by them. The funds of helpless and dependent persons, of widows and orphan children, had been used to pay fictitious dividends and advance schemes in which the people had no confidence. An amount of recklessness and demoralization in the financial interests of the country was revealed that startled the most hardened; and in all of it the hand of the railroad gambler could be clearly and unmistakably traced.

The lesson has been severe, but it was needed. The people of this country now see what recklessness and greed in the management of our railroad interests can do, and what they dare attempt; and the nation will richly merit all the evil that will come to it if it does not profit by the lesson.

CHAPTER XI.

WILD CAT RAILROADS.

False Assertions respecting Railroad Property—Railroad Building a profitable Work—Useless Railroads—Why they are built—Theory of Wild Cat Railroad Constructors—Forming the Company—A Specimen Enterprise—A Share of the Public Lands—How to raise Money to build a Railroad—Disposing of the Bonds—Where the Money comes from—"Judicious Advertising"—Bribing the Press—The Road in Operation—What becomes of the Stock—Where the Profit lies—The Crime of the Bankers—A Confidence Game—How to stop Wild Cat Railroad Building.

ONE hears a great deal now-a-days of the risk assumed by men who undertake the construction of a railway; and we are told, with a great array of figures in support of the assertion, that railway property is among the least profitable of all the investments open to capitalists. But, nevertheless, the work of building roads goes on, and it is a fact that the incorporators of these enterprises, in spite of their assertions respecting their risk and the uncertainty of their investments, make large sums out of their connection with their respective schemes.

The truth is that railroad making is a very profitable undertaking to men who understand their business; and it is for this reason that so many useless roads are built. That railroads are a necessity to the community no one will deny; but it is a fact that a very large number of the roads constructed during the last four years, or at present under construction, are useless.

Many of them are built through sections of country which cannot yet support them, and which will be unable to do so for many years to come; and others are located in regions where in strict truth they are not wanted. These roads must of necessity languish for a considerable period, if, indeed, they are ever profitable; and yet they constitute a majority of the roads now being built.

The reader will then ask why are such roads, to which the term "wild cat" has been popularly, and not inaptly, applied, ever entered upon? The reason will be apparent when we have investigated the matter.

In order that the originators of a railroad scheme may make money out of it, it is not necessary that there shall be a real demand for the road by the community in which it is to be located. That is a consideration that does not enter into the scheme.

A number of shrewd men, with an ample supply of "brass," and very little money often, combine for the purpose of constructing a road from a certain point in Missouri, let us say, to a certain point in Kansas. Their object is purely and simply to make money. The country through which the road is to be located is new and unsettled. There is no trade there to make the road profitable when constructed, and it is evident that many years must elapse before it will possess a population large enough, or business interests sufficient, to enable the road to do a paying business. These facts are well known to the originators of the road, but they are of no consequence. They are looking to immediate and not to ultimate profits.

The first step after locating the road is to obtain the necessary charters and a share of the public land. We

SCENE ON THE TRUCKEE—CENTRAL PACIFIC RAILROAD.

have seen how easy it is for the roads to manage the land grab system, and how docile the Congress of the United States can be at the bidding of a railroad director.

The charters are obtained, the lands are "donated," but the treasury of the road is empty. The work cannot be done without money. That must be had, and the incorporators of the scheme have no idea of advancing their own funds for this purpose. They have no intention of building the road with their own money. Their plan is to build it at the cost of the public, or with other people's money. In order to accomplish this, they proceed to issue mortgage bonds for an amount sufficient to pay the cost of construction, pledging the lands given to them by Congress as security. These bonds are to be sold, and the road built with the money thus received.

The sale of the bonds is the only really difficult part of the undertaking, and this is managed with consummate ability. Arrangements are made with some prominent banking-house in one of the principal cities of the country for the sale of the bonds, a heavy premium being paid to said house for its services in negotiating the loan. Or it may be that the banker will advance the company a certain sum for the commencement of the work, and receive, as security, a sufficient number of the bonds, which he is to sell at the highest price, and at a large profit for himself.

None of the great banking firms are able to hold the securities thus placed in their hands, nor do they take them for purposes of investment. They buy them to sell again, or sell them on commission. The persons who are expected to buy the bonds for investment are private individuals, who take them as safe investments paying a large interest; and the difficulty of the whole matter lies in persuading the public that the bonds are a safe investment, and that the interest will be promptly paid. That is the task of the banker.

The process is as follows: The bonds are advertised in the most prominent newspapers, and large sums are thus expended which must be made good out of the money paid by the unsuspecting public. The papers advertising these bonds are often induced to recommend them, or, in plainer words, to "puff" them in their editorial columns, editorial virtue being overcome by the temptation of a large advertising contract. The religious papers are a favorite medium for the advertisement of wild-cat railroad bonds, and it is a singular fact that the most enthusiastic endorsements of these securities have been from this source. In effect, the press is bribed to puff the securities of the road. The banking-house negotiating the loan pledges its faith to the public that the bonds are a valuable commodity, and it is principally upon its representations that the bonds are sold. So common has this practice become, that bankers seem to lose sight of the responsibility they incur towards the public in thus recommending doubtful securities.

Well, the bonds are sold. Who are the purchasers? Not the great capitalists, who are shrewd enough to see through the whole transaction, but people of moderate means, persons who cannot afford to take risks or to lose any of their income, and who buy the bonds because of their confidence in the representations of the house disposing of them.

With the money received for the bonds the road is built, and put in operation. The stock is then issued and divided among the originators of the scheme. The people have paid the cost of the road, and also the large commissions paid to the bankers who sold the bonds, and as they hold a mortgage on the road, they are its real owners. But the stock, which represents the ownership of the property, is divided among the originators of the scheme, who are also charged with its management. It has not cost them anything, but now that the road is built and in running order, it is a valuable commodity. The holders of it are shrewd men,

WILD-CAT RAILROAD LANDS.

and, foreseeing that the road will not be able to pay the interest on its bonds for some time, take advantage of the *éclat* which attaches to all new enterprises, and sell their stock at the earliest possible moment. The amount received is clear gain to them, for the stock has cost them nothing. Should it sell for only fifty cents on the dollar, it is fifty cents clear profit, for they paid nothing for it. The people built the road, and the stock is the reward of the originators of the scheme. Herein lies the profit of railroad building. What does the incorpo-

VIEW OF THE COUNTRY TO BE OPENED BY THE SOUTHERN PACIFIC RAILROAD.

rator care for the future of the road, so long as he is safe? What does he care whether the interest on its bonds is paid or not? The men who advanced the money to build the road are nothing to him. He has received and sold his stock, and has pocketed the money, and the future of the road is a matter of indifference to him.

The case would be very different were the road constructed at his expense. He would be very careful to locate it in the most favor-

IN THE TUNNEL—SIERRA NEVADA. CENTRAL PACIFIC RAILROAD.

able section, and only where it is actually needed; but since it costs him nothing, and affords him the means of obtaining a large sum of money for his stock, it matters little what becomes of the road after the stock is sold.

And so, year after year, the practice of building useless roads, and roads of doubtful profit, goes on. The country is flooded with worthless railroad bonds, which are advertised and puffed by bankers and newspapers, and which unwary people purchase, only to find their investment entirely the reverse of what was represented to them by the firm disposing of the bonds. The legitimate business of the country is crippled by the withdrawal of the sums that go to purchase these wretched securities, and an element of unsoundness is introduced into the finances of the country that may yet involve us in a disaster infinitely worse than the panic from which we have just emerged.

Unquestionably the people are to blame for much of the evil that is upon us, for they are free to refuse to purchase these securities when placed in the market; but they are not altogether to blame. There are always numbers of people of moderate means who are very naturally and properly seeking some simple and safe investment for their money, and they are not fitted to judge very accurately of the value of the different schemes proposed to them. They must of necessity depend upon the representations of the house offering the securities for sale, and it is but natural that when a prominent house, enjoying the confidence of the public, presents such securities, and endorses them as both safe and profitable, such securities should be eagerly taken by the class we have described. The banker who undertakes a negotiation of this kind assumes a responsibility to the purchasers of the bonds, of which no amount of "Wall street logic" can relieve him. It is a melancholy fact that one of the first banking houses in the country, men whose names have hitherto commanded the confidence of the public, has been among the foremost in flooding the market with wild-cat railroad securities. The recent panic revealed some ugly facts of this kind, and the people of the United States, it is to be hoped, will not be unmindful of them.

The people owe it to themselves to put a stop to this reckless system of railroad-building, which is simply gambling under another form. They should refuse to countenance the securities of such roads as those which have produced our recent troubles; and, above all, should demand of Congress, and see that the demand is complied with, that there be no more subsidies of the public lands. The national domain should no longer be at the mercy of railroad gamblers. Let the land-grab system be once overturned, and we shall have more caution in the construction of new railroads. Let it be understood that the people will no longer pay for roads which are built only for the profit of the incorporators, and the era of Wild-Cat Railroads will have passed away forever.

CHAPTER XII.

THE CASE OF THE NORTHERN PACIFIC RAILROAD.

The Road chartered by Congress—An Imperial Gift of Land—The Nation robbed of Fifty Millions of Acres—Route of the Road—Character of the Country through which the Road is to be constructed—A Wilderness—Popular Doubts respecting the Success of the Road—The Capital of the Company—How it was to be raised—The People to pay for the Road—The Stock-Holders to receive all the Profits—The Bonds of the Northern Pacific Railroad declined in Europe—A "Popular Loan" inaugurated—Jay Cooke & Co. undertake its Negotiation—A Terrible Blunder—The Loan does not command the Public Confidence—The True Character of the Scheme—What Might Have Been—The Sequel—Report of the German Commissioners—A Capitalist's View of the Scheme—The Risks too great to warrant the Investment of German Capital—A Remarkable Statement of the Character and Prospects of the Northern Pacific Railroad.

The Northern Pacific Railroad has lately been brought very prominently before the public by the failure of the great house that went down weighted with the bonds of this road, which it could not sell. It will be well to examine its history. It will be found full of food for reflection, and overrunning with instruction.

In 1864, Congress granted a charter to the Northern Pacific Company, and by this charter and subsequent acts authorized this company to build a railroad from Lake Superior, through the State of Minnesota, and the Territories of Dakota, Montana, Idaho, and Washington, to Puget Sound, by the valley of the Columbia river, through Portland, in the State of Oregon. In

DULUTH—EASTERN TERMINUS OF NORTHERN PACIFIC RAILROAD.

aid of this road, Congress made large grants of land, the amount now being about 50,000,000 acres.

The charter was granted during the last year of the war, and matters were too unsettled then to allow the company to commence the work at once, and it was not until long after the close of the Rebellion that the construction of the road was fairly begun.

It was proposed to construct this road through the most northern portion of the United States, and from Lake Superior to the Pacific, a distance of 2000 miles. The expense of the undertaking was enormous, and the road was to be built through a section of country that was simply a wilderness. There were scarcely any settlements along its line, a great portion of which lay through the territory of hostile Indians. Much of the region through which it was to pass was barren and unfit for settlements, and a large part of the proposed route lay through the sterile region of the Yel-

14

lowstone. The entire route lay through the extreme northern portion of the Republic, a country which, it was popularly believed, would long remain unsettled, by reason of the severity of the climate and the inhospitable nature of the country. The best informed men expressed grave doubts of the practicability of the scheme. They did not believe that this region would be sufficiently settled to warrant the construction of such a railroad for many years, and they based this belief upon the fact that the region offered scarcely any inducements to settlers. Consequently people regarded the road with doubt, and when its bonds were offered, held aloof from them.

By the terms of the company's charter, a share capital of $100,000,000 was authorized; but of this amount only $2,000,000 was required to be subscribed in advance, and but $200,000 to be paid in. The last-named sum perhaps covered the preliminary expenses of the scheme, such as the cost of surveys, of legislation, and such other operations as were necessary for the commencement of the enterprise. The cost of building the road was to be paid by the people. Congress had given a criminally large area of land to the company, and the proceeds of the bonds which were issued were to constitute the capital with which the road was to be built. The $200,000, subscribed and paid in by the stockholders, was the only contribution they seem to have expected to make to the road. "This was a slender provision, it would seem, for a road 2000 miles long, through an uninhabited country, without commerce at either terminus, and without an important town on its whole route. But the projectors intended that Congress should build the road and put them in possession of it.

They secured a grant of nearly 50,000,000 acres of public land, and on the security of this magnificent estate they proposed to negotiate a loan of $100,000,000. As the estimated cost of the road was only $85,000,000, this loan would pay for the whole work and leave a handsome surplus for contingencies. The land is now worth, say $125,000,000, or perhaps more; the portions thus far sold have brought, on an average, over five dollars an acre. As the country becomes developed it will of course rise in value; and it was calculated that the sales would be sufficient to pay whatever of the interest on the bonds the road might fail to earn, and to pay the principal likewise at maturity. Anything that remained would be the property of the stockholders."

When the bonds were issued, they were offered in Europe, but were declined. European capitalists considered the risks assumed by the road too great to render the bonds a safe investment, and they declined to have anything to do with them. The company, thus driven back upon a home market, resolved to make the people of the United States pay for the road in another sense. They made a popular loan of their scheme. They succeeded in enlisting the house of Jay Cooke & Co. in it, and Messrs. Cooke & Co. agreed to place the loan in the market, using in its behalf much the same system that they had found so successful in their management of the great war loans of the General Government.

The people of the United States were somewhat surprised when they found Messrs. Cooke & Co. in charge of the Northern Pacific loan, and there were many that did not hesitate to assert their belief that the Cookes had committed a serious error in connecting themselves with the scheme.

The truth is, the Cookes had committed a very serious error in undertaking the management of this loan. The Northern Pacific scheme never possessed the confidence of the people of the country, and Jay Cooke & Co. undertook too much in endeavoring to carry it through. Their judgment may have been, and doubtless was, satisfied, but the people were not so easily deceived. They did not see the necessity for the road, in the first place, nor could they understand how the road was to earn the money needed to pay the interest on its loan after discharging its ordinary expenses.

The loan was extensively advertised, and the more complaisant section of the newspaper press puffed it liberally. It was declared to be equal to any of the loans of the United States. The bonds were pronounced by some papers better than Five-Twenties. The advantges of the scheme, as they appeared to those in charge of it, were set forth in glowing terms, and the Cookes lent the whole force of their reputation and their great popularity to the task of popularizing the loan. But without avail. Many there were who purchased the bonds, and thus placed their funds at the mercy of the road, but the loan, though it absorbed immense sums, never became popular.

The mass of the people could not forget that this loan was in behalf of a road that was twenty years in advance of the demand for it. They could not forget that the road was to be constructed through an uninhabited region, much of which never would be fit for the dwellings of human beings; and much of which was exposed to the fury of hostile savages, who would

PUGET SOUND. PROPOSED WESTERN TERMINUS OF THE NORTHERN PACIFIC RAILROAD.

resist any effort to settle the region. There was no local business to furnish the road with ready money; and the through business, it was clear to thinking men, would amount to nothing; for whatever Duluth or Puget Sound might offer in years to come in the way of inducements to commerce, their advantages at present were too insignificant to be taken into serious consideration. Messrs. Jay Cooke & Co. assumed a most serious responsibility in venturing to assure the purchasers of the bonds that the interest on them would be paid. The amount of money necessary for this purpose could be derived only from the earnings of the road, and it was clear to men of as profound knowledge and as great financial skill as the Cookes, that it would be impossible for the road, for the first years of its existence at least, to earn this amount.

It is true that matters *might be* different; true that the country along the road *might be* settled with a rapidity that would surpass all previous western growths; true that the hostile Sioux *might* offer no resistance; true that there *might* spring up along the line a local and a through business that might enable the company to earn $20,000,000 a year, the amount needed for their wants; but all these things were, and still are, uncertain. They might be, but it was very improbable. There was a terrible doubt hanging over the whole matter. The road could pay its interest and other expenses only in the event of an unusually brilliant success, but the presumption was against it. The risk was too great. People shrank from the loan, and the Cookes, who had made heavy advances to the road, found themselves encumbered with a mass of bonds which they could not sell, and when the first serious disturbance of the market arose, they were among the first to sink under the weight of their unpopular loan. The result was in accordance with the general opinion of men who knew the true nature of the scheme, and it was a blessing to the country at large, however unfortunate it may have been to a few individuals.

"If the bonds had been duly negotiated according to programme," says the New York *Tribune*, "the case would have stood just thus: A few speculators would have subscribed $200,000 and persuaded Congress to build a railroad for them worth fifty times that amount out of the national estate. In a short time they would get back their original investment in dividends. Then they would be the absolute owners of 2000 miles of road, for which they had paid nothing, and probably they would still have also a large quantity of unsold land to divide among themselves. Whether we should have had a repetition of the Crédit Mobilier Building-Rings, and a rapid absorption of the profits and estate of the company by a little coterie of inside managers, railway Congressmen, and Christian statesmen, we leave our readers to conjecture.

"The bonds were offered in Europe and declined. Then the house of Jay Cooke & Co. undertook to place them among the multitude as a popular investment—in other words, to persuade the middle classes to advance the money which the nation was ultimately to repay with interest.

"Probably it was a combination of accidents, rather than any intrinsic defect in the arrangements, which brought this scheme to grief. Similar methods have succeeded before, and, if we are not cautious, will be attempted again. But what we particularly wish to call attention to is the bearing of this case upon the transportation question. If the Northern Pacific Railroad were in running order to-day from Duluth to Puget Sound, the directors would undoubtedly claim the right to put their tolls high enough to yield eight or ten per cent. on the cost of the work. But the cost of the work would have been paid wholly, or almost wholly, out of the national estate. Congress has made a grant rich enough to cover the entire expense, and leave the stockholders handsomely provided for likewise. For more than twenty years the Government has given away land in reckless prodigality to aid in the construction of railways.

"In 1871, the total amount of the public domain thus appropriated reached the stupendous total of 217,847,375 acres. It is true that a large part of this grant will prove inoperative, as the quantity of vacant land within the designated limits will fall short of the appropriation; but probably over 100,000,000 acres has been or will be deeded to the favored companies. These concessions represent, at the very lowest computation, a money value of $300,000,000, and an area considerably greater than the whole of the British Isles, and greater than New York, New Jersey, Pennsylvania, and Illinois combined. But when it is proposed that the roads for which the nation has done so much should be required to do something for the people, we are met with the objection, 'Oh, you must not interfere with the vested rights of railroad stockholders.'

"Whatever may be said of the railroad question generally, it must be evident that the land-grant roads hold a peculiar relation toward the people, and may

THE FALLS OF THE YELLOWSTONE, ON THE ROUTE OF THE NORTHERN PACIFIC RAILROAD.

fairly be compelled to do much more than roads built entirely by private capital. 'Reasonable' profits on roads of the former class cannot be measured by a percentage on their cost, because their cost was defrayed in a great degree, if not entirely, out of the public fund. It cannot be measured by the capital, because that is largely fictitious. The limit of their right to take toll must be fixed by a general review of all the circumstances of the roads and the history of their construction; and common sense will demand that in coming to a determination on this point the share which the public took in building them shall be fully considered."

When the bonds of the Northern Pacific road were offered in the European market, the German capitalists sent two commissions of experts to this country to examine into the affairs of the road, and upon receiving the reports of these commissioners, they declined to take part in the loan. One of these reports, that of Herr Haas, of Berlin, has been recently published by the New York *Tribune*. We give it as sustaining our view of the matter, and as a queer commentary upon the assertions and reassertions with which our press has overflowed of late years, that this was the safest and most profitable loan of the day.

After describing the location of the road, Herr Haas says: "Accordingly the estimates prepared for building the Northern Pacific Railway are limited to the present project, the cost of the branch line not being included.

ESTIMATES OF COST.

"7. These estimates calculate the cost of construction of the main line as follows:

1. Grading, masonry, bridges, rails, and entire surface works of the line	$60,320,000
2. Sidings	4,200,000
3. Sundry expenses inclusive of engineering	5,000,000
4. Telegraph lines	600,000
5. Buildings	2,312,000
6. Working capital	3,615,000
7. Small branch line	1,200,000
8. Extra expenses	800,000
9. Interest on capital during construction, minus the income derived from the working of already finished lengths during that time	7,230,000
Total	$85,277,000

"8. To raise this sum the Northern Pacific Railway Company intends issuing bonds to the amount of $100-000,000, and pledges itself to pay interest at the rate of 7.30 per cent. per annum in gold out of the surplus revenue from the traffic of the line, and to redeem the bonds within thirty years.

"9. As security for the payment of interest and the redemption of the bonds, the whole of the property of the Northern Pacific Railway Company, the line and buildings, as well as the land grants, have been made over to the trustees, as representing the bondholders, by a general mortgage deed, registered July 1st, 1870, in the office of the Secretary of the Interior of the United States.

"By these data the extent and aims of the Northern Pacific Railway enterprise, and the means by which it is to be accomplished, are clearly set forth. Adding to this the fact that up to August, 1871, a length of line of 140 miles, extending from Duluth, Lake Superior, to twenty miles beyond the Mississippi, was already completed and in working order; that 120 miles additional, as far as the border of the State of Dakota, are, save little interruptions, almost complete, and that lastly, in the Western division of the road, in the vicinity of Portland, Oregon, twenty-five miles of the line, through Washington Territory, in the direction of Puget Sound, are so far advanced that they can be opened for traffic by the end of this year, we have all that can be stated about the present condition of the Northern Pacific Railway, and we can turn to the consideration of the future.

"10. The consideration of the future, in order to keep within bounds, should be limited to answering the following three questions: *First:* Are the means provided for the construction of the Northern Pacific Railway adequate to complete the line ready for traffic? *Second:* Does the finished line offer the necessary guarantee that the net profit of its income will yield the sums required for the half yearly payment of the stipulated rate of interest on the bonds? *Third:* Will the sums realized from the sale of lands suffice to redeem the bonds within thirty years?

"11. Respecting the first of these questions, whether the building capital is adequate to the completion of the line, we must revert to the detailed estimates. The first item of these estimates, which provides $60,320,000 for the construction of the line, or $30,000 a mile, leaves no cause for uneasiness, inasmuch as the contracts already disposed of afford proof that the lengths contracted for can not only be completed for the amount, but that savings are made so considerable that by means of them the more expensive mountainous parts can be undertaken. The chief engineer of the company is a well-tried man, his honesty, experience, and capacity are beyond question, and he has positively declared that the item in question will not be exceeded.

"12. As little can be said against items 2, 3, 4, and 7; but 5, 6, and 9 produce serious misgivings. Item 5 of the estimate, providing $2,312,000 for buildings, includes $850,000 for repair-shops for machinery and cars, 134 stations at $2000 each, or $268,000; lastly ten principal stations at $25,000 each, or $250,000. These figures are out of all proportion low; for a length of line of 2000 miles, workshops at the collective amount of $850,000, stations the whole arrangements of which are put down at $2000 only, and principal stations at $25,000 each, cannot be looked upon as adequate to the requirements. This item, therefore, will have to be increased. The same remark applies to item 6, which provides $3,615,000 as the working capital, and out of this are to be procured 120 locomotives, 100 first-class passenger cars, 50 second-class passenger cars, 30 smoking cars, 30 mail and baggage cars, and 1500 freight and cattle cars. This working capital is so small, and stands in such glaring contrast with the length of line, that much more will be required than has been provided by the estimates. In North Germany a line of similar length would require more than 25,000,000 thalers; and though it may not be quite fair to measure American expenses by a German standard—a maxim which underlies this report—still the most superficial critic must perceive that here a very considerable augmentation is needed. With regard to item 9, providing $7,230,000 for the payment of interest during construction, a similar claim will have to be put forward. Suppose the time of building to be four years—a supposition based upon exact information obtained—the works would consume a quarter of the building capital in every year during the building period. This interest will have to be paid at the end of the first year upon one-quarter; at the end of the second year upon one-half: at the end of the third year upon three-quarters, and at the end of the fourth year upon the whole of the building capital. Hence the interest during the four years will be two and a half times the annual interest.

"13. With an emission of bonds to the amount of $100,000,000, at a rate of interest of 7.30 gold, the interest during the building period will foot up $18,250,-000 gold, and though the building fund need not be burdened with that, since the money obtained for bonds and not immediately required for building purposes may be otherwise employed with advantage, and since also the intermediate finished portions of the line will yield a revenue before the whole line is opened, nevertheless, what may be gained in this way must not be overestimated. The estimate of $3,250,000 is sufficiently high, so that $15,000,000 interest will have to be paid out of the capital. Item 9, therefore, will have to be increased in proportion.

"14. If, according to these calculations, the estimates require manifold augmentations on the one hand, we must not lose sight of the fact that on the other hand the fixed building capital, $100,000, 000, is not reached by the sum total of $85,000,000 of the estimates, but exceeds these estimates by $15,000,000. In case, therefore, the bonds are not issued too much below par, the respective items may be augmented by this surplus, and this may be the more easily effected, as, according to the communications of Mr. Jay Cooke, in a conversation on the subject, that gentleman is prepared to consider the matter with a view to such augmentation.

"15. With regard to this last point, but only in view of the possibility of these anticipations being realized, there is no cause for anxiety about the estimates, and after such augmentation no occasion for an unfavorable judgment.

"16. Turning to the second question, whether the Northern Pacific Railway, after its completion, offers the necessary guarantee that the surplus accruing out of the traffic revenue will suffice to pay the half-yearly rates of interest on the bonds, it is to be observed in the first place that the interest at the rate of 7.30 per cent. on a capital of $100,000,000 amounts to $7,300,000. To obtain a net profit of a similar amount requires a gross income of $20,000,000 a year. It is proved by official data that the net profits of the American railways are equal to 35 per cent. of the gross income, 65 per cent. of the total being consumed by the working expenses.

PROBABLE TRAFFIC.

"17. Whether the completed Northern Pacific Railway will be able to count on a traffic that will yield an income of $20,000,000 a year can only be ascertained by an inquiry into the state of the population and its industrial and commercial relations, and it will be important to keep the actual state of these relations very carefully in view. The tract of country traversed by the Northern Pacific Railway upon which at the outset the line depends for acquiring and securing a local traffic, is situated in the States of Minnesota, Dakota, Montana, Idaho and Oregon, and in Washington Territory. These have, according to the census of 1870:

	Area in square miles.	Inhabitants.
Minnesota	83,531	435,511
Dakota	147,490	14,181
Montana	143,776	20,594
Idaho	90,932	14,998
Oregon	95.244	90,922
Washington Territory	69,994	23,901

Or an aggregate area of 630,917 square miles and an aggregate population of 598,147, while in the year 1860 they had a population of only 250,000 persons. According to this the population has increased 350,000 persons within a period of ten years, but this increase belongs for the most part to the State of Minnesota only, since the population of that State has grown from 172,000 to 435,500 during the period in question—an increase of 263,000 persons. There is hardly any room for doubt that 600,000 people, scattered over an area of 30,000 German square miles, even if all are taken as contributing to the success of the Northern Pacific Railway, will not be able to insure a traffic that will produce an annual income of $20,000,000.

"18. The enthusiastic adherents of the undertakers of the Northern Pacific Railway will not deny the truth of this assertion, and they are only able to hold out hopes for the future by pointing to a possible rapid increase of population in the adjacent districts consequent upon the completion of the line, and a corresponding increase of the income of the company. Willing as I am to acknowledge, respecting America, the well approved fact that, contrary to what we see in Germany, where railways are the product of already cultivated and well-populated regions, the railways in America have hitherto drawn culture and population after them into uncultivated regions, and thereby drawn an income to themselves, yet the deductions from these facts must always be made with a certain reserve. We must not forget that though a growth of the population in regions newly traversed by railways is certain, it is not sufficiently rapid to cover thinly inhabited regions within a few years with numerous and densely peopled settle-

CAMP OF RAILROAD BUILDERS ON THE NORTHERN PACIFIC RAILROAD.

ments of a commercially agricultural or industrial character, particularly when it is a question of filling a region of 2000 miles with such settlements.

GROWTH OF POPULATION.

"We learn, by way of example, from a comparison of the population of the United States in 1870 and 1860, that in that decennial period the population increased from 31,500,000 to 38,500,000, an increase of 7,000,000,—very considerable in itself, but not of very great importance compared with the proportional increase of railways, when we bear in mind that the railways in the United States increased during the same period from 31,286 miles to 53,400 miles. Even the most favorable rate of increase of the population during the last ten years, namely, that of Minnesota, or about 150 per cent., would, extended over the whole region of the Northern Pacific Railway, during another ten years, only result in an aggregate population of 1,500,000, or one-third of the present population of the State of New-York. That such a population—which, by the way, would not exist at the time of the completion of the line, nor till ten years afterward—will not yield the required income to pay out of the net profits the interest on the capital invested in the enterprise, I think I am, according to my conviction, bound to maintain.

"Of course an increase of population beyond the percentage mentioned is possible, in consequence of accelerated exertions and efforts to direct immigration toward the hitherto neglected States of Dakota, Idaho, and Oregon and Washington Territory; but we must not ignore the fact that such exertions and efforts will always have to contend with great difficulties. The stream of immigration has hitherto flowed notoriously and preferably in the direction of the more Southern States, its diversion northward, particularly in the States of Dakota, Idaho, and Montana, where the Indian tribes are still hostile, will not succeed until after years of struggle and perseverance. Take, for instance, Minnesota, which has only just succeeded in attracting the influx of population already mentioned, though its emigration agents have for years traversed the length and breadth of Europe, and though the State has for a considerable time been provided with railways. To rely so confidently and strongly upon the exertions and efforts in favor of an immigration, the results of which must be reserved for a future day, and to deduce with certainty that they will necessarily produce a considerable local traffic for the Northern Pacific Railway from the time of the completion of the line, I consider hazardous.

TRANS-CONTINENTAL TRADE.

"19. If thus, according to my conviction, the prospects for an advantageous local traffic on the Northern Pacific railway during the first years of its operation do not exist, the through traffic will hardly offer any better chances. The Northern Pacific Railway, one terminus of which is situated at Duluth, on Lake Superior, in Minnesota, and the other at Puget Sound, will hardly be able to reckon on a through traffic in the course of a series of years, so far as it relates solely to American products, since the termini do not furnish any basis for such a traffic. Puget Sound, favorably as it may be situated for shipping, has, for the time being, nothing in the shape of industrial establishments except a few embryo collieries, as yet insignificant, and some important saw-mills, which procure their raw material by water, and whose selling markets are not on American soil but abroad, requiring ships to export their product, while Duluth, a town of about 4000 inhabitants, is still in embryo, so that it is hard to tell what traffic it may afford hereafter.

"20. The Asiatic through trade, so far as it affects the existing Pacific lines, and the importance of which must not be too highly estimated, since it consists of two articles only, tea and silk, will only be attracted with difficulty to the Northern Pacific road, because, on the one hand, until the branch line provided for in the charter is completed, which will reduce the distance from Puget Sound to Duluth from 2000 miles on the main line to 1775 miles on the branch line, the distance on the existing Pacific lines is about equal to that of the Northern Pacific, which offers no shortening of the journey by land, and because, on the other hand, the commercial relations between New York and Chicago and San Francisco are so closely tied that the removal of the agency of the San Francisco houses concerned in this commerce can hardly be thought of.

"21. After this exposition, though I readily acknowledge *that the Northern Pacific road will come in for* something at the opening—for instance, the important consignments to supply the military forts with provisions, the transport of provisions for the Hudson's Bay Company, the products of the mines of Montana, Idaho, and Washington, insignificant at present—I must incline to the opinion that after the completion of the line and the simultaneous cessation of paying interest out of the capital, a longer or shorter period of time will intervene during which the working of the road will not produce the required income to pay an interest of 7.30 per cent. on the capital invested after deducting the working expenses. As the means provided in the statutory regulations, of which I shall speak further on, by which supplies are to be raised in cases when the regular income does not suffice to pay the interest, will fail to afford a remedy, as I shall endeavor to prove, I have no hesitation in saying that the obligations undertaken by the promoters of the enterprise respecting the payment of interest cannot be fulfilled during the period immediately following the opening of the line.

VALUE OF THE LAND GRANT.

"22. With regard to the third question, whether the sale of land will realize the amounts by which the redemption of the entire bonded debt can be effected within the thirty years specified, I must, first of all, contradict an opinion which has been widely circulated, representing the regions traversed by the Northern Pacific Railway as unfavorable to civilization, agriculture and industry The experience I have acquired from personal inspection, as well as the most trustworthy information from official sources I have everywhere gathered on the spot, has convinced me that the region through which the Northern Pacific Railway passes is one of the most fertile on the American Continent, and is in every respect suitable for colonization. Minnesota and Dakota belong to the grain-growing region; Montana and Idaho are rich in minerals and pastures, and

Oregon and Washington Territory belong to the region of minerals, furs, timber and agriculture.

"The Government land grant to the Northern Pacific Railway Company, representing, as already mentioned, 50,000,000 acres for the main line, has an appreciable value, the money price of which, at the rate at which the Illinois Central sold its land, would amount to $550,000,000; at the rate at which Minnesota disposed of her school land, it would amount to $350,000,000; and at the rate at which the Kansas Pacific realized, it would amount to $165,000,000. Granted that even the last figure is put too high, we may fall back on the minimum price below which the United States Government has bound itself to the Northern Pacific Railway Company not to sell land in the alternating sections adjoining the line. At $2.50 an acre the land grant of the company is worth $125,000,000—an amount still more than sufficient to redeem the entire $100,000,000 of bonds; even the redemption at ten per cent. above par permitted by the statutory regulations requiring $110,000,000 to redeem the bonds, would still leave a handsome surplus. However, another question is whether the land can really be disposed of by way of sale within the thirty years fixed for the redemption of the bonds, and whether the demand for these lands will so nearly equal the supply that sales can be made at the given prices and during the specified period of time? The answer to this question is essentially dependent on the same points which have been already considered in connection with the security for the payment of interest on the bonds, namely, whether in this case the population of the respective regions will increase at a rate sufficiently rapid to insure that after the lapse of thirty years the whole land of the company shall have changed hands and been transferred to private owners. So far as human calculations can at all forecast what may be accomplished in periods of such duration, this question may be answered in the affirmative. The experience of the past in the United States, as well as what is seen every day, gives ample proof that, as a rule, after a number of years have passed subsequent to the opening of a line in new and uncultivated regions, and new settlements have been established in consequence of the railway, such land can be sold well and easier than any other land. Thus, though a period of several years following immediately upon the opening of a line may have few sales to show, yet in the long run all will be right, and the period of redemption is long enough to warrant a confidence, based upon experience, that within the period of redemption the area of the Northern Pacific Railway Company will have changed hands for the benefit of the company. However, that during the first few years after the opening of the line the sales will assume but small dimensions, can hardly be doubted, since, as has been already observed, the immigration turning toward the newly-opened regions of the Northern Pacific Railway will not be of sufficient magnitude to create an active demand, and as, furthermore, a great many immigrants who may turn in that direction to look out for new settlements will prefer to avail themselves of the facilities offered by the Homestead law, according to which every man on American soil, above twenty-one years of age, and declaring that he will remain, may secure on easy terms the possession of eighty acres of public land by simply paying a fee. Now, as in the regions crossed by the

Northern Pacific Railway the public land lies adjacent in alternating sections, and is consequently without any difference of quality, the immigrants will, at the outset, prefer the public land, and the railway company will only be able to effect exceptional sales.

"23. To what extent public land is disposed of under the Homestead law may be gathered from the fact that during the year 1869, the Government of the United States sold for cash 2,900,000 acres, and allotted under the Homestead law 2,737,000 acres; during the year 1870, it sold for cash 2,150,000 acres, and allotted under the Homestead law 3,700,000 acres. If thus the land sales of the railway company should be of little account for several years, one of the means already alluded to, of supplementing the fund out of which the interest on the bonds is to be paid in years of deficiency, will simultaneously fail. The statutory regulations stipulate that the bonds shall be successively bought up with the proceeds of the land sales and cancelled. At the same time, however, it is permitted, in case of the treasury of the company being exhausted, and consequently without the necessary means for paying the interest on the bonds, to make up the deficiency out of the proceeds of the land sales. Of course the company is bound to make restitution to the land fund of the amount taken out, by handing over the first net profits of the line. But if the land sales are only of a limited extent at first, and the proceeds correspondingly small, it will not be possible to make any substantial advances toward paying the interest, and simultaneously, as I have previously asserted, the line will not produce the requisite amount for the payment of the interest on the bonds; and thus, in my opinion, it is certain that a longer or shorter period will ensue, immediately upon the opening of the line, during which the bondholders will have to forego interest.

INSUFFICIENT SECURITY FOR BONDHOLDERS.

"24. In continuation of the foregoing, I have yet to mention in what manner the rights of the bondholders are to be guarded under such critical circumstances. The mortgage of July 1st, 1870, executed by the railway company to the trustees, as representatives of the bondholders on the other side—to which the bondholders have to submit—stipulates that, in case the company should not be able to fulfil their obligations respecting the payment of interest on the bonds, and (1) the delay in the payment has lasted three months, the trustees shall have the power to sell so much land out of the area of the company as will be necessary to realize the amount required to pay the interest; (2) when the delay in the payment has lasted six months the trustees shall be empowered to take charge of the line and work it themselves, and to make all the necessary arrangements for that purpose; (3) when the delay in the payment has lasted three years, the trustees shall be empowered to sell the line, and all the possessions pertaining to it, for the benefit of the bondholders. These stipulations, however, if acted upon, would do the bondholders little good. With reference to the powers conferred upon the trustees by Article 1, it is difficult to see what advantage would be gained for the bondholders by making use of them. If the land were saleable the company could and would sell to pay the interest on the bonds, and it will only be on account of the land not being saleable, shortly after the opening of the line, that no proceeds of the land sale will be available to supplement the interest-paying fund, and that the circumstances will arise for which the sale of land is to furnish a remedy. How the trustees can realize money by the sale of unsaleable land is not very clear. The same objection applies to the measures provided for in Article 2, empowering the trustees to take charge of the line and work it themselves. What benefits are the bondholders likely to derive from that? The trustees will hardly be able to convert a non-paying line into a paying line, and it might even be hazardous to take the management out of the hands of people acquainted with local and other conditions, simply because a crisis had ensued which, by no exertions and foresight, could by any human possibility have been averted. Finally, the powers conferred upon the trustees by Article 3 ought to have been given, in my opinion, by Article 1. If the interest on the bonds should not be paid, the bondholders—the mortgagees of the Northern Pacific Railway—must, according to my view of the matter, have the power to bring the object mortgaged, in the case before us the property of the Northern Pacific Railway Company, under the hammer of the auctioneer, in order to realize out of the proceeds a pro rata dividend upon their claims, as in any other case of insolvency; but this power must not be withheld for a space of three years, and only conceded when other and questionable measures have proved fruitless.

"I find in the mortgage of July 1st, 1870, which determines the rights of the bondholders, less a security for the exercise of those rights than a troublesome obstacle to the execution of rights guaranteed to the bondholders by the general laws of the United States of America.

CONCLUSION.

"25. To sum up, I cannot deem it advisable that European capitalists should be encouraged to participate in the enterprize of the Northern Pacific Railway, as, in my opinion, after the completion of the line a period will ensue during which the company will not be able to fulfil the obligation it is under respecting the payment of interest on the bonds. It is certainly possible that this period will not ensue immediately upon the completion of the main line, inasmuch as the branch line, equally provided by the charter, will issue new mortgage bonds, out of which the interest may be paid for a while, but this will only postpone, not avert, the crisis.

"That the Northern Pacific Railway may be a good and profitable enterprise after the years of its childhood and troubles have been survived, will not be enough to commend it to the European money market; it will have to be proved that the company will be able during the early years of the enterprise to fulfil all the engagements entered into. In the full consciousness of the responsibility incumbent on me as a member of the European Commission of Experts, I cannot consider that this proof is forthcoming. HAAS.

"BERLIN, *Nov.* 30*th*, 1871."

CHAPTER XIII.

DANGER AHEAD.

Evils resulting to the Country from Railroad Mismanagement—The Danger of Monopolies—Disregard of Individual and Public Rights—Efforts to corrupt the Legislative and Judicial Powers of the Country—How the Corporations menace the Public Liberties—Mistakes of the People—Helplessness of the Community—Mr. Thomas Scott's Boast justified—A Railroad King—Contrast between Vanderbilt and Drew—Immense Power of Commodore Vanderbilt—A Gigantic Monopoly—A Real Danger—An Unsafe Power in the Hands of an Interested Man—Danger Ahead—The Way to meet it.

We have now examined hastily some of the evils of the present system of railroad management, and have pointed out some of the troubles likely to arise therefrom. Our purpose in doing so is not to excite unnecessary or ill-advised hostility to the railroad system of the country, but to arouse the people to a sense of the danger with which the mismanagement of this system threatens them. That there is danger, we presume no one will deny.

Looking back over what we have been considering, we find:

I. That the railroad system of the United States, which was intended to give the people rapid and cheap communication and transportation, and which was designed as the servant of the people, has grown into a powerful combination of monopolies, each and all animated by a common object.

II. That the object of these monopolies is to compel the people to pay whatever rates they may see fit to establish for the service rendered them, and to keep these rates at the highest possible point.

III. That the corporations have a decided advantage over the public in this struggle, and that they are determined to resist, and do resist, all efforts on the part of the latter to obtain cheap transportation.

IV. That they are utterly regardless of the rights of the people, either as individuals, or as a community, and that they resent and punish to the extent of their power, any attempt on the part of an individual to dispute their regulations, however arbitrary and unjust the said rules may be.

V. That they are practically irresponsible for their action, and resist any and all efforts to render them amenable to the law.

VI. That they pursue a systematic course of plunder, robbing the nation of its property, and levying exorbitant rates upon individuals and freight, to pay "fancy dividends" upon their fictitious stock.

VII. That in order to secure the success of their schemes, they do not hesitate to resort to the most corrupt practices. They have done what they could to debauch the men placed in positions of public trust by the people, bribing legislators, and taking them into their pay, literally purchasing courts of justice, and thus closing the means of obtaining justice once open to the people.

VIII. That they are directly responsible for a large share of the corruption that is fast demoralizing our public service, and are seeking to render themselves the masters of the National and State governments.

IX. That they have introduced an element of reckless gambling in stocks into the monetary affairs of the

INHABITANTS OF THE COUNTRY THROUGH WHICH THE NORTHERN PACIFIC RAILROAD WAS BEING BUILT.

country, which is utterly destructive of all sound business management, and have succeeded in demoralizing this portion of our financial system to such an extent that great evils must follow unless they are compelled to desist.

X. That they are growing bolder and more audacious in their designs upon the people, caring for nothing but an increase in their gains, and that the liberties, the free institutions, the property, and the national existence of the American people are seriously endangered by the unlawful designs and the insolent acts of the railroad corporations.

There is danger in all this, and it would be worse than folly to shut our eyes and profess not to see it. It is a danger that must be met and turned aside. The power of a railroad corporation is not an imaginary thing. The corporation employs many hundred men, and disburses large sums of money; and it does these things for the avowed purpose of "earning" as much money from the people who are compelled to use the road, as they will pay. It is carrying out a system of operations opposed to the interests of the people, and it is a compact, solid body, under the direction and control of one vigorous mind, and it possesses every chance of success against the people, who are generally divided and indisposed to assert their rights, though sensible that they are being injured. It is almost absolute master of the market of the region it supplies. It can benefit or injure a community by liberal tariffs or extortionate rates, as it pleases, and the managers are free to decide which policy shall be pursued. It is subject to no control. It can do as it pleases. It controls hundreds of votes along its line, not one of which will be cast against it in any contest with the public, and the lobby it maintains at the centres of government takes care that no adverse legislation shall stop its encroachments upon the rights of the people. Relying upon its wealth and power, it insolently defies the community to protect itself, and pursues its course of extortion unchecked.

Now, if this be the power of a single road, what shall we say of the vast combinations of roads which are being organized and are in operation throughout the country? Does any one for an instant imagine that these combinations, whose sole object is to enrich themselves, are careful of the rights of the public? The very essence of their system is to make charges as high as possible, and, by combining, prevent competition. They know their roads are a necessity to the public, and that persons using them must pay whatever rates they see fit to impose. They have combined for the purpose of compelling the public to submit to their extortions, and they have no intention of abandoning their design. They are masters of the situation thus far, and they know it.

As for the people, they have no redress as matters are conducted at present. "With packed legislatures," says the *Atlantic Monthly*, in a recent issue, "with paid or intimidated judges, and with a civil service consisting of several thousand cunning clerks and able-bodied brakemen, conductors, and switch-tenders, they would be in just that position most dreaded by all lovers of liberty—a powerful and enormously rich corporation, surrounded by a timid, weak, and hopeless public. While we were still engaged in singing pæans over the glorious institutions of our happy country, we should

EVENING RECREATIONS OF RAILROAD GAMBLERS.

suddenly find that our institutions had disappeared, and that we had riveted round our necks the chains of a worse despotism than any we ever lamented for our fellow-creatures. This is really no imaginary picture, as any one will admit who recollects the stronghold, absolutely inaccessible to the law, which Fisk and Gould erected and for a time maintained in New York, or the military operations of the employés of the Erie and the Susquehanna railroads during the 'Susquehanna War,' and who has followed with any attention the helpless struggles of the Government of the United States—formerly supposed to be quite able to take care of itself—in the foul toils of the Union Pacific Railroad."

Mr. Thomas Scott, the vice-president of the Pennsylvania Railway, is credited with the remark that he preferred his position to the Presidency of the United States, as it was a post of more real power, and offered greater opportunities to a man of ambition and talent. This was no idle boast. Colonel Scott has shown, upon more occasions than one, that he feels that his road is the master of the vast communities dependent upon it.

Few monarchs enjoy as much substantial power as is vested in the hands of Cornelius Vanderbilt, of New York, who is often called the "Railroad King." A man of unbounded ambition, and with every quality for the successful organization and management of great monopolies, he has, by his genius and daring, placed himself at the head of the railroad interest of the United States. Naturally arrogant, his continued success has made him a true king of the absolute school, in the execution of his plans. Mr. Adams draws the following contrast between Vanderbilt and his rival,

CROSSING THE PLAINS ON THE UNION PACIFIC RAILROAD.

Daniel Drew :—"Drew is astute, and full of resources, and at all times a dangerous opponent; but Vanderbilt takes larger, more comprehensive views, and his mind has a vigorous grasp which that of Drew seems to want. While, in short, in a wider field, the one might have made himself a great and successful despot, the other would hardly have aspired beyond the control of the jobbing department of some corrupt government. Accordingly, while in Drew's connection with the railroad system his operations and manipulations evince no qualities calculated to excite even a vulgar admiration or respect, it is impossible to regard Vanderbilt's methods or aims without recognizing the magnitude of the man's ideas and conceding his abilities. He involuntarily excites feelings of admiration for himself and alarm for the public. His ambition is a great one. It seems to be nothing less than to make himself master in his own right of the great channels of communication which connect the city of New York with the interior of the continent, and to control them as his private property. Drew sought to carry to a mean perfection the old system of operating successfully from the confidential position of director, neither knowing anything nor caring anything for the railroad system, except in its connection with the movements of the Stock Exchange, and he succeeded in his object. Vanderbilt, on the other hand, as selfish, harder, and more dangerous, though less subtle, has by instinct, rather than by intellectual effort, seen the full magnitude of the system, and through it has sought to make himself a dictator in modern civilization, moving forward to this end step by step with a sort of pitiless energy which has seemed to have in it an element of fate. As trade now dominates the world, and railways dominate trade, his object has been to make himself the virtual master of all by making himself absolute lord of the railways. Had he begun his railroad operations with this end in view, complete failure would have been almost certainly his reward. Commencing as he did, however, with a comparatively insignificant objective point—the cheap purchase of a bankrupt stock—and developing his ideas as he advanced, his power and his reputation grew, until an end which at first it would have seemed madness to entertain, became at last both natural and feasible."

Not long since, the Presidency of the Lake Shore & Michigan Southern Railroad became vacant by reason of the death of the president, Horace F. Clark. Mr. Clark was the son-in-law and a valuable ally of

Commodore Vanderbilt, and at his death the directors transferred the presidency to Vanderbilt. Their action elicited considerable comment from the press at the time. The editorial remarks of *Harper's Weekly* are significant and worthy of preservation, as showing the view of the matter held by an influential portion of the American people. Said that journal:

"The election of Commodore Vanderbilt to the presidency of the Lake Shore & Michigan Southern Railway marks another step in the gradual consolidation of our great railroad and financial enterprises. Starting from Chicago, the metropolis of the northwest, and the greatest grain depot in the world, the Lake Shore Railway, running through a country which has been settled for two generations of men, drains the rich peninsula between the lakes, and connects the populous towns of the south shore of Lake Erie with the railway system of the east. At Buffalo it connects with the extensive system of railroads which Mr. Vanderbilt has consolidated within the past few years under the title of the New York Central & Hudson River Railway, and which drain the best counties of New York from Lake Erie to Westchester. Adding to the Central & Hudson the Harlem, which is now operated under a perpetual lease, Mr. Vanderbilt thus controls 2150 miles of railway, constituting the main line between the west and the sea-board, and the chief outlet of such cities as Chicago, Toledo, Cleveland, Buffalo, Rochester, Syracuse, Utica, Lockport, Schenectady, Troy, Albany, Hudson, Poughkeepsie, and the other river-side towns. The property which he thus administers is represented on the Stock Exchange by securities equal to $215,000,000, and its gross income last year was not less than forty-five millions of dollars—more than the whole income of the United States Government a few years ago.

"It is impossible to contemplate this vast aggregation of money power and commercial control in the hands of one man without feeling concern for the result. Neither military, nor political nor commercial supremacy can be pushed beyond certain limits without danger. It would seem as though the limit in this case had been reached. Yet, not content with the mastery of 2150 miles of railway, involving in a large degree the control of the internal trade of the States of Illinois, Indiana, Ohio, and New York, it is well understood that, in October next, at the annual election of the Western Union Telegraph Company, the Commodore will enter into possession of that great property likewise, with its sixty or seventy thousand miles of wires, its forty millions of capital, and its eight or nine millions of revenue. When this occurs, not only will the commerce of the four chief States of the North be subject to Mr. Vanderbilt—under such feeble restrictions as our Legislatures may impose—but the whole telegraphic correspondence of the country will obey his law. He may prescribe, not only what shall be the price of a barrel of flour in New York, but also when, how, and at what cost citizens may communicate with each other by telegraph.

"Of course he will be subject to legislative control. What that will amount to we all know. In the past no Legislature in this State has ever dared to beard him. He will be a bold man, indeed, who attempts to do so now, when his resources are so unbounded, and his power so far-reaching. It was said that the late James Fisk, Jr., who controlled a paltry 450 miles of

Erie, running through a half-settled country, could, on an emergency, bring 25,000 votes into the field. At how many votes, then, must we reckon the master of 2150 miles of railway through a thickly settled country, and 70,000 miles of telegraph? It is, moreover, one thing to pass laws, and quite another to execute them against a man fertile in resource, energetic in action, obstinate in combat, and inexhaustible in purse. We have some fine laws prescribing the rate of fare on the Central & Hudson, but every traveller knows that if he would be comfortable he must pay thirty to forty per cent. extra for a seat in a drawing-room car.

"If, again, the concentration of these great enterprises in one grasp were likely to be attended with a reduction of the cost of travel and the burdens of trade, or if it insured improved facilities to keep pace with the development of the country, these would be redeeming features in the Vanderbilt régime. But great as Mr. Vanderbilt undoubtedly is as a railway manager, his greatness shows itself not in increased facilities for travel and trade, but wholly and altogether in economy of administration. He makes money by saving it. Economy is his watchword, his motto. It was by new economies, he said in his letter accepting the presidency of the Lake Shore, that he hoped to set that company on its legs. It is by economy that he makes the Central pay four times more than Erastus Corning or Henry Keep could ever make it yield. Now economy in railway administration is admirable from the stockholders' point of view. It is not so good for the traveller. If it means, as some evil-disposed critics of the Vanderbilt régime pretend, filthy cars, wretched stations, general discomfort, and decreased instead of increased accomodation, the prospect of its indefinite extension is not likely to overwhelm the public with delight. It would, of course, be childish to expect Commodore Vanderbilt or any one else to run railroads from philanthropic motives or as charitable institutions. He runs them to make money, and for no other purpose. This being the case, and such being his policy, it is not surprising, considering how extensively railways control commerce, govern prices, and influence our closest interests, that people should feel nervous at the news of this Great Economist capturing another thousand miles of railway, and stretching out his long hand to grasp all the telegraph wires in the country. It is probably unfair to grudge the Central stockholders their dividends. But people who are not so fortunate as to belong to that happy class cannot be blamed for remembering that the Central & Hudson property, which is now made to pay dividends on $115,000,000, was represented in 1862 by only $50,000,000 of stock and bonds, and really cost about $35,000,000, the difference between this sum and $115,000,000 being mostly what is called, in the jargon of the street, 'water;' and that if there had been no water mixed with the good old Central wine, the road could have carried passengers at one cent a mile in clean, well-ventilated cars, and have paid the same dividends that it does now.

"In truth, however, it is a small matter to a prosperous people like ourselves whether we pay one or three cents a mile to go to Albany, or whether dust and discomfort do or do not drive us into drawing-room cars. A much graver matter is the inevitable tendency of these grasping monopolies that are springing up around us, and the inexorable law which punishes corporate

SNOW SHEDS ON THE CENTRAL PACIFIC.

greed with confiscation. No student of history, especially of the history of the great corporations of the last century—we mean, of course, the ecclesiastical bodies—can fail to discern the fate of many of our great railway companies. One set of men after another growls and submits. One Legislature after another threatens and is cajoled or bought off. But the intolerable oppression continues, and year by year the instinct of rebellion grows stronger and stronger, until it has coherence enough for demagogues to make it a plank in their platforms. Then no man can stem the tide or set a limit to party fury or popular injustice. You can hear the first mutterings of the storm in the proceedings of the 'granges' at the West. Already, monstrous as it may seem, judiciary elections turn on railway against anti-railway. Even Congress the other day, in its fury over the Crédit Mobilier, stopped but little short of confiscation in its proceedings against the Union Pacific. How long will it be before some party in this State seriously proposes to tax the Central all it makes on its watered stock? Is there no Ben Butler in the Western counties to ride this magnificent hobby-horse into Albany and Washington? When the Central was borrowing money to pay three per cent. half yearly, it seemed mean to tax it. But now, with a revenue of twenty-five millions, Aristides himself might vote to confiscate. Idle to talk about measures being unconstitutional. Constitutions can be changed as well as laws, and when the day comes for the spoliation of the railways, neither vested rights nor common honesty are likely to obtain a hearing."

Now Commodore Vanderbilt is by no means a bad man personally, but he is the representative of one of the most perfect despotisms in existence, and he is human enough to regard the interests of his roads before those of the public. It is not safe to lodge such power as he holds in the hands of any individual, and it is for the people to decide how long such a state of affairs shall continue. His power is exerted against them. His despotism, in common with that of the other monopolies, threatens them in every relation of their national life. He exacts tribute from them in every act of his official existence, and his great wealth is made up of the aggregation of the sums he has wrung from them as a successful leader of a grinding monopoly.

Nor is he the only "dangerous character" before the

public. Each of the great railroad monopolies has its representative, who does on a small scale that which Vanderbilt accomplishes in his regal style, and each is dangerous to the community, as being engaged in a struggle against its best interests and most cherished rights and privileges.

Such vast power as these men possess would be a source of danger in the most disinterested hands. It is doubly so in the hands of men who are engaged in such a warfare against the public as the railroads are now carrying on. The people owe it to themselves to curtail their powers, and to render them harmless by subjecting them to a series of regulations which shall compel them to respect the rights of the community to whom they are indebted for the very existence of their roads.

The people have the right to do this, and it should be done promptly. There is no necessity for placing burdens upon the roads heavier than they can bear. They have a right to a fair return for their investments, but they have no right to plunder the public. A series of wise and liberal regulations will protect the people against railroad tyranny and extortion, and at the same time enable the roads to do a profitable business.

UNITED STATES TREASURY—THE MEDIUM THROUGH WHICH THE NATIONAL LEGISLATURE ROBS THE PEOPLE.

PART II.

THE COAL MONOPOLY.

CHAPTER XIV.

OPERATIONS OF THE COAL RING.

Character and Extent of the Coal Deposits of the United States—The Supremacy of Anthracite—Enormous Coal Wealth of the Country—This should be a Land of Cheap Fuel—Coal one of the Costliest Articles of Consumption—The Cause of this—The Anthracite Fields—Their Location and Value—The Pennsylvania Coal Ring—A Crushing Monopoly—Efforts of the Corporations to keep up the Price of Coal—Condition of the Companies kept secret—History and Present Condition of the Reading Railroad—A Dangerous Monopoly—Immense Wealth and Power of this Corporation—Ten per cent. on Watered Stock—How Money is extorted from the People by the Coal Ring—An Inside View of the Scranton Coal Sales—Amount of the Tax paid by the People to the Coal Ring—An Imperial Tribute—Who are the Sufferers—The Poor driven to Despair—How a Scarcity of Coal is brought about—The People at the Mercy of the Coal Ring—Popular *vs.* Corporate Rights—The Remedy for the Great Evil—How to bring down the Price of Coal and destroy the Power of the Monopoly—The Remedy in the Hands of the People—The Future of the Country at the Mercy of the Coal Ring—The Duty of Congress—Will Congress stand by the People or yield to the Monopoly?

We have seen how the railroad monopolists have inaugurated and are carrying out a systematic warfare upon the people of the United States, the result of which has been to increase the cost of food. We shall now consider the efforts of these monopolists to raise the price of coal, next to food the chief necessity of the people of this country. We shall find the same disregard of the rights of the community, the same heartless greed, on the part of the corporations, that we have seen in the management of the railroads.

Coal is found in twenty of the States and Territories of the Union. According to the census of 1870, the product for that year was as follows:

	Tons of Bituminous Coal.	Tons of Anthracite Coal.
Alabama,	11,000	——
Colorado,	4,500	——
Illinois,	2,624,163	——
Indiana,	437,870	——
Iowa,	263,487	——
Kansas,	32,938	——
Kentucky,	150,582	——
Maryland,	1,819,824	——
Michigan,	28,150	——
Missouri,	621,930	——
Nebraska,	1,425	——
Ohio,	2,527,285	——
Pennsylvania,	7,798,518	15,650,275
Rhode Island,	——	14,000
Tennessee,	133,418	——
Utah,	5,800	——
Virginia,	61,803	——
Washington,	17,814	——
West Virginia,	608,878	——
Wyoming,	50,000	——
Total Product of the United States...	17,199,415	15,664,275

The capital employed in the production of coal amounted, in 1870, to $110,008,029.

By the above table it will be seen that the amount of anthracite coal is but 1,535,140 tons less than the total product of bituminous coal. The latter is divided among nineteen States and Territories; the former is almost exclusively confined to the State of Pennsylvania. The bituminous coal is used principally in the West, but the anthracite is the fuel of the Middle and

Eastern States. The bituminous coal deposits of the United States are supposed to be inexhaustible. As yet they have been scarcely touched. They are capable of supplying the wants of the entire continent, and of contributing towards the needs of other lands. The single State of West Virginia contains coal enough to supply the entire Union. There is no reason to believe that the supply open to the American people will ever be exhausted, and, with our facilities for mining, coal should be produced at a comparatively small cost. It never should reach a high figure, but should be so moderate in price that no one in this land of plenty should ever suffer from the lack of it.

Yet scarcely a winter passes that we do not see, in the Eastern States at least, a scarcity of coal, which is the cause of great and unnecessary suffering to the poorer classes. In spite of the immense qantities of fuel within reach; in spite of the fact that this should be a land of cheap fuel, we find, in all the Eastern States, that coal rises in price so rapidly upon the approach of winter that the poor can scarcely obtain it, and people of moderate means are compelled to make serious sacrifices of other comforts in order to obtain the necessary supply.

The cause of this unnatural state of affairs will appear as we examine into the manner in which New England and the Middle States are supplied with fuel.

The anthracite coal used in the Eastern States, and preferred by Eastern people to the bituminous coal of the West, comes almost entirely from a small district in the State of Pennsylvania, in the Eastern portion of the State. This region is included within an area of about five hundred square miles, and within this

MINING VILLAGE IN PENNSYLVANIA.

area is contained the richest, and indeed almost the only bed of anthracite coal in the world. "In the beautiful valley of the Wyoming, in the wild and desolate regions of the Schuylkill and the Lehigh, strewn in great basins, or sometimes upheaved to the tops of the mountains, the rich stores of coal have been concentrated by nature in an unequalled profusion. The supply is at present boundless. Many beds reach to an unknown depth. Some have as yet been only partially explored; and notwithstanding the extraordinary waste practised in getting out the coal, there seems enough to satisfy for generations the wants of the nation."

Upon this small, but immensely valuable region, the people of the Eastern States are dependent for their supply of fuel. The product of the mines can find its way to market only by the railroads which lead from the coal regions to the principal cities of the East, and by the canals which communicate with the Delaware and the Hudson rivers. These are few in number, repeated consolidations having confined them to the corporations known as the Delaware, Lackawana & Western, the Delaware & Hudson, the Lehigh Valley, and the Philadelphia & Reading Companies. The other roads leading to the coal regions are either leased or controlled by these companies, and these constitute a monopoly of the carrying trade. It is a well understood fact that these corporations will not permit the construction of any road which may endanger the monopoly they enjoy, and they are well known to be powerful enough to enforce this decision. If the coal of the Pennsylvania mines reaches the markets of the East at all, it must pass over the routes controlled by these corporations, and pay such tolls as they may see fit to levy upon it. They take care that these tolls shall be high enough to return them enormous dividends on the stock, which stock is popularly believed to be perhaps the most liberally watered of any in the market.

The companies mentioned are not only the sole controllers of the transportation from the mines to the market, but, together with the Pennsylvania Coal Company, they own nearly the whole of the anthracite fields of Pennsylvania. Their policy has been, and is, to buy up the coal lands as fast as possible, in order that they may obtain control of every branch of the coal trade of the State, and thus be the absolute masters of the fuel of the East. By becoming the owners of the mines and of the lines of transportation by which the coal must reach the markets of the Eastern States, the corporations are enabled to control the amount of production and to regulate it in such a manner that coal shall always command such prices as they see fit to ask for it. So great has been their success that at present they are able to name the price at which coal shall be sold in Philadelphia, New York and Boston. The people have no choice in the matter; they must pay the rates demanded, or do without fuel.

The majority of the companies manage their affairs in such a manner that the public are able to learn very little concerning them. Some refuse to publish any statements at all; others make public the most meagre and unsatisfactory returns. Enough is known, however, to make it evident that a very heavy profit is reaped by the stockholders upon their investment, in spite of the efforts of the companies to keep secret the exact ratio of this profit.

The principal member of the great coal monopoly is the Philadelphia & Reading Railroad, one of the richest and most powerful corporations in the Union, and, beyond a doubt, "the life and soul" of the monopoly we are discussing. Poor's *Railroad Manual of the United States*, for 1873–74, sums up the history and present condition of this company as follows:

"This company was chartered by the Legislature of Pennsylvania, April 4th, 1833, to build a road from Philadelphia to Reading, in Berks county, 58 miles from Philadelphia. Work was commenced early in the Spring of 1835, and portions of the road were opened for travel in July, 1838. By Act of March 20th, 1838, authority was given to extend the road to Mount Carbon, or to Pottsville, one mile above Mount Carbon. As these two points were already connected by a road called the Mount Carbon Railroad, it was decided to extend the road to Mount Carbon and connect therewith. The Mount Carbon Railroad was leased, and on May 13th, 1872, was merged into and became part of the main line of the Philadelphia & Reading Railroad.

"The first through trains between Philadelphia and Pottsville, 93 miles, were run in January 1842, although local trains were run in 1838.

"The branch from the falls of the Schuylkill to Port Richmond, from which the shipments of coal are made, was completed in 1842; since then over forty six millions of tons of coal have been shipped from that point, principally for consumption in the Eastern and Middle States.

"In 1850, the company bought that portion of the Commonwealth's improvements extending from Broad and Vine streets in Philadelphia, to and including the inclined plane on the Schuylkill, and the Columbia bridge over the river.

"In 1858, the Lebanon Valley Railroad, 54 miles long, extending from Reading to Harrisburg, was merged into the main line.

"Within the past two years, the following railroads and branches have been merged into the company's railroad proper. The length of these roads is given in the tabular statement on page 263:

"The Mahanoy & Broad Mountain Railroad, wholly within the coal region.

"The Lebanon & Tremont Railroad, partly in Lebanon and Schuylkill counties.

"The Northern Liberties & Penn Township Railroad (commonly called the Willow Street Railroad), from Broad street to the Delaware River, Philadelphia.

"Port Kennedy Railroad, Montgomery county.

"Schuylkill & Susquehanna Railroad, extending from Rockville, on the Susquehanna river, 5 miles above Harrisburg, to Auburn, on the Schuylkill river.

"Shamokin & Trevorton Railroad, wholly in the coal region.

"Zerbe Valley Railroad, from Port Trevorton, Snyder county, on the Susquehanna river, to a point near Shamokin.

"The Mount Carbon Railroad, from Mount Carbon to a point above Pottsville.

"The following roads are leased to the company, generally in perpetuity:

"Catawissa Railroad, to Williamsport.

"Chester Valley Railroad, Bridgeport to Downingtown.

"Perkiomen Railroad, Montgomery county.

"East Pennsylvania Railroad, Reading to Allentown.

"Little Schuylkill Railroad, from Port Clinton to Junction with Catawissa Railroad.

"Mount Carbon & Port Carbon Railroad, wholly in coal region.

"Mill Creek Railroad, wholly in coal region.

"Schuylkill Valley Railroad, wholly in coal region.

"Mine Hill & Schuylkill Haven Railroad, wholly in coal region.

"East Mahanoy Railroad, wholly in coal region.

"Philadelphia, Germantown & Norristown Railroad.

"In addition to the above, the company controls and operates the Reading & Columbia Railroad and the

Allentown Railroad, from Topton to Port Clinton, completed to Kutztown.

"The company has also leased in perpetuity the canal of the Schuylkill Navigation Company, extending from Port Carbon, Schuylkill county, to Philadelphia, a distance of 108 miles; also the Susquehanna and tide-water canals, extending from Columbia to Havre de Grace, on the Susquehanna river. The chief business of the Philadelphia & Reading Railroad Company is the transportation of coal from the first and second anthracite coal-fields of Pennsylvania to tidewater in the Delaware river, at Port Richmond, Philadelphia. At this eastern terminus, extensive wharves, 23 in number, and extending from 300 to 800 feet into the river Delaware, have been erected with trestlework and chutes, allowing a direct discharge of coal from the cars into vessels. To accommodate this immense shipping business, 35 miles of track are distributed on the wharves or their immediate neighborhood. The main line of the road winds through the Schuylkill valley, extending its numerous branches east and west, draining completely the two southern coal-fields, and making them tributaries to the main stem.

"The heavy freight of this road, being generally in one direction, that is, from the coal region to the seaboard, the grades of the road have been adapted to its economical working, by establishing exclusively down grades and levels in the direction of the main traffic, and the heaviest grades admit of a locomotive taking back the same numbers of empty cars she is able to move downward loaded.

"At Lebanon, 28 miles west of Reading, a connection is made with the Cornwall Railroad, contributing

INTERIOR OF A COAL MINE.

the products of the immense magnetic iron ore deposits of Cornwall, the largest unbroken mass of ore known, to the business of this branch.

"The Broad Mountain, dividing the two coal-fields, is crossed by the different branches at four different points, three of which lead directly into the Mahanoy coal-field and one into the Wiconisco basin. The ascent from the southern side is by steep but practicable gradients, but the descent of the above points is by means of inclined planes, the steepest of which, the Mahanoy plane, has a gradient of 22 feet per 100 feet and is $\frac{5}{8}$ mile in length.

"The following is a tabular statement of the several lines, owned, leased and controlled by the Philadelphia & Reading Railroad Company, November 30, 1872:

NAME OF ROAD.	Single Track.	Double Track.	Length of Road.	Sidings and Later'ls.	Tot Length of Tracks and Sidings.
MAIN LINE.—Philadelphia & Reading Railroad	....	98.4	98.4	143.0	339.8
Northern Liberties & Penn Township	....	1.4	1.4	.8	3.6
Port Kennedy Branch	1.2	....	1.2	.4	1.6
Lebanon Valley Branch	13.0	40.7	53.7	18.9	113.3
Lebanon & Tremont Branch	42.2	....	42.2	24.0	62.7
Schuylkill & Susquehanna Branch	53.4	....	53.4	9.3	15.6
Mount Carbon Branch	8.5	....	8.5	7.1	15.6
Mahanoy & Shamokin Branch	53.8	10.8	64.6	66.9	142.3
Total roads owned	172.1	151.3	323.4	270.4	745.1
Chester Valley Railroad	21.5	....	21.5	2.3	23.8
Perkiomen Railroad	18.4	....	18.4	2.8	21.2
Colebrookdale Railroad	12.8	....	12.8	1.7	14.5
Pickering Valley Railroad	11.3	....	11.3	.4	11.7
East Pennsylvania Railroad	36.0	....	36.0	15.5	51.5
Allentown Railroad	4.5	....	4.5	.3	4.8
Little Schuylkill Railroad	28.2	....	28.2	25.5	53.7
Mine Hill Railroad	31.0	21.8	52.8	63.2	137.8
Mount Carbon & Port Carbon Railroad	....	2.5	2.5	9.8	14.8
Mill Creek Railroad	....	3.8	3.8	18.1	25.7
Schuylkill Valley Railroad	5.7	5.3	11.0	16.1	32.4
East Mahanoy Railroad	10.7	....	10.7	3.7	14.4
Philadelphia, Germantown & Norristown R. R.	13.5	20.2	33.7	13.7	67.6
Catawissa & Williamsport Railroad	92.6	....	92.6	13.5	106.1
Total roads leased	286.2	53.6	339.8	186.6	580.0
West Reading Railroad	1.9	....	1.9	1.8	3.7
Reading & Columbia Railroad	39.5	....	39.5	13.6	53.1
Lebanon Branch Reading & Columbia Railroad	3.4	....	3.4	....	3.4
Total roads controlled	44.8	....	44.8	15.4	60.2
RECAPITULATION.					
Roads owned	172.1	151.3	323.4	270.4	745.1
Roads leased	286.2	53.6	339.8	186.6	580.0
Roads controlled	44.8	....	44.8	15.4	60.2
Total	503.1	204.9	708.0	472.4	1385.3

"Statement of the length of tracks (equated to single track) in use by the company, yearly, November 30, 1863–72:

	1863.	1864.	1865.	1866.	1867.	1868.	1869.	1870.	1871.	1872.
Main line	266.1	283.3	289.0	306.7	315.8	320.2	326.0	334.3	336.8	339.8
Branch lines	248.0	402.8	428.5	441.9	454.8	486.5	815.9	833.7	929.5	1,045.5
Total	514.1	686.1	717.5	748.6	770.6	806.7	1,141.9	1,168.0	1,266.3	1,385.3

"For terms of leases of Schuylkill Navigation and Philadelphia, Germantown & Norristown Railroad and branches, *see* 'Manual' for 1871–72.

"For nature of the consolidation of the Mahanoy & Shamokin Railroad, the Northern Liberties & Penn Township Railroad, and the Lebanon & Tremont Railroad, with the Philadelphia & Reading Railroad Company, *see* 'Manual' for 1872–73.

"On the 1st of January, 1872, a formal lease and contract was entered into, by which the possession of the Susquehanna Canal, extending from Columbia to the tidewater on Chesapeake Bay, passed to this company, at an annual rent equal to the interest upon the debt of the Canal Company, and one-half of the net profits of operating the canal, after deducting rents and the cost of all improvements; provided that during and after the year 1880 the annual rent to be paid in addition to the amount of interest shall not be less than a sum equal to *three* per cent. upon the present capital of the Canal Company.

"On the 10th of October, 1872, the Philadelphia & Reading Railroad Company entered into a lease with the Catawissa Railroad Company for its entire line from the junction with the Little Schuylkill Railroad to Williamsport, including the contracts of the latter company with the Sunbury & Erie and the Lehigh & Mahanoy Railroads, the Empire Transportation Company, and the Lehigh Coal & Navigation Company, for 999 years, from November 1, 1872; the Philadelphia & Reading Railroad Company agreeing to pay the interest on the bonds and such other sum as shall equal 30 per cent. of the gross earnings, being not less than $77,000 on the 1st days of May and November, 1873; $89,000 on the same days of 1874; $101,000 on the same days of 1875; and $113,000 on the same days in 1876 and thereafter. Provision is also made for preserving the separate organization of the Catawissa Railroad Company.

"Notwithstanding the largely increased production of coal and its transportation over this road, the receipts per ton were less than for the previous year, having been $1.54.4 per ton, against $1.80.8 for 1871, and $1.94 for the average of the previous ten years.

"The following table shows the product of anthracite coal in Pennsylvania for ten years:

Year.	Tons of Coal.	Over previous years.			
		Increase.	Decrease.	Increase per cent.	Decrease per cent.
1863	9,566,066	1,696,599	....	21.56	....
1864	10,177,475	611,469	....	6.39	....
1865	9,652,391	....	525,084	....	5.16
1866	12,703,882	3,051,491	....	31.61	....
1867	12,988,725	284,843	....	2.24	....
1868	13,834,132	845,407	....	6.51	....
1869	13,723,030	....	111,102	....	.80
1870	15,849,899	2,126,869	....	15.49	....
1871	15,113,407	....	736,492	....	4.64
1872	18,929,263	3,815,856	....	25.25	....

"It will be seen by the above that the increase in the production of anthracite coal in Pennsylvania in ten years has been a fraction less than 100 per cent., and that one-third of the entire production of last year has been transported over the roads of this company. The business of the canals has involved a loss of $467,-755.60, but in view of the annual increase of the production, it is believed that it will not be long before both road and canals will be taxed to their utmost limits.

"During the year various connecting lines heretofore controlled by this company have become merged in it.

The Mount Carbon Railroad, of which this company owned 3202 shares, was absorbed by exchanging share for share, the investment amounting to $178,229.25; the Schuylkill & Susquehanna Railroad, of which this company owned 21,702 shares, was merged by the issue of one share for three, representing a cost of $404,388.34; and the stock of the Port Kennedy Railroad, all the stock of which, amounting to $26,893.98, was owned by this company, transferred to the item of railroad in the balance-sheet. Stock of the company to the amount of 536 shares was also issued and exchanged for stock in the following named leased companies:

Mount Carbon & Port Carbon Railroad	195 shares for	163
Schuylkill Valley Navigation Railroad	177 " "	354
Mill Creek & Mine Hill Navigation & Railroad	164 " "	328

"The basis of the exchange was such as to secure to stockholders the same income as was derived from their stock in the above-named companies.

"There were also created and issued 50,190 shares of stock in exchange for a similar amount of convertible bonds, as follows:

Loan of 1857–86	$17,500
Loan of 1870–90	2,492,000
Total amount converted	$2,509,500

"There remain outstanding only $124,000 of convertible bonds, of which $96,000 are of the 6 per cent. loan of 1886, and $28,000 of the 7 per cent. loan of 1890. The sterling loan of 1872, amounting to $110,400, was retired, and $110,000 consolidated mortgage bonds of 1911 issued.

"The company took the express business of its lines

into its own hands on the 1st of September, 1872, and the results for the three months show that, while in September the receipts were 36.8 per cent. less than the amount received from the express company during the same month of the previous year, in October they were only one-half of 1 per cent. less, and in November 58.87 per cent. more than for the corresponding month of 1871.

"The Philadelphia & Reading Coal and Iron Company, of which the Philadelphia & Reading Railroad Company is the sole stockholder, and a full account of whose organization is given in the 'Manual' for 1872–73, now controls 80,000 acres of land, on which are 98 collieries, 27 of the largest of which will be worked by the company, and the remainder leased. The tonnage of these lands amounted for the year to 3,030,881 tons, and the rents were $946,774.69. Almost the entire issue of $19,000,000 of the consolidated loan was applied to the purchase and development of these lands, in addition to which the Coal and Iron Company have issued bonds amounting to $11,131,000, guaranteed by the Philadelphia & Reading Railroad Company. It is believed that the entire production of coal on its estates during the current year will reach 4,100,000 tons, of which the company will mine over 2,000,000. For the still further development of this industry, a new debenture loan of $10,500,000 has been placed upon the market, consisting of 7 per cent. coupon bonds, payable in 1893, convertible after July 1, 1876, and before January 1, 1892, each stockholder having the right to subscribe, at par, in proportion to his stock, payments to be made in instalments up to April 15, 1875, with the privilege of anticipating any or all instalments. The subscrip-

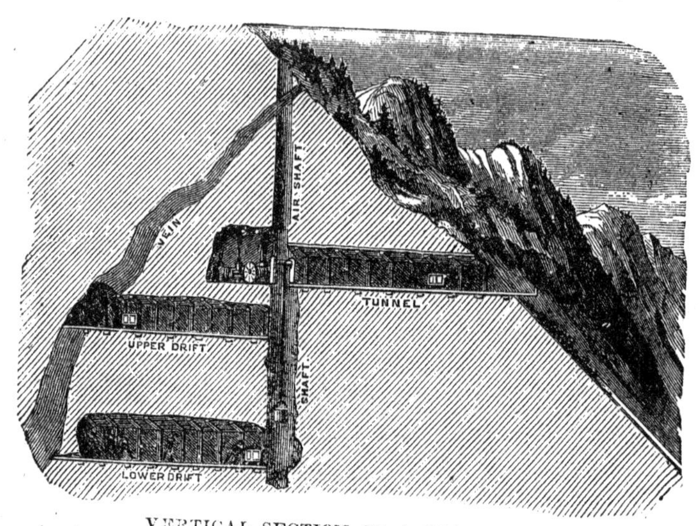

VERTICAL SECTION OF A COAL MINE.

tions, up to January 13, reached $12,857,400, of which $8,543,000 were pro rata, and $2,305,630 has been paid in.

"The company have determined to construct a fleet of additional steam colliers, some of 600 and some of 1200 tons capacity, the machinery to be built at the company's shops. A contract has been made for the hulls and boilers for two vessels for a trial. If they are successful, a large fleet will at once be completed."

Now this is a remarkable history of prosperity, and it is significant as showing the rapidity with which the Coal Monopoly is fastening its yoke upon the people of the United States. In the year 1872, one-third of the total production of anthracite coal in Pennsylvania was carried to market over the roads of the Philadelphia & Reading Company, or, in other words, the people of the Eastern and Middle States were at the mercy of the Philadelphia & Reading Railroad for one-third of their fuel, and paid to this company a tax amounting to large sums during the year. When we remember this we shall have no difficulty in understanding why the stock of this road is such a favorite in the market.

But the Philadelphia & Reading Railroad is not the only corporation engaged in the effort to run up the price of coal. It is only the principal sinner. All the corporations engaged in the coal carrying trade and in mining have a common interest, which is to extort from the public the highest price it will pay for coal. Their profits are large, excessively large, and it is believed by the public that in order to conceal them the companies have watered their stock to a very great extent. The exact point to which the watering process has been carried no one but those in charge of the affairs of the companies can tell, but the popular belief places it at a very high figure.

"Since 1864 the minimum dividend paid in any year by either the Reading & Delaware, Lackawanna & Western, or Lehigh Valley Railroad companies, or the Delaware & Hudson Canal Company, is ten per cent. The Reading has paid as high as fifteen per cent., and each of the others as high as twenty per cent. It cannot be said that the business has not been profitable to them. Ten per cent. per annum is a good return for a railroad investment, and is certainly far above the average profits of individuals who have embarked in the business of coal mining.

"It would be interesting to know how much of the round two hundred million dollars representing the capital and debt of the corporations composing the great coal combination is watered stock. A great deal of it is fictitious capital, in our opinion. It is matter of history that the Reading & Delaware, Lackawanna & Western railroads, and the best part of the Lehigh Valley, and also the Delaware & Hudson Canal, were completed works in 1864. They were all bringing coal to tide-water in that year, and the quantity they brought was more than half the quantity they are bringing now. The total anthracite production that year exceeded ten million tons, while for the last twelve months it has been about eighteen millions. With completed roads and canals, with the mines in their possession, and about the same terminal facilities in 1864 as they now enjoy, these corporations certainly did not require double the capital to do double the business. We should judge that an increase of capital of one-third would be amply sufficient to double the production. As an instance, the capital stock and debt of the Boston & Albany Railroad were sixteen million dollars in 1861 and twenty-one millions in 1870, and in the interval the business more than doubled, a great deal of double track was laid, and other expensive improvements were made.

"The figures of the coal corporations, without further explanations, will speak for themselves. The capital stock of the four companies we have named has been increased from $43,000,000 in 1864 to $90,000,000 in 1872, or considerably more than doubled, and the debt has grown from $17,000,000 in 1864 to $43,000,000 in 1872. Taking the stock and debt together, the increase has been from $60,000,000 in 1864 to $133,000,000 in 1872, or 120 per cent., while the increase in the production of coal is only 80 per cent. In making the calculation we have deducted from the debt of the Reading Railroad the bonds of the

Schuylkill Canal, and also $19,000,000 more, being the amount of its recent investment in coal mines. The Delaware & Hudson Canal Company has doubled its stock and debt since 1864, and the Delaware, Lackawanna & Western and Lehigh Valley companies have multiplied the nominal amount of their investment by three, the stock of the two companies last named having been raised from $13,500,000 to $41,000,000, and the debt from $5,000,000 to $19,000,000. (See Poor's *Manual of the Railroads of the United States.*) We suspect the managers of these corporations think it looks better to distribute among the shareholders ten per cent. on twenty millions of nominal capital than twenty per cent. on half the amount.

"In those parts of the country where anthracite coal is used, the consumption is reckoned at one ton a year for each inhabitant. A family of six will burn six tons, and will pay this winter a special tax of from ten to twelve dollars in order that these great coal corporations may do a little better than paying only ten per cent. on their watered stocks. 'The exceptionally low price in America during the past year,' say the managers of the Reading Railroad in their report, 'has introduced coal in competition with wood into districts where it never had been sent before; and it is well-known that when the appliances for burning anthracite coal are once introduced, and the advantages of that fuel once understood, it is never displaced by any other.' We are wedded to the glowing anthracite as the Englishman is wedded to his beer. The corporations have found it out, and have fixed the excise accordingly."

At the time these pages are passing through the press, coal is selling in New York for seven dollars a

SCENE IN THE COAL REGIONS.

ton, and is rising in price. Probably, when the mid-winter arrives, the rate will have advanced to ten dollars per ton, and this will be the average price throughout the Middle and Eastern States. And yet it costs but from two dollars to two dollars and fifty cents to produce this coal at the mines. The balance is made up by the enormous rates charged for transportation and by the profits of the retail dealers. Mr. Franklin B. Gowen, president of the Philadelphia & Reading Railroad Company, in his report to the stockholders in January, 1871, made an estimate of the natural average price of coal at Port Carbon, a point on the Reading Railroad, about ninety miles from Philadelphia. He declared "that if all the coal fields were producing largely, the average price of coal at Port Carbon would not exceed from $2.25 to $2.50 per ton."

It is the custom of the retail dealers in the East to lay in their supplies of coal for the winter during the summer and early fall. At this period of the year large quantities of coal are sold by the miners, and the prices then obtained to a great degree govern the retail prices for the remainder of the year. The companies exert themselves to keep the prices at these sales at the highest possible figure, and as they are masters of the situation, they succeed in obtaining whatever they may choose to ask. Some idea of their high-handed manner of conducting their business may be gained from the following comments of the New York *Tribune* upon the August (1873) sales of Scranton coal:

"The poor, starving coal companies have just given another turn to the screw, just 'to steady the price' of the fuel by which ten millions of people are to cook their food and keep themselves warm the coming winter. The Scranton auction sale was held yesterday, and the average price obtained for the 90,000 tons sold was $5.17 per ton, against an average of $3.46 for 100,000 tons sold August 28th, 1872, and of $5.03 for 140,000 tons sold August 30th, 1871, the famine year. In the year 1870, the production was also interrupted by strikes, the suspension of mining in the Schuylkill region for four months, terminating August 1st, enabling, as President Gowen tells us, the Wyoming and Lackawanna Companies to obtain high prices. Yet the average price for 80,000 tons of Scranton coal, sold by auction, August 31st, 1870, was only $4.83. We bring these prices together in a line in order that the public may compare them·

Average price, August 31st, 1870....................$4.83
Average price, August 30th, 1871.................... 5.03
Average price, August 28th, 1872.................... 3.46
Average price, August 27th, 1873.................... 5.17

"We confess our inability to do justice to the facts which the above figures so eloquently proclaim. In 1870 and 1871, the supply of coal was restricted by strikes of the laborers, which for magnitude and obstinate persistency are without a parallel in the history of Trades Unions. Last year is the only one of the four when the price of coal was regulated by natural laws, with both supply and demand unimpeded. This year we have a conspiracy of the coal corporations, and as the result of that conspiracy we have prices already far above the point to which strikes and suspensions carried them in 1870 and 1871. The State of Pennsylvania might have levied a tax on anthracite coal of the difference between this and last year's prices, and paid her entire bonded debt in one year. The annual product of such a tax would extinguish the whole debt of the State of New York, Canal bonds and all, even with the present deficiencies in the Sinking Funds. Yet people are expected to pay over these millions to the wealthiest corporations in the world without a murmur. We are told by Mr. Mead, the secretary of the little twenty per cent. Pennsylvania Coal Company, that it is 'beneficial' to us to submit to this extortion; that it is 'right and reasonable' that we should pay this tax. Here have these companies for five years been fighting each other and their private competitors, and distributing all the while ten, fifteen, and twenty per cent. dividends; and now they come forward and extort a coal tax of more than a dollar per ton over and above what would give them a fair profit, and they claim to be public benefactors! What colossal impudence!

"Another man, Mr. R. G. Moulton, the general agent of the Delaware & Hudson Canal Company, gives his views: 'If the low rates had continued, the result would have been a general stoppage of the business and a tremendous inflation of prices.' As well talk of a general stoppage of agriculture! In the opinion of these gentlemen, prices have now reached a 'healthy level.' To the shallow twaddle of these tax-gatherers, to which, four weeks ago, we appropriated a column of our space, we oppose the eminently sound and conservative views of President Gowen. 'High prices and uncertainty of supply,' says Mr. Gowen, 'will drive away buyers, force manufacturers to turn to other fuels, and prevent the natural increase of demand which would result from low prices, and which would soon supply a certain market for any temporary over-production.' ('Reading Railroad Report,' 1871, page 18.) It is also certain that Mr. Gowen regarded the over-production of last year as only temporary, for he states his belief that the stock on hand at the close of the season was no greater than in the year preceding; argues that the market will require three million tons more in 1873 than in 1872, and concludes that it is a reasonable supposition 'that it will be difficult to produce this year any quantity so greatly in excess of the demand as to depress the market to any considerable degree.' ('Reading Railroad Report,' 1873, pages 20 and 21.) We might quote further from Mr. Gowen, but have we not quoted enough to show that, as a disinterested observer, he would have pronounced the work of this Anthracite Coal Combination to be one of the most inexcusable

WESTERN COAL MINE.

displays of rapacious greed that the world has witnessed since Christianity became the religion of civilized man?"

There is no good reason why coal that costs two dollars per ton at the mines should cost seven dollars to the Eastern consumer. The actual cost of transportation could be paid and a fair profit returned to the railroad companies, and yet the price of coal be reduced very much. But a fair profit is not to the taste of the monopolist. He must have an exorbitant return for his capital, or he is not satisfied. He cares nothing for the thousands of suffering poor who are unable to procure fuel at the prices at which he holds it. He is deaf to the voice of humanity; he thinks only of his gains. So the curse of the railroad monopolist hangs like a black pall over the Republic, growing darker and darker as the time wears on. His greed and heartlessness have made bread dear in a land capable of feeding the world, and the same accursed spirit of avarice has put the next great necessity of human life almost beyond the reach of the poor. "The cheapness of food and coal most concerns the comfort of the people; to lower their price must be the aim of every popular government. Yet both with us have become the subjects of monopolies, and are dealt out to the people by the great companies in such quantities as they think will aid them best in paying their dividends. We are threatened with a dearth of coal, because the companies have resolved that there shall be one. There is no complaint of any deficiency in the supply, of any failure of the mines, or any want of labor, but the coal companies and the railroads combine to stop the production of coal in order to raise the price, and are willing to starve the miner and the consumer to enlarge their own profits. It is of no consequence to them that the poor must suffer or perish, that every honest working family must be pinched and straitened, that manufactures are impeded and commerce checked, so long as their dividends are maintained and their wasteful extravagance supplied. The price of coal has already risen one-third; by December, if the companies choose, it may be doubled. The mines are to be left unworked, the coal retained, and the all-powerful companies' rule unchecked over the helpless people. It has long been their custom to produce these periods of unnecessary dearth; they scarcely seem conscious of the cruelty of their policy, or of the painful consequences of their avarice."

It is an ugly state of affairs, but there is no denying

the facts in the case. The people of the New England and Middle States are utterly at the mercy of the great corporations controlling the Pennsylvania coal fields. They must pay for their fuel just what these companies choose to ask for it, and they have no means of escaping from their dilemma if the companies are left in their present condition. The companies have been steadily increasing their exactions, and unless something is done to check them, they will no doubt increase them to a point at which anthracite coal will become accessible only to the rich.

Doubtless these companies have rights, among which is the right to earn a fair return for the labor and capital invested in their business. But the people, the consumers, have a right in the matter, which they will yet be driven to assert. The Almighty did not create the coal beds of Pennsylvania for the sole benefit of the railroads and coal corporations that have secured the control of them. He placed this magnificent gift in a region easily accessible, for the benefit of the millions who people the vast region it is intended to supply. The people have a right to obtain it at moderate rates, and they have a right to compel the great monopoly that is bleeding them so unmercifully to respect this claim; and the day may come, and ought to come, if there is not a change for the better, when the coal monopolist will find that vested rights, and charters, and stocks and bonds, are powerless to restrain the wrath of a defrauded people bent upon supplying one of the chief necessities of their existence.

It has been suggested that a remedy could be found in the introduction of the bituminous coal of the Western States into the market of the East. This fuel, however, would be subject to a heavy transportation tax which would be levied upon it by the railroads over which it would have to pass, and this would go far towards preventing it from becoming a formidable rival to Pennsylvania coal.

Another remedy that has been proposed is the construction of a freight railroad by the General Government from New York to Chicago. It is believed by the advocates of this scheme that such a road, on which minimum rates of transportation would be charged, would bring Western coal to the Eastern markets at rates which would effectually break down the anthracite monopoly. There can be no doubt that if the Western coal could be laid down in New York and Boston at reasonable rates, it would prove a formidable rival to anthracite; but the railroad rates, as at present managed, preserve the supremacy of the kindred monopoly. Should the proposed Government road touch the Pennsylvania coal regions on its way to the West, another blow would be struck at the monopoly. By affording cheap transportation to the Eastern markets, it would reduce the price of coal at least one-half. It is the cost of transportation alone that raises the price of fuel, and not the cost of producing it.

A recent number of *Harper's Weekly*, commenting upon the extortions of the coal monopoly, thus speaks of the necessity for such a railroad:

"But the future of the coal trade is still more suggestive. Soon the mines of Pennsylvania and the West must supply the fuel of the world. The mines of England, yielding 120,000,000 tons a year, already show signs of exhaustion. Coal has doubled in price in England within three years. English iron manufacturers are turning to the United States as the scene of their future successes. It is rumored in Philadelphia that a prominent English firm engaged in building steamers has resolved to remove with all its capital and labor to the banks of the Delaware. It is not unlikely that Western Virginia and Pennsylvania may soon supply the factories of Europe with fuel, that the great iron-works of the world will follow the line of coal from Pittsburg to the Tennessee; and it seems more than ever the duty of the National Government to prevent this great trade from falling into the hands of monopolists. A Government railway penetrating the West from New York to Chicago seems the only means of opening to the world the immense masses of coal that lie every where scattered through the inaccessible country. Our export of fuel and the growth of our iron factories depend upon the cheapness of coal. It seems the duty of the National Government to provide at least economical transportation; and the best mode of tempting the steam-ship builders from the banks of the Clyde to the Hudson or the Delaware would be to provide a sufficient communication between the mines and the sea. Nor with a Government route would the great monopolies ever be able to prey upon the famishing people."

But will Congress build such a road? Will that honorable body condescend to consider the rights of the people against the interests of the railroads? We confess we have little hope of it. The Congress of the United States, so far from showing any disposition to check the extortions of the monopolists, has aided the coal monopoly in its robberies of the people by compelling the people of the Eastern States to purchase

Pennsylvania coal. It has, in its eagerness to serve the men who are plundering its masters—the people—levied a prohibitory duty upon coal imported from the British Provinces. The rich coal fields of Nova Scotia lie at the very doors of New England, and a fine quality of coal could be delivered at the New England ports at prices lower than the Pennsylvania dealers are demanding. It is to the interest of New England to purchase this coal. With its markets thus supplied, it would be relieved of its dependence upon Pennsylvania, and the competition thus introduced would result in a scale of moderate prices. The anthracite monopolist would be deprived of his power to rob the people, and would be obliged to sell his coal for its actual value. But Congress has joined the coalition against the people, and in order that the Pennsylvania coal ring may rob and plunder the community, by charging unfair prices for its coal, it has levied upon foreign coal a duty which keeps it out of the market.

Let the duty be taken off foreign coal; let there be free trade in this great necessity of life, and a very different state of affairs will ensue. Let the people enjoy the benefit of a free market, and let them be rid of their slavery to an insolent and unscrupulous monopoly.

Think of this, farmers and workmen of New England, as you sit by your costly coal fires, and reckon the value of each lump of the precious fuel. Demand of your servants in Congress that justice shall be done you, and that you shall be able to buy your coal cheap in a land which is the richest of all lands in that mineral. You have a right to a free market. You have a right to your hard earnings, and it is a shame to your manhood to submit to such infamous extortion as is being practised upon you with the assistance of your servants in Congress. The remedy lies in your own hands. Members of Congress will do much for the monopolists, but they will not dare to resist the determined expression of your will, for do you not hold their official lives in your hands? and can you not make or unmake them with a ballot? Organize, combine for the protection of your interests. Join hands with the farmers of the West in their courageous war upon monopolies of all kinds. There is not a monopoly in the land but can be broken to atoms by the combined and determined action of an indignant and wronged people.

PART III.

THE FARMERS' WRONGS.

CHAPTER XV.

THE AGRICULTURAL CLASSES AND THEIR WRONGS.

Detailed Statement of the Agricultural Wealth of the United States, and of the Strength of the Agricultural Class—The American Farmer—His Defects and Virtues—His Character as a Man and a Citizen—The Superior of the Old World Farmer—He should be the most independent and contented Man on Earth—The actual State of Affairs—Hard Lot of the American Farmer—Difficulty of making the Farm pay—A real Grievance—Wrongs of the Farmer—The Effect upon the Young Men—Driven from Home—Sad Story of a Farmer's Daughter—Not an isolated Case—Cause for Apprehension—A Remedy needed.

In the Ninth Census of the United States, taken in the year 1870, the number of persons engaged in all classes of occupations was 12,505,923. Of these 5,922,471 were persons engaged in agricultural pursuits. The rest of the great army of workers was divided as follows: Persons engaged in personal and professional services, 2,684,793. Persons engaged in trade and transportation, 1,191,238. Persons engaged in manufactures and mechanical and mining industries, 2,707,421. Thus it will be seen that nearly one-half of the industrial class of the United States is engaged in agricultural pursuits. Taking the number of inhabitants of the Union of ten years of age and over, which is

28,228,945, we gain a still clearer conception of the strength and importance of the agricultural class of the country.

The agricultural workers are thus subdivided, according to the Census quoted:

	Total.	Males.	Females.
Agricultural laborers	2,885,996	2,512,664	373,332
Apiarists	136	136	——
Dairymen and dairywomen	3,550	3,133	417
Farm and plantation overseers	3,609	3,609	——
Farmers and planters	2,977,711	2,955,030	22,681
Florists	1,085	1,046	39
Gardeners and nurserymen	31,435	31,202	233
Stock-drovers	3,181	3,181	——
Stock-herders	5,590	5,545	45
Stock-raisers	6,588	6,558	30
Turpentine farmers	361	361	——
Turpentine laborers	2,117	1,933	184
Vinegrowers	1,112	1,105	7

This being the force employed, let us glance at the field in which its operations are performed, and the results accomplished by it.

In 1870, the total area of land in farms amounted to 407,735,041 acres. This was divided as follows:

Acres of improved land	188,921,099
" " woodland	159,310,177
" " other unimproved land	59,503,765

The total number of farms was 2,659,985. These were reported as follows:

Under 5 acres	6,875
5 acres and under 10	172,021
10 acres and under 20	294,607
20 acres and under 50	847,614
50 acres and under 100	754,221
100 acres and under 500	565,054
500 acres and under 1000	15,873
1000 acres and over	3,720

Average size of farms.........................153 acres.

The capital invested in the farms and their products was stated as follows:

Cash value of farms	$9,262,803,861
Cash value of farming implements and machinery	336,878,429
Total amount of wages paid during the year, including the value of board	310,286,285
Total (estimated) value of all farm productions, including betterments and additions to stock	2,447,538,658
Value of orchard products	47,335,189
Value of market-garden products	20,719,229
Value of forest products	36,808,277
Value of homemade manufactures	23,423,332
Value of animals slaughtered, or sold for slaughter	398,956,376
Value of all live stock	1,525,276,457

The live stock was stated as follows:

Number of horses	7,145,370
Number of mules and asses	1,125,415
Number of milch cows	8,935,332
Number of working oxen	1,319,271
Number of other cattle	13,566,005
Number of sheep	28,477,951
Number of swine	25,134,569

The farm products were as follows:

Bushels of Spring wheat	112,549,733
Bushels of Winter wheat	175,195,893
Bushels of sweet potatoes	21,709,824
Bushels of Irish potatoes	143,337,473
Pounds of butter	514,092,683
Pounds of cheese	53,492,153
Gallons of milk sold	235,500,599
Bushels of clover seed	639,657
Bushels of grass seed	583,188
Hogsheads of cane sugar	87,043
Hogsheads of sorghum sugar	24
Pounds of maple sugar	28,443,645
Gallons of cane molasses	66,593,323
Gallons of sorghum molasses	10,050,089
Gallons of maple molasses	921,057
Pounds of beeswax	631,129
Pounds of honey	14,702,815
Bushels of rye	19,918,795
Bushels of Indian corn	769,944,549
Bushels of oats	282,107,157
Bushels of barley	29,761,305

Bushels of buckwheat	9,821,721
Pounds of tobacco	262,735,341
Bales of cotton	3,011,996
Pounds of rice	73,635,021
Pounds of wool	100,102,386
Bushels of peas and beans	5,746,027
Gallons of wine	3,092,330
Tons of hay	27,316,048
Pounds of hops	25,456,669
Tons of hemp	12,746
Pounds of flax	27,133,034
Bushels of flaxseed	1,730,444
Pounds of silk cocoons	3,937

These figures present to the reader a fair average of the condition of the farming interest of the country. They show that nearly one-half of the whole toiling class, the men and women who use hands and brain to procure their daily bread, are engaged in the labors of the farm and in pursuits kindred to it. These constitute the real backbone and sinew of the country, and are its mainstay in times of trial and danger.

Whatever may be the merits of other classes, it cannot be denied that our agricultural community comprises a population of which any country may be proud. The American farmer is, as a rule, an intelligent, clear-headed, practical man. He is the possessor of, at least, a common school education. A reader of the newspapers and a lover of books, he manages to keep himself abreast of the questions of the day, and has definite and intelligent opinions concerning them, which he is able to express vigorously when occasion demands. He is strong-armed as well as strong-minded, and his out-door life keeps him in robust health. He is industrious, ambitious to improve his temporal condition, and attentive to his duties. As a citizen he is faithful to the obligations imposed upon him by the

WAGONING GRAIN TO MARKET.

laws of the land, and deeply interested in the welfare of his country. As the head of a family he is kind, affectionate, earnestly striving to advance the welfare of those dependent on him. In short, his ambition extends to two things chiefly—to provide for his family in such a way that his children may have a comfortable and happy home, and enter upon life prepared for its struggles, possessing vigorous bodies and well-trained minds; and to make his farm the best in the county.

No one who has travelled through agricultural districts of the Old World, can withhold from the American farmer his meed of praise. The farmer of England is, at the best, the inferior of the American. He is coarser, worse in morals and general habits, and inferior in education; while on the continent, the farmer is little better than a slave.

Now, in the natural order of things, and especially in this happy land of freedom, the farmer should be the most fortunate man in the community. Owning his land, with good health, dependent only upon the bounty of the Giver of all Good, his should be a life of absolute independence. Not that this should free him from the great necessity of earning these blessings by the sweat of his brow, but that his hard labor should be blessed with a fair reward, and that he should be able to look forward with confidence to a comfortable provision for his old age, and a fair start in life for his children. Yet, looking around upon the community, we find matters very different. The condition of affairs which meets us at every turn is not that which should exist in a well-regulated state of society.

Instead of the lighthearted, independent, contented owner of a domain sufficient to support him in comfort, we find the American farmer a toiling, overworked man from the beginning to the close of his life. However intelligent he may be, however determined to succeed, we find him, as a rule, doing little more than providing food and clothing for those dependent on him, often struggling under a load of debt, and conscious that he is not receiving as fair a return for his labor as the merchant or mechanic receives for his. There is something wrong in the system which thus dooms him to perpetual slavery, and it will be interesting to endeavor to ascertain the causes of the evil which has come upon him.

It is a common saying that "no man makes a fortune on a farm." It is true there are instances of great wealth amassed by farmers, but they are the exceptions. These fortunate persons have been possessed of large capital, which has enabled them to conduct their operations on a scale which compelled large returns. The average farmer, however, is a man of limited means, and it taxes all his energy to make "both ends meet" at the end of the year. He may dream of great wealth, and, in his mind's eye, may see himself a capitalist, able to carry out his theories upon a proper scale; but the golden harvest never comes to him, and he ends his days where he began, a struggling man, vainly grappling with a difficulty that is too great for him. Yet he labors as hard, as honestly, and as intelligently as the merchant or manufacturer. He has, perhaps, quite as much capital invested as they, and yet in spite of his efforts, in spite of his intelligence, in spite of his honesty, he cannot keep up with them in the race for wealth. He is distanced by men of less ability, and he cannot help himself. The most that the majority of farmers are capable of achieving is to become the absolute masters of their property. He is a lucky man who can do this, who can keep the farm clear of mortgages and himself free from debt, and earn enough to educate his children and afford his family the refinements and comforts of our present civilization.

Now here is a real grievance; here is a most singular state of affairs. The most useful class in the land, that which should be the most fortunate and independent, is becoming the most oppressed, the hardest worked, and the poorest paid. There is no necessity for this state of affairs. It should not exist. But it is a reality, and the farmers know it, and their efforts to remedy it have thus far been without success. Every year the trouble is becoming more serious. The young

19

ST. PAUL, MINNESOTA.

men, seeing the fate of their fathers, their lives of constant toil, and the hopelessness of realizing a proper return for their industry, are leaving the farm and flocking to the great cities, to seek in other pursuits the rewards denied them in the calling they prefer, and to which they are best suited. They leave behind them the sweet restraints of home, and the happy innocence of a country life, and come to the cities to meet trials and temptations to which they too often fall victims.

The young women, dismayed by the hard lot of their mothers, and wishing to escape from the drudgery which is the inevitable doom of the farmer's wife, follow the example of their brothers. They come to the city, where there is no room for them, and where snares and dangers lie thick along their paths. Too often they yield to them, and go down into the dreadful abyss from which no woman ever comes back. Oh! the sad stories of farmers' daughters that one hears in the great city of New York. How they come crowding there, year after year, frightened by the hard life at home, and, in their eagerness to escape from it, rushing upon a doom so terrible that even the hardest lot of honest labor would be joyfully embraced could they but see the end when taking the first step.

Not long ago, a gray haired old man came to New York from his farm in New England, in search of his daughter. He was accompanied by his son. He told a sad tale to the police. Owning his farm, and nominally well to do in the world, he was still barely able to provide a support for his family. One of his daughters —the eldest—appreciating his difficulties, and wishing to help him by relieving him of her support, had left home with his consent, and had come to the great city to obtain work. For awhile she succeeded, and not only earned enough to keep her in comfort, but managed to send an occasional remittance to the old people at home. By and by there was a change. Her letters became rarer, and at length ceased. Weeks passed away, but no letter came, and, in alarm, the old man had come to the city to find his child. He was sure she was not dead, and he dreaded to find that she had gone the way that so many go. He applied to the police, and in the official to whom he confided his story he found a sympathizing friend. Inquiries among the members of the force enabled the officers to recognize the girl by her father's description as one who had but recently become known to them as a lost woman, and they at once led the father and brother to the house in which she was living. As they entered the brightly lighted parlors, the girl recognized her father, and, with a wild cry of joy, sprang into his arms. There was no need of explanation, and the father had no reproaches for the poor, sinful wanderer. Lifting her tenderly in his arms, he bore her to the street, sobbing out joyfully as they passed the portals of the house of shame:

"We've saved her, thank God! We've saved Lizzie!"

That night all three left for their distant home.

This is not an isolated case. There is many an unfortunate who has trod the hard streets of the metropolis, with no eye but the pitying one above to look kindly on her, who could tell of a home from which the hard hand of poverty drove her.

It is a most lamentable condition of affairs that makes these things possible. It is not so with other callings. The labor of the merchant, the manufacturer, the small trader, the professional man, enables him to do better for his children, and to give them a home of refinement and happiness, which they enjoy, and which makes them eager to provide homes of their own when they arrive at man's and woman's estate. But the reverse is too often the case with the farmer. His sons have no wish to toil as their father, and they seek to escape from the farm, and change their pursuits as soon as they can. The farmer's daughter, seeing her mother's daily toil and care, and the constant struggle of both parents to earn a decent living, has no wish to become a farmer's wife, and she too seeks to leave home at the earliest possible moment.

Now these evils, like most of those from which the farmer suffers, are traceable to the cause we have been considering. The farmer does not receive a proper return for his labor, and is therefore unable to provide for his family as he should. The same amount of labor, of energy, intelligence, and capital in other pursuits would produce different and happier results. Here they do not produce even fair results.

The whole farming community will bear testimony to the truth of our assertion. We are stating no imaginary case. It is a sad history of facts we are relating. We assert that the farmer's grievances are real, and not fictitious. They deserve from the entire community a patient hearing. He who can find a remedy for them will deserve and receive the thanks of a grateful country.

CHAPTER XVI.

THE MIDDLE-MEN.

A Leading Cause of the Distress of Farmers—Working at Starvation Prices—High Price of Bread—Who is responsible for it—How the Middle-Men grow Rich at the Expense of the Farmer—An Unequal Division of Profits—The Farmer receives too Little—Comparison between Agricultural and Manufacturing Profits—The Story of Two Brothers—A Lesson for Farmers—Profitable and Unprofitable Labor—Contrast between the Middle-Men and the Farmers—Where the Profit on Grain goes—A Palace and a Farm House—Who pay for the Splendors of the Large Cities—Need of the Farmer for Ready Money—How this Necessity is taken Advantage of—The Local Grain Dealers—How they Plunder the Farmers—The Excess of Western Production—The Real Cause of it.

ONE of the principal causes of the great distress prevailing among the farming interest to-day is the low price which the farmer receives for his products. There can be no question that his labor is repaid at too low a rate, and that until he receives a fairer price for his toil and skill he must be content to struggle on as he is forced to do at present.

Now, we are aware that there will be many who will take alarm at the idea of an increase in the price of breadstuffs, and who will meet our assertion with the old argument that the people pay enough for their bread already. We are fully aware of this. We believe that it would be a great evil to the country at large to increase the price of bread, and that that article is already sufficiently high; but we maintain that this state of affairs is not due to the farmer.

The farmer, the producer of the food of the country, is compelled to make a heavy outlay of money and time in order to make a fair crop of grain. To this outlay he adds his labor, which is long and severe. When his crop is harvested and sent to market, he receives a small profit upon his investment in the price paid for his grain, and this is still further reduced by the iniquitous freights levied upon him by the railroads for transporting his produce to market. The grain upon reaching market passes into other hands, and sells at an advance upon the price paid the farmer. The next step is to convert it into flour or meal, and here another profit is added to its cost, and one very much larger than that received by the farmer. Millers are fortunate men, and they are experts in the art of making money. The flour or meal next passes into the hands of the commission merchant, or the retail dealer, and by the time it reaches the final consumer the cost is vastly increased by the number of profits it has to pay. As flour or bread a price is paid for it which the whole community pronounce too high, and the result is that the farmer is universally denounced for his rapacity. He is believed by the average consumer of bread to be enjoying the proceeds of an extortionate price for the chief article of human consumption.

Now the truth is, that of all the profits we have enumerated, that of the farmer is the smallest and the most unfair. It is not in proportion to that of the merchant or the miller. He is robbed by the railroads in the first instance, and in the next place his price is kept down in order that the grain merchant and the miller may enlarge their profits. These worthy gentlemen are shrewd enough to join in the general demand for

THE MIDDLE-MAN'S DREAM OF HIS PLUNDER.

cheaper food, and in denouncing the farmer for keeping up the price of bread, when it is their heavy profits that keep up the high prices. They know where the trouble lies, but by denouncing the farmer, they seek to screen themselves.

The merchant and the miller make too much upon their investment of labor and capital, and the farmer makes too little upon his. Matters should be adjusted upon a different basis. The middle-men should be content with a smaller profit, and the profit of the farmer should be increased. It will be difficult to do this, for the middle-men, being masters of the situation, will not be satisfied without a certain return for their labor, which return they place at a very high figure, and the farmer, who is entirely dependent upon them, must be satisfied with what he can get. Yet the farmer's labor is as much entitled to reward as that of the middle-man, and it is fair that his profit should be as large in proportion as that of any one through whose hands his grain passes. But, instead of this, instead of receiving the reward to which his honest labor is entitled, he is compelled to accept a miserably small return for that labor, in order that the middle-men may make large profits and grow rich. They grow rich at the expense of the farmer and the community. The farmer either remains where he started or grows poorer.

Some time ago Mr. John T. Campbell, of Rockville, Indiana, undertook to compare the average profits upon the capital of the country invested in manufactures and that invested in farming, as reported by the last Census of the United States:

"Let us take," says he, "a general survey of the statistics as shown by the census reports of the three last decades, and carefully note the comparison between the general manufacturing interests of the nation, much of which is protected, and the general agricultural interests, and see if it teaches us a lesson.

"But we would be discouraged from this comparison by General Walker's foot note to the compendium of the Ninth Census, page 797, in which he says in substance that capital reported as invested in manufactures is entirely untrustworthy and delusive, and that it is the one inquiry which manufacturers resent as uselessly obtrusive, and that they could not tell what they were worth or how much of their wealth was invested as capital if they were disposed to. He thinks the capital that yields a gross product annually of four and a quarter billions should be reported at *eight billions* instead of two billions, as shown by the Ninth Census, and recommends the abandonment of that inquiry in the future.

"Perhaps the following showing will explain why the manufacturers are so reluctant about showing their capital, and also that generally the amount of their capital is *overstated* rather than *understated*.

"While the statistics may not be strictly accurate, they are a safe guide by which to compare one industry with another, or the same industry at different periods.

"In 1850 the value of the manufactured product, after deducting the cost of wages and raw material, was 42½ per cent. on the capital invested. This, of course, was not the clear profit on capital, for doubtless the owners have greatly augmented its profits by the assistance of their own labor.

"In 1860 the profits on manufacturing capital (assisted by the owners' labor, as before), after deducting the cost of wages and raw material, was 47 per cent. per annum.

"In 1870 the manufacturing capital, put under like conditions as before, yielded a profit of nearly 46 per cent. per annum, or including mining and fishing, as per census of 1850 and 1860, 44⅓ per cent.

"However 'untrustworthy' the census reports may be as to the amount of capital engaged in manufacture, here is a uniform percentage for three decades.

"The value of the total agricultural product not being given for any but the census of 1870, I cannot compare the profits of capital invested in agricultural industry with the profits of capital invested in manufacturing industry for any but that year, but it will doubtless be a good indication of the results in 1850 and 1860.

"In 1870 the value of the entire agricultural product, after deducting the cost of wages, as in the case of manufacturers, was 19⅕ per cent. per annum on the capital invested in agriculture, assisted by the owners' labor.

"In this case I assume the value of the farm, farming implements, and live stock (less its annual increase) to be the farmer's capital, to which *should* be added the value of the seed for the crops and provender for the stock.

"From 1850 to 1860 manufacturing capital increased 89⅓ per cent., while the number of establishments increased only 14⅙ per cent., and the number of operatives increased only 37 per cent.

"This leads *one* hereditary protectionist to the conclusion that manufacturing capital has but little tendency to diffuse itself among the people; but on the contrary, that those who possessed it in 1850 also possessed about 85 per cent. of the increase from 1850 to 1860.

"From 1860 to 1870 manufacturing capital (including mining and fishing, as these branches of industry were classed with and included in manufacturing statistics in 1850–60) increased 132½ per cent. The number of establishments increased nearly 87 per cent.; and the number of operatives increased only 70 per cent.

"While this showing is much better than the previous one, it still shows that the fortunate few who had possession of the capital still get the lion's share of the increase.

"The per cent. on the dutiable imports from 1850

to 1860 averaged about 25 per cent., and from 1860 to 1870 about 40 per cent. Possibly, or even *probably* this high duty had much or *most* to do with increasing the capital—the number of establishments and the number of operatives during the last decade—but must they continue to be propped up by protection at the expense of the greatly overworked agricultural laborers? For it really appears this way. Perhaps some one better versed in statistics can see some other and greater reason why the agricultural capital of the nation, consisting of farm implements and live stock, increased 101½ per cent. from 1850 to 1860, during which decade the average duties on imports was about 25 per cent., and during which period manufacturing capital increased only 89⅓ per cent. And why from 1860 to 1870 agricultural capital increased only 39½ per cent., with an average duty of 40 per cent., while manufacturing capital increased 132½ per cent.

"It is not sufficient to answer that the ravages of war have depreciated the value of farming capital, for other things being equal, manufacturing capital would suffer as much as farming capital, and would be as difficult to resuscitate.

"The State of Louisiana has quite as bad a showing of agricultural capital as any Southern State, and the following is the comparison of increase in agricultural and manufacturing capital for the last two decades in that State: Agricultural capital increased from 1850 to 1860 102 per cent., and from 1860 to 1870 it decreased 55 per cent. Manufacturing capital increased from 1850 to 1860 42 per cent., and from 1860 to 1870 156 per cent. Perhaps the people are lazy in Louisiana and the other late rebel States, and the manufacturing capital has been taken there by the Yankees. Let us therefore take Ohio, a first-class agricultural and manufacturing State with an industrious population.

"From 1850 to 1860 agricultural capital in Ohio increased 87 per cent., and from 1860 to 1870 it increased only 54⅔ per cent. Manufacturing capital increased from 1850 to 1860 97 per cent., and from 1860 to 1870 it increased 164 per cent.

"Perhaps it may be urged that agricultural capital in Ohio has about reached its climax and cannot expand much more, while manufacturing capital may expand illimitably. If so to any extent in Ohio it would be doubly or trebly so in the older States. Let us examine Connecticut. Agricultural capital increased from 1850 to 1860 27¼ per cent., and from 1860 to 1870 it increased nearly 39 per cent., showing that it is possible for agricultural capital to still increase greatly in value even in the oldest States. The manufacturing capital (in Connecticut) increased from 1850 to 1860 76⅕ per cent., and from 1860 to 1870 it increased 109 per cent. In this old State, as elsewhere, to him that had it was given, for the number of establishments increased from 1850 to 1860 19½ per cent., and from 1860 to 1870 they increased 69½ per cent.

"How about the operatives? They increased in number from 1850 to 1860 27 per cent., and from 1860 to 1870 they increased 39 per cent.

"The number of operatives have not maintained their former ratio to the capital on which they are employed. In 1850 there was an operative in the United States to every $557 of manufacturing capital, in 1860 one to every $770 of capital, and in 1870 one to every $1031 of capital.

"In Pennsylvania there was in 1850 an operative for every $644 of manufacturing capital, in 1860 an operative to every $855, and in 1870 an operative to every $1270.

"In the State of Massachusetts in 1850 there was an operative to every $501 of manufacturing capital; in 1860 an operative to every $611 of capital, and in 1870 an operative to every $829 of capital. These examples are a fair type of the rest. Protection in this country is fast coming to mean Protection for improved machinery with a Chinaman to operate it, against cheap labor in Europe.

"Let us see if the agriculturists of Pennsylvania make enough on their capital and labor to pay for protecting the manufacturers of that State. In 1870 the agricultural product of Pennsylvania, less the cost of wages, was 13½ per cent. on the agricultural capital assisted by the owners' labor, while the manufacturing product, less cost of wages and raw material, was 40 per cent. assisted by the owners' labor. But possibly agricultural capital in Pennsylvania is rated too high, for there is a limit to what the land may be made to produce, though there is no well-defined limit to the price which the presence of a large landless population may be the innocent cause of putting on it. Let us examine Illinois, where land is cheap and productive and markets easy of access. If the Illinois farmer can't keep pace with the manufacturer none other can.

"In 1870 the value of the agricultural product of Illinois, less the cost of wages, was 17 per cent. on the agricultural capital assisted by the owners' labor; while the manufactured product, less the cost of wages and raw material, was 49⅞ per cent. on the manufacturing capital of that State for the same year assisted by the owners' labor.

"In short, 5,500,000 laborers, operating with $11,000,000,000 of capital, can give a product of only $2,500,000,000 a year in agriculture; while 2,000,000 of laborers, operating on $2,000,000,000 capital in manufacturing, can give a product, less the cost of the raw material, of $1,500,000,000.

"While we may have to concede that certain specified industries will languish and fail if not protected, we also insist that the field in general manufacturing is wide enough and profitable enough to fairly employ all that may fail in special industries."

These figures are eloquent, and go far to sustain the assertions we have made.

Thirty years ago, two brothers started out in life. The elder, loving the country, its pure air and free life, invested the sum of five thousand dollars, which he had inherited from his father, in a snug farm in the West. His farm lay in a growing section of country, within easy access of the markets, and the land was as good as any in the State. The young man, whom we shall call David Dean, had been brought up on a farm in New England, and was by nature in every way fitted for an agricultural proprietor. Under his management his farm prospered, and year by year it grew until he became one of the best to do farmers in his State. A man of intelligence and education, he kept himself abreast of the times, adopting every improvement in cultivation that his judgment sanctioned, and rendering his homestead in all respects a model farm. In due time he married. Children grew up around him, and these he was enabled to keep in comfort and to provide each with a fair education as an equipment for the battle of life. He counted himself a fortunate man, as, indeed, he was, for he was an exceptionally prosperous farmer. Thirty years of honest and intelligent labor brought him their reward, and when at last he was gathered to his fathers, he was found to have earned for his family, besides the support he had given them, a fine farm of several hundred acres, and about twenty thousand dollars in other investments. His whole estate was worth, perhaps $60,000, representing, apart from the amount expended in maintaining his family, an average gain of about $2000 per annum during the thirty years of his manhood.

The younger brother also inherited five thousand dollars from his father, but, being more ambitious than David, he obtained employment in New York. In those days $5000 was a good round sum, for we were then a nation of small dealers; and the young man, being provident and temperate in his habits, invested this in such a manner that it nearly supported him, leaving him a considerable portion of his salary, which he carefully invested. In the course of a few years he was enabled to purchase a minor interest in the business of the house in which he was employed, and being a steady, industrious and frugal man, his condition was bettered every year. In ten years from the day he entered the house he was the owner of half the business. In ten years more he had bought out the interests of the other partners, and was the sole owner of the business. He too had married, and children had grown up around him. At the end of the twenty years he was in possession of a splendid and lucrative business, which was increasing his wealth rapidly from year to year, and his home was

LIFE AMONG THE MIDDLE-MEN.

an elegant mansion replete with every luxury. At the end of the thirty years men called him a millionaire; and having occasion about that time to make an inventory of his possessions, he found that he was the master of the handsome fortune of two millions of dollars, invested in various forms. Apart from the cost of maintaining his family, his wealth represented an average gain of nearly $67,000 for every year of his working career. As he stood by the grave of his farmer brother, to whom he was deeply attached, and to whom he had often lent a helping hand, he inwardly thanked Heaven that he had chosen at the first to abandon the farm and adopt a mercantile life; and he was very clearly of the opinion that his brother David, whom men called fortunate, had received a very imperfect return for the capital and intelligent labor he had expended upon his calling.

And if this was his opinion, as a shrewd, practical man of business, of one who had been really fortunate in his pursuits, what must he have thought of the recompense of the vast army of farmers who are doomed to perpetual servitude in order that others not of their blood may grow fat upon their exertions?

It would be an interesting, if a very saddening journey of inspection, for one to visit the counting-houses of the fat, sleek dealers in grain, the forwarders and middle-men of our great cities, where every evidence of wealth and prosperity is visible, and where the huge ledgers contain the summary of the gains which have flowed into the house from the results of its sales and speculations in grain. It would be interesting to talk with the heads of the houses, the well-fed, well-clothed, soft-handed and portly members of the Produce Exchange, and learn how rosy are their views of life, how bright their anticipations of the future. It would be still more interesting to ride up with the head of the house, in his elegant coach with liveried servant, to his superb mansion in the fashionable quarter of the city; to enter with him and see how luxurious are all the appointments of the establishment; how soft and deep the carpets into which one's foot sinks noiselessly; how rich the frescoes, how superb the furniture, and how beautiful the general appearance of the place. It would be interesting, we say, to see all these, the evidences of wealth earned in the successful prosecution of the business of buying and selling grain; and it would be even more interesting, and should be deeply instructive, to go from these scenes of splendor, and visit in succession the homes of the farmers in whose grain the house that has been so successful has been operating. One might take the names from the accounts in the merchant's ledger. What a contrast there would be! It would be like passing from the master's mansion to the cabins of the slaves on a Southern plantation in the *ante-bellum* days. What a succession of plain and often unattractive homes we should find! Instead of the elegant mistress of the city mansion, we should find the worn, anxious, prematurely old farmer's wife, whose dreams of an easier lot have faded before the unceasing toil and care demanded of her. Instead of the fat, rosy, well-dressed middle-man, we should find the farmer in homespun, with hard, brown hand, and a man worn down in body and soul with care and toil in the present and anxiety for the future. Instead of the cheery views of the middle-man, we should find heavy-hearted broodings over the unsatisfactoriness of his position and the injustice with which his labors are repaid. We should find scores of honest, industrious, and deserving men toiling early and late, enduring hardships and privations, only to see their just reward taken from them to enrich the middle-man. Is it a wonder that they regard the grain dealer as their worst enemy?

The farmer is not a capitalist. He has very little ready money as a rule, and his need for cash is very great. Often when the harvest has been gathered in, and the crop is ready for the market, there are numerous expenses devolving upon the farmer which must be met in cash. There is but one way to obtain ready money, and that is to sell a portion of the grain that has been harvested to the local grain dealer, numbers of whom abound in every agricultural district. The farmer cannot wait for the slow process of sending his crop to a distant market, and awaiting the remittances of his commission merchant; and he often argues that the grain will bring him but little more after the expenses of transportation, selling, etc., are paid. His need of money is urgent, and a portion of his crop goes to the nearest local grain merchant; and when once sales of this kind have begun, the entire crop is usually disposed of in the same manner.

The local dealer is perfectly acquainted with the necessities of the farmer. He knows that nothing but the need of money has driven him to sell his crop in such a manner, and his offers are governed by this knowledge. He takes advantage of the farmer's necessity to offer him a price below the actual value of his grain, and the latter is obliged, or feels himself obliged, to accept it; and the crop that has cost so much labor and care and capital to produce, goes to swell the profits of the local grain dealer, and the farmer is doomed to another disappointment.

Said a gentleman, writing from Dubuque, Iowa, not long since, referring to the hardships endured by the farmers of that State:

"While every other interest seems to 'bear down' upon the farmer, he has been, until recently, entirely powerless to resist or to retaliate by putting up the price of what he has to sell. In a former letter I have explained how the railroads often play into the hands of the grain speculators, to the injury of the farmers. But these speculators have a way of combining against the farmers, independently of the railroads. The great bulk of the grain of this State has been sold by the producers at the nearest railroad station, and it has been very rare for a farmer to ship his own crop. The reasons for this are numerous: very few farmers know much about the methods of doing business in the cities; they are not acquainted with the commission merchants, and if they were would often be afraid to trust them. Again, it often happens that the farmer desires to sell a load of grain and get the money for it immediately without waiting for it to be sold in Chicago, while he may not desire to sell enough at the present state of the market to load one or more cars. The result of these circumstances is that the local wheat buyers have had almost a monopoly of the shipping business, and, as I have already remarked, the bulk of the crop is sold at the nearest railroad station. But even here there is little or no competition between rival buyers, because they either agree each morning to pay a certain fixed price that day, or else agree not to outbid each other, and divide their profits in proportion to the amount each man has bought."

Thus the farmer is robbed year after year, and when he utters his complaint, there are loud outcries from the robbers that his grievances are purely imaginary, and that he has brought all his woes upon himself by an excessive production of grain, and is experiencing only the evils which attend upon an overstocked market.

But this outcry about a surplus of grain in the West is mere folly. The agriculture of the Western and especially the North-western States is of necessity confined to grain. "The North-west not only must produce cereals, but must produce a surplus. The hope that growth of manufactures may create a sufficient 'home market' in the farming States is cherished by many, in complete disregard of necessary conditions of manufacture, or the ratio of production to consumption of agricultural products. The average consumption of wheat is four and seventy-six hundreths bushels *per capita;* and of all cereals, including the quantity fed to animals, thirty-six bushels *per capita.* If it were possible to gather up all the hands employed in all the cotton mills of the United States and deposit them in a single county in Iowa, either one of the fourteen counties in that State now produces more wheat than all those hands could consume. All the hands employed in the factories and shops of the United States, if added to the present population of Illinois, would consume less than half the surplus of cereals now produced by that State. A mill of 273 hands on every farm of 100 acres of wheat would only suffice to consume the wheat which that farm would produce. Until hands in manufacturing establishments eat very much more than they are able to do at present, and manufacturers establish themselves without regard to natural facilities and resources, the great agricultural States will continue to produce a surplus of cereals. The costly exchange of products between farms and factories widely separated, supports a class which consumes nearly half as much as do the hands employed in manufactures. In the year 1871, about thirty-nine per cent. of the wheat grown in this country was consumed by farmers and those dependent upon them; about eighteen per cent. by persons employed in manufactures and those dependent upon them; about eighteen per cent. by those engaged in personal and professional services and others dependent upon them; about eight per cent. by persons engaged in trade and transportation and their dependents; and the rest, about seventeen per cent., was exported. The surplus of cereals in the North-western States, therefore, is not the result of accident or mistaken whim, but the inevitable consequence of fixed laws. Increasing density of population and cost of land steadily drive the larger operations of agriculture to regions more remote from the great centres of population, manufactures, and commerce, and to fresher and cheaper lands. New York produces less wheat and less corn than it did twenty years ago. The cost of moving the ever-increasing surplus of agricultural States, over a steadily increasing distance, to points where it is needed to supply an ever-increasing deficit in production, is a condition of the growth and prosperity of agriculture in this country which it cannot escape." *

* W. M. Grosvenor.

It is idle, then, to speak of the Western surplus of cereals as the fault of the farmer or the cause of the evils from which he suffers. The wrongs of which he complains spring from no such source. We have pointed them out, and in doing so have expressed the sense of the entire farming community.

CHAPTER XVII.

THE RAILROADS AND THE FARMERS.

Opportunity of the Railroads to plunder the Farmers—Extent of the Wheat Production of the United States—Amount consumed at Home—The Western Surplus—Amount of Corn produced—The System of High Freights—The West shut out from Market—Effect of the Civil War—Burning Corn for Fuel—Greed of the Railroad Companies—The Cost of getting Grain to Market—Facts for Farmers—Combination of the Railroads and the Middle-men—The Story of a Car Load of Corn—Mr. Walker's Views—The Farmers' Complaint—Railroads disregard the Law—Futile Efforts of the Western States to protect their Citizens—How High Freights are arranged—The Dependence of the Farmers upon the Railroads—The Effect of High Freights upon the Value of the Farm—A Startling Exhibit.

We have endeavored to familiarize the reader with the greed and tyranny of the railroad corporations. We come now to consider their dealings with the farmers, and to show how the latter are plundered by the corporations of their earnings by the iniquitous rates levied upon them. The farmers of the entire country are sufferers at the hands of the railroads, but the farmers of the Western States are their principal victims.

The reason of this will be evident when we consider the relative production and consumption of the various sections of the country. In 1870 the total wheat crop of the United States was 287,745,626 bushels. Of this the Western States, not including the Pacific States, or the Territories, produced over 202,000,000. About one-third of this vast amount is used for home consumption, and the rest, or about 130,000,000 of bushels, is shipped away to other States and to foreign countries. Exactly how much is exported to Europe depends upon the condition of the harvest in the Old World, but about 50,000,000 bushels are annually sent over the ocean. The remainder is sold in the American markets. The New England States are large consumers of Western wheat, requiring about 35,000,000 bushels more than they produce themselves. About 33,000,000 bushels are received in New York, and though some of this is finally exported, the bulk remains in the Empire State. Pennsylvania, New Jersey and Maryland require about 5,000,000 bushels more than they produce. The Southern States do not produce all they need, and are consequently large buyers of Western wheat.

All the surplus of the West, in order to find a market, must, to some extent, at least, pass over some railroad. The bulk of the wheat is sent to Chicago and the lake ports and to St. Louis, whence it finds its way during the season of navigation to the Erie Canal at Buffalo and the cities communicating with the canal, thence by the canal to Albany, whence all that is intended for the East and for export is sent to New York or Boston. But the season of navigation occupies only a portion of the year, and during the remainder the grain of the West must find its way eastward over one of the great railways. The cost of transportation to the East eats up about one-half of the value of the wheat, and the farmer's profit is made small in order that the heavy freights may be paid and the large profits of the middle-men gained.

But wheat is not the only grain crop that seeks a

CAMP OF WAGONERS HAULING GRAIN TO MARKET.

market. The production of Indian corn in the United States in 1870 amounted to 760,944,549 bushels. Of this product 492,664,536 bushels were grown in the Western States. Very much of this is used at home, but an immense quantity is shipped to the East for consumption there and for exportation.

Thus it will be seen that the Western States, being the principal grain producers of the country, are the most dependent upon the railroads, and, therefore, the greatest sufferers by their extortions.

Every farmer will doubtless remember the expectations that were held out to him, when the road upon which he is principally dependent was seeking the right of way and subscriptions towards defraying the cost of its construction. How he dreamed of the days which were to come when he could send his grain cheaply and rapidly to market. But alas, those dreams were never realized. The road once built, the rates were based upon a scale to meet the financial needs of its managers, and the farmers' interests were never considered. The road was built to make money for its stockholders, and the farmer must pay its tolls or his crops could not reach the market. The inducements and promises made by the corporations at the outset were never fulfilled.

When the civil war burst upon the country in 1861, the Western farmers were among the first sufferers. Their Southern market was closed entirely against them, and their great avenue to the sea, the Mississippi river, was lost to them. Corn fell rapidly in price until it reached ten cents per bushel. At this rate no one could afford to pay the cost of sending it to market, and it being cheaper than coal, vast quantities were consumed in the West for fuel. But as the war progressed, the needs of the Government and the Eastern States created a strong demand for Western corn, and the price of that grain rose in the Eastern markets. The Western farmers at once began to ship their corn East, hoping to make up some of the losses they had sustained. No sooner had the demand sprung up, however, than the railroads at once advanced their tolls, and the money that should have gone to repay the farmers' losses was secured by them by reason of their exorbitant freight charges. So burdensome were the exactions of the roads that there was a general outcry from the entire community against the outrage. Said the Secretary of the State Agricultural Society of Iowa, in his report for 1862: "Our great national highway to the ocean for two years has been closed, and we have been left to the tender mercies of relentless gamblers in railroad stocks. With facilities altogether inadequate to carry the marketable products of the teeming West, they have taken advantage of the necessities of the people to make one advance after another in their tariff of charges, until it now costs, in some instances, three times as much to carry our grain to market as it does to produce it."

In the ten years that have gone by since this complaint was made, neither Iowa nor any of the States have experienced any relief from the evils referred to. New roads have been built, but the rates have remained high. They have even grown higher. In a season of scarcity they are sufficient to throw a gloom over the entire farming interest; and when such a magnificent yield as that of 1872 blesses the country, the general joy of the agricultural community is embittered by a strong advance in freights. The roads combine against the farmers, be the season good or bad. The grain of the West must go to market, and the roads combine to demand what they please for its transportation. The farmers find that to get the product of one acre of corn to market, they must pay the railroad the product of three acres. The reader can easily calculate the result. A prominent farmer in Iowa recently declared to a correspondent of the New York *Tribune* that if the cost of producing grain was as great in Iowa as in the States farther east, the corn crop could not be sent to market at all.

The farmer, unlike other producers of a valuable commodity, is not allowed to fix the price of his produce. That is decided for him by the railroads and the middle-men, who generally work together, as they can well afford to do, the farmer paying the cost of their iniquitous combination.

Said an Iowa farmer recently:

"The railroads of this State discriminate unjustly against the farmers in the transportation of crops; that is, give other men advantages which they deny to the farmers. Let me explain: Here is a wheat or corn buyer who makes a living by purchasing grain of the farmers and shipping it to Chicago. Of course he makes a profit by it—grows rich, in fact. Now the farmers think that if they ship their own grain directly to Chicago they might save the profit that this middle-man makes. They engage a lot of cars, load them, and send them forward, but they find when they have paid the freight and the other expenses which the middle-man must necessarily also incur, they don't have as much left for their grain as he offered them. Now how is that explained? The railroad company gives the grain trader a drawback on the grain he ships, which it refuses to the farmers; and in some instances, at least, these traders are in partnership with railway officials. I thought, when the idea of cooperative shipments was first proposed, that these favors were given solely on account of the amount of business that these men brought to the railroads. I supposed that the deductions were simply those that would be naturally made to wholesale trade, and in speeches to the farmers

I told them so. But we have learned differently, for when our farmers have combined and offered freight in large quantities to the railroad companies, they have refused to give us the advantages which they give to the favorites.

"The terms of these contracts are secret. But we know that they must be considerable, or these men who have them could not make so much money. You see what this kind of railroad management amounts to. The company comes in and says: 'You shall sell your corn to a certain man and for a certain price, which we will fix.' That's one thing we complain of, and we will not long submit to it. But I haven't told you all. In certain cases the roads have fixed the rates of freight very high, and then men have appeared among the farmers, offering to buy our produce at prices just a shade higher than what it would net us to ship it ourselves, but at rates much below what it ought to bring us. We have often suspected that those men were the agents of the railroad companies or of the railroad managers. If our suspicions were correct, you see what an outrage upon the farmers it was. The railroad people knowing our necessities, and that many of us are obliged to sell, even at a loss, for the purpose of obtaining money, first arbitrarily fix the price of our produce and then force us to sell to them.

"Nor are these discriminations confined to our shipments east. They discriminate in favor of certain men in bringing freight westward, and in that way force us to trade with those men. Take salt, for instance, and let an association of farmers and a local trader purchase the same amount at the same price in Chicago. When that salt is in Iowa, the local trader, if there is strong competition, will retail it to the farmers cheaper than what their own cost them with the freight added. Now there must be some cat in that meal (or salt). It may be that in some cases the wholesale dealer may give the Iowa trader a drawback; but in others we know that he is favored with special rates by the railroads which they refuse to give to others shipping the same goods in like amount."

Said another:

"We fare worse than the man who fell among thieves between Jerusalem and Jericho. The great railroad corporations first extort from us everything they possibly can, and then they turn us over to Chicago to be still further plundered. Why, they don't allow us to say which elevator our grain shall go into when it reaches Chicago; we have no redress if the railroad don't deliver as much grain as we ship from here, and it is utterly impossible for us to have any of our grain passed as 'No. 1.' We may ship the best wheat that ever went to Chicago, and the probabilities are that they will mix it up with their 'imperial' wheat and make a 'No. 2' that will bring a higher price, and the increase that we ought to have goes to the owner of the elevator. We have no particular interest in Chicago's prosperity; indeed, if our grain could go forward without going into Chicago to be taxed for the benefit of her speculators, we should be much better off."

Some time ago a Philadelphia merchant stated that a car load of corn had been recently shipped to him from the interior of Iowa. The freight charges, commissions and other expenses amounted to $233.70, and the grain sold for $233.07, leaving a deficit to the shippers in addition to the value of the grain at the point of shipment. Another consignment of corn netted the shippers five cents a bushel, from which, however, was to be deducted the price of the corn at the place of production, which would entail a loss of ten or fifteen cents a bushel.

In some cases the farmers within twenty-five or thirty miles of a convenient market in the West, have found it cheaper to haul their grain to market in their own wagons than to ship it by the railroads.

Said the Hon. Amasa Walker, of Brookline, Mass., in a speech delivered in Boston on the 1st of September, 1873: "The farmer's products must be transported a great distance, and they are heavy products, and it takes a large part of what he can raise to pay the freight and get his goods to market. You have, perhaps, heard of the man who went out to Iowa and bought a lot of corn for thirteen cents, and, selling it in Springfield, Mass., for sixty-nine cents, made just one cent a bushel. Now that is a very startling illustration, but no more startling than true, of the manner in which the whole thing appears. Now you will see why the farmers are the first to move, because they are made so much more interested than others by its taking such a larger portion of what they can raise. For instance, a manufacturer will send on his goods West and pay not more than five per cent.—a twentieth or thirtieth part of what the farmer pays on his products, and the difference is a very wide one."

Said Mr. Stephen Smith, the Illinois farmer, speaking for the men of his calling in his own State, recognizing and pointing out the evils of railroad extortion: "For the past three or four years the conviction has been gradually forcing itself upon us that something was wrong in our affairs; for while every other industry was being fairly remunerated, we have been steadily going behind, until poverty, if not bankruptcy, stared us in the face. We found that, while we labored harder and more hours than the artisan and workman in other pursuits, we were forced to content ourselves with poorer food and clothing, with fewer social privileges, and less opportunities for mental cultivation than they. We could not help seeing that if they were as steady and industrious as we, they were able to live in better houses, and had more money to spend in their adornment than we had; that if they had the taste for such things, as most of them had, they had more pictures, books and newspapers, and more leisure to enjoy them, than we, and that they often indulged in such luxuries as lectures, concerts, excursions, and festivals, while it was rare, indeed, that we could afford to give wife and children one of these treats. Then we began to see that the men who did nothing but handle the products of our labor were still better off, and were getting rich while we were growing poor; that those who supplied us with the implements for our work added from twenty to fifty per cent. to the original cost, and charged it over to us; that the merchant and grocer who supplied us with necessaries in their line never forgot their profits; that the lawyer, who spent half an hour in drawing up the mortgage upon our farm, charged us what would be equal to four days of our labor; that to the doctor who came five or six miles into the country to cheer the coming or speed the departing member of our family, we paid the price of an acre of corn or five days' labor with our team; that the teacher, for whose education we had paid, earned as much in six hours as we could in six days of sixteen hours each; and so on through all the branches of trade, professions or productions, we found all getting a fair, and some an exorbitant, profit on their commodities and services with which our own would bear no comparison."

"Is it any wonder," says the New York *Tribune*, commenting upon this declaration, "that the men who turned from their hard labor and profitless crops to see these features of their surroundings should put up the cry, 'There's something wrong about all this'? And the story is not much exaggerated; from the farmer's point of view not at all, but on the contrary very mildly stated. You may say some of these things that seem so unjust and harsh are but the natural and inevitable accompaniments of the profession of agriculture; that men take up and follow farming knowing all the disadvantages and risks of the business; that they go into it with their eyes open, and that even with these drawbacks the business is overdone, and low prices are brought about by over-production. But with all that, you do not remove or explain the patent injustice which always stares the farmer in the face, that all his neighbors in other pursuits and occupations are get-

WHAT IS LEFT OF A CROP AFTER PAYING RAILROAD CHARGES.

ting rich and living in comfort upon the profits of his business and his labor. For many of the discomforts and privations of their lot there are compensations, of course. They do not deny this, though they could hardly be expected to enumerate them in the recital of their complaints, for they belong to the other side of the case. On the other side of the case, too, are considerations that pertain to the kind of crops they raise, whether they could not make their business more profitable by the exercise of sounder judgment in the choice of crops to be produced, and other similar suggestions. But underlying all this is a grievance actual and tangible, and that is their present and immediate objective point, to wit: the absolute power over them and their business of the railroad corporations which have been created by their votes. They have seen the railroads discriminating against them in freight tariffs, and paying no heed to remonstrance or protest. They have appealed to legislatures and to courts, and found themselves met with the money and power of great moneyed corporations; and finally they have betaken themselves to organization and to trying the force of numbers for the acquisition of what they believe to be their rights. They may be striking out in some cases blindly and in a hasty, unreasoning way; but what they mean to do is to agitate the subject till it gets some attention and some thought from men competent to devise a remedy, or at least a relief."

Efforts have been made by some of the Western States to protect their farmers against the extortions of the roads; but little has been accomplished. The railroad corporations, relying upon their great wealth and their immense patronage, insolently defy the States that seek to control them, and disregard the laws concerning them. They do not mean to submit to control. They have reaped a rich harvest from the farmers. Railroad directors have fattened too long upon the plunder of the farm to give it up without a struggle. They mean to make it so terrible for the farmers to fight the road that the latter will be forced to cease to defend their rights, and submit to any exaction that may be levied upon them. They mean that the world shall continue to witness the unequal state of affairs now existing, in which the farmer is growing poorer and the railroad director richer.

It certainly is a very unfortunate state of affairs which makes the bread-producing interest of the country the chief sufferer from the rapacity of railroads. Yet there is no denying the fact that the cereals of the agricultural States are taxed more heavily by the railroad corporations than any other products of the land.

Mr. W. M. Grosvenor, from whom we have quoted in a previous chapter, has recently published an article of unusual ability in the November number of the *Atlantic Monthly* (which we commend to the perusal of the readers of this work), in which, though differing widely from the views expressed in these pages, he sets forth in a masterly manner the reasons why grain is the principal sufferer from railroad extortion. Let us hear him:

"Three men meet in a room in New York. They are not called kings, wear no crowns, and bear no sceptres. They merely represent trunk lines of railway from the Mississippi to New York. Other points settled, one says, 'As to the grain rate; shall we make it fifty from Chicago?'

"'Agreed; crops are heavy, and we shall have enough to do.'

"Business finished, the three enjoy sundry bottles of good wine. The daily papers presently announce that 'the trunk lines have agreed upon a new schedule of rates for freight, which is, in effect, a trifling increase; on grain, from forty-five to fifty cents from Chicago to New York, with rates to other points in the usual proportion.'—The conversation was insignificant, the increase 'trifling.' But to the farmers of the Northwest, it means that the will of three men has taken over thirty millions from the cash value of their products for that year, and five hundred millions from the actual value of their farms.

RAISING THE RATES OF RAILROAD FREIGHTS.

"The conversation is imaginary; but the startling facts upon which it is based are terribly real, as Western farmers have learned. The few men who control the great railway lines have it in their power to strip Western agriculture of all its earnings,—not after the manner of ancient highwaymen, by high-handed defiance of society and law, the rush of swift steeds, the clash of steel, and the stern 'Stand and deliver!' The bandits of modern civilization, who enrich themselves by the plunder of others, come with chests full of charters; judges are their friends, if not their tools; and they wield no weapon more alarming than the little pencil with which they calculate differences of rate, apparently so insignificant that public opinion wonders why the farmer should complain about such trifles. Yet the farmers have complained, and, complaining in vain, have got angry. When large bodies of men get angry, the results are likely to be important, though they may not always prove beneficent. The farmers' movement threatens a revolution in the business of transportation, if not in the laws which protect investments of capital. It seems strange, no doubt, to those who do not know that a change of one-twentieth of a mill per one hundred pounds, in the charge for transportation per mile, may take hundreds of millions from the actual value of farms. It can neither be comprehended nor intelligently directed, without a full understanding of the conditions under which agriculture exists in the Northwestern States, and of the power which the railway has exerted and still wields for the development or destruction of that great industry.

"About 150,000,000 bushels of wheat, 11,000,000 tons of hay, and 1,012,000,000 bushels of cereals are annually produced by eleven States, having in 1870 a population of 14,283,000. In this statement, as in the term 'Northwestern States,' when used in this article, Kentucky and Missouri are included with the former free States of the Mississippi Valley. Had these States consumed in proportion to their population, there would have remained a surplus of eighty-one million bushels of wheat and five hundred million bushels of cereals. Though the consumption *per capita* is greater of cereals other than wheat in this region than in the whole country, over two hundred million bushels of grain are received yearly at seven chief points of shipment from the West, while a large quantity besides goes directly to consumers at the East and South, without passing through either of these cities. Probably eight million tons of grain, besides hay and other products of the farm, go forth from this fertile region each year in search of distant markets.

"Because the surplus is so enormous, distant markets control in a great degree the price of the whole crop. As the water behind a dam never rises far above the level of the overflowing sheet, so the prices of products largely exported do not rise much above the export price, less cost of transportation to the port of shipment. That this is true of wheat, of which we export about one-sixth, is well known; of other grain and of hay we export comparatively little, and yet the surplus at the West is so large, and the demand at the East for consumption or shipment so essential to a profitable sale of the crop, that the Eastern markets rule prices, not only of the quantity forwarded, but of the entire product. A very large proportion of the corn crop is consumed at the West. Yet the average of monthly quotations for the three years 1869, 1870, and 1871, at New York, Chicago, and Cincinnati, the difference per bushel and per cental, and the summer rate for freight per cental from Chicago and Cincinnati to New York, all in cents, compare as follows:

	Price.	Difference. Per bu.	Per ct.	Freight. Rates.
New York	87 2-3			
Cincinnati	64 2-3	23	41	41
Chicago	62 1-3	25 1-3	45 1-5	45

"Thus even in corn, the average rate for three years at these three markets corresponds exactly with the summer rate of transportation between them.

"In spite of wide fluctuations, 'corners,' and local disturbances, the tendency of Western markets is to approximate closely during any term of years to the rates at which the surplus of products of the farm can be shipped to and sold in Eastern markets.

"Consequently, an increase of one cent per bushel in cost of transportation ordinarily costs the Western farmer one cent per bushel in the selling price of his crop. Neighborhood consumers, millers, produce merchants, cattle-feeders, do not ordinarily pay more than the price fixed by Eastern quotations less the rate of transportation, because they know that millions of bushels all around them must find a market at the East, or be wholly lost.

"Cotton was 'king,' only because it could bear transportation, its value being great in proportion to its bulk. Hay would wear the crown, if, instead of one cent, it was worth twenty cents a pound. Crops differ very widely in their dependence upon cost of transportation, and hence the question of transportation affects the Southern and Southwestern States much less than the States of the Northwest. Wool excepted, Northern crops vary in value, from about two cents per pound for wheat to less than one cent for potatoes. But tobacco is worth eight, sugar ten, and cotton nineteen cents a pound. Transportation of cotton one hundred miles by wagon (at twenty cents per ton per mile) would cost only one-nineteenth of its value. Carriage a like distance would cost about half the value of wheat, and more than the whole value of potatoes or hay. At three cents per ton per mile, by railroad, the entire value of potatoes (at fifty-four cents) would pay for transportation 600 miles; of hay (at twenty-two dollars a ton), 733 miles; of wheat ($1.24 a bushel), 1377 miles; of tobacco (eight cents), 5533 miles; of sugar (ten cents), 6666 miles; and of cotton (nineteen cents), 12,666 miles. Even in the palmy days of Southern agriculture, the building of railroads was regarded with comparative indifference by the people of that section; and, for the same reason, the contest between the farm and the rail is mainly confined to the Northwest.

"Unable to raise Southern crops, the farmers of the Northwest must raise products peculiarly affected in value by the cost of transportation, or relapse into a patriarchal form of industry, and derive their only profit from flocks and herds. The value of animals for food is limited by the demand for consumption. All the animal food required by States which do not produce enough for their own use—in value about forty millions, or one-tenth of the entire consumption—could be supplied by a single State. Texas now has one-seventh of all the neat cattle in the country, and the difference in cost of transportation from Texas and from Northwestern States is more than compensated by the difference in the cost of land. No large increase in the production of animals at the Northwest could be profitable, unless the people of this country should continue to eat very much more animal food. Wool bears transportation a long distance, but, again, the demand is limited; the entire value of wool consumed, not of our own production, is less than that of the wheat alone exported. Meanwhile, Texas, New Mexico, and California will soon supply wool in such quantity that the growing of sheep for wool alone must become even less profitable than it now is in the Northwestern States. . . .

"An increase of five cents per one hundred pounds in the cost of transportation from Western States to New York or other Eastern markets is equivalent to three cents a bushel on wheat, two cents and eight-tenths on rye and corn, one cent and six-tenths a bushel on oats, and (allowing for convenience forty-nine pounds to the bushel of barley and buckwheat, laws of different States varying widely) two cents and four-tenths to the bushel of barley or buckwheat. At these rates, supposing the change in rates to affect the whole crop of all the Northwestern States alike, the loss in value to the farmer upon the crop of 1871, as given in the latest agricultural report, may be thus stated:

	Quantity produced.		Loss in Value.
Wheat	149,600,000	bushels	$4,488,000
Corn	690,900,000	"	19,345,000
Hay	10,915,000	tons	10,915,000
Oats	153,789,000	bushels	2,460,000
Rye	6,625,000	"	185,000
Barley	10,019,000	"	245,000
Buckwheat	1,694,000	"	41,000
			$37,679,000

"This loss of over $37,000,000 in the selling price of the products of 44,375,100 acres cultivated is about eighty-four cents an acre. It is a loss not of valuation, but of the yearly income or profit upon which valuation is based. The actual value of land for farming purposes is that sum upon which the net profit of the yearly crop will yield a fair interest. At seven per cent. interest, whatever reduces the net profit seventy cents per acre reduces the actual value of the land $10 per acre. Hence the loss of $37,679,000, in the yearly income from certain lands, is equivalent to a loss of $538,271,000 in their actual value. Such is the result of a 'trifling' change of five cents per cental in rates of freight!

"It is true, the farmers of the eleven States are not affected in the same degree by a general change in rates. For the effect upon the value of land depends upon the number of bushels produced to the acre. The average yield of different crops to the acre in different States, for the four years 1868–1871 inclusive, and the effect of a change of one cent. per one hundred pounds upon the average value of land employed in growing each crop in each State, are stated in the following tables:

AVERAGE YIELD OF DIFFERENT CROPS.

"(Hay in tons and hundredths; other crops in bushels and tenths.)

	Wheat.	Corn.	Rye.	Oats.	Barley.	Buck't.	Potato.	Hay.
Ohio	14.0	40.4	14.2	31.9	24.4	15.1	90	1.32
Kentucky	8.9	29.2	11.0	22.1	19.1	17.6	74	1.28
Michigan	13.9	32.8	16.7	33.7	28.9	17.3	103	1.31
Indiana	12.1	33.1	14.5	28.2	23.7	16.1	76	1.32
Illinois	11.7	32.7	16.2	30.9	23.0	16.1	79	1.37
Wisconsin	13.5	33.8	15.8	33.7	26.6	18.6	86	1.36
Minnesota	14.8	33.2	18.5	35.0	25.1	19.4	109	1.49
Iowa	12.7	36.2	18.1	35.3	27.0	20.5	111	1.58
Missouri	13.6	32.5	17.0	29.9	25.4	21.0	96	1.50
Kansas	16.2	33.8	21.6	32.6	24.8	17.9	134	1.56
Nebraska	14.7	34.1	19.8	36.3	28.4	19.6	106	1.56

"Effect upon value of land per acre of a change of one cent per one hundred pounds in value of crop:

	Wheat.	Corn.	Rye.	Oats.	Barley.	Buck't.	Potato.	Hay.
Ohio	$1.20	3.20	1.13	1.46	1.05	1.70	7.71	3.71
Kentucky	.76	2.33	.88	1.01	1.23	1.33	6.34	3.57
Michigan	1.19	2.62	1.33	1.54	1.21	2.02	8.83	3.71
Indiana	1.03	2.65	1.16	1.29	1.12	1.62	6.51	3.71
Illinois	1.00	2.61	1.29	1.41	1.12	1.61	6.77	3.85
Wisconsin	1.16	2.70	1.26	1.54	1.30	1.66	7.37	3.85
Minnesota	1.21	2.65	1.48	1.60	1.35	1.75	9.34	4.14
Iowa	1.09	2.89	1.45	1.61	1.48	1.89	9.51	4.43
Missouri	1.02	2.60	1.36	1.37	1.47	1.77	8.23	4.28
Kansas	1.39	2.70	1.73	1.49	1.25	1.73	11.48	4.43
Nebraska	1.68	2.73	1.56	1.66	1.37	1.98	9.08	4.43

"From these tables, the effect of any change of rates of transportation or of price, upon lands employed in growing either crop in either State, may be readily calculated; also the profit on every one hundred pounds of each crop necessary to yield seven per cent. interest on any value of land per acre; also, the effect of any change in rates of freight per ton per mile, the distance to the controlling market being known. Thus, from a farm nine hundred miles from New York, a change in freight rates of one nine-hundredth of one cent per one hundred pounds, or one forty-fifth of one cent per ton, per mile, will affect the value of wheat land in Illinois one dollar an acre. It is easy to see, also, that there are limits within which only these effects follow, fixed, on the one hand, by the lowest cost of raising any crop compared with its value at a consuming market, and on the other hand by the cost of land. But within those limits, as far as the price of crops is controlled by distant markets, all the profits and even the very existence of agriculture depend upon the rate charged for transporting its products. It is not strange that the owners of land and producers of grain regard with constant apprehension a power which may at any moment affect the value of a thousand million bushels of cereals, and of forty-four million acres of cultivated land. Even if a change of five cents per cental does not affect the whole crop so much as three cents per bushel in price, it may take away all the profit—all the reward of a year's labor. And the same power may also raise rates even more at pleasure. The farmers have been taught that the cost of transportation depends upon the will of a few men, and varies with their agreements or quarrels. The *quondam* pedler of Vermont fell out with Vanderbilt, and their quarrel was worth, during the year 1870, one-fifth of a cent per ton per mile to the farmers; $9,000,000 on the crop of wheat alone, if it had all been shipped at the reduced rate. In July, 1872, somebody raised the rates from the West five cents per cental. His act cost the farmers millions of dollars. Is it strange that our greatest industry grows restive under fluctuations which it can neither foresee nor comprehend? Elsewhere the world moves. The beneficent progress of civilization in other lands is toward cheaper transportation and better wages for the producer. Russia pushes railroads through her vast territory, in order that her subjects may obtain at the Baltic and Black Seas better pay for their industry. We cannot maintain sufficient private markets of our own, nor force upward prices in those great markets of the world upon which ours depend. If, while the world makes transportation cheaper, we make it more costly, the loss will be our own.

"This the farmer believes we are doing. He declares that others who stand between him and the consumer, amass great wealth, while pinching economy barely saves him subsistence and does not keep him from debt. His beliefs, as to the cause of existing evils, and the best remedy, whether correct or not, will soon take the shape of laws. He has the votes. Before that power, legislators drop like leaves shaken by the autumn wind. Governors, politicians of all grades, crush each other in their hurry to seize the new standard. Lawyers who do not forget the Dartmouth College case, already find themselves ineligible to the judiciary. Has not this same generation set its heel upon the Dred Scott decision? Reverence for judicial precedents is a dam which floods have carried away. Restraints devised by founders of our Government no longer bar the people from their will. We have trusted all power to the majority. If its opinion is in error, we have but one remedy: that freedom of discussion which remains the only safeguard of our institutions."

CHAPTER XVIII.

THE STORY OF FARMER GREEN'S REAPER.

A Common Fault with Farmers—Not in a Condition to incur Risks—The Danger of running into Debt—The Curse of Mortgages—Labor Saving Machines—What they are worth—Unfair Prices demanded for them—Farmers paying twenty per cent. Interest.—An iniquitous Business—Danger of Indiscriminate Purchases of Machinery—A few Words of Sober Counsel—Farmer Green and his Farm—Getting on in the World—Farmer Green buys a Reaper—How he paid for it—The first false Step—Beautiful Calculations—An Iron-clad Note—In the Toils—Arrival of the Reaper—Disappointment—Second Visit of the Agent—The Theory of Deferred Payments—How it works—Deeper in Debt—The Farm mortgaged—New Misfortunes—Selling the Homestead—Beginning anew—What Farmer Green's Reaper cost him—A Lesson for Farmers.

We have spoken of the evils from which the farmers of the country have suffered, for which they are not responsible. We come now to consider one for which the farmer is entirely responsible, and which should receive the earnest and candid consideration of every farmer in the land. We mean the recklessness with which they incur debts which they afterwards find themselves powerless to pay. We do not mean that the farmer is the only one who thus hampers himself, or that he is any more given to incurring debts than other people, but we do mean that his carelessness in this respect is a source of serious trouble to him. He, of all men, should avoid debt, for, as a general rule, his means of discharging such obligations are too limited to permit him to incur risks that are but trifles to other men. He must look to his land for the means of discharging his debts, and he should shun as a pestilence anything that tends to increase the burdens upon that land. To keep the farm clear of mortgages should be the farmer's chief effort, and nothing but the direst necessity should ever tempt him to peril the safety of his home.

Of late years it has become a habit with the farming class to depend upon the labor-saving machines to do the work of the farm. No one will deny that improved machinery is a great advantage to the farmer, or that a good planter, cultivator, or reaper is worth making a sacrifice to obtain; but the labor is saved by the machine at a high cost when the farmer goes in debt to purchase it. If a machine is wanted, wait until the money is in hand to pay for it.

In the first place, the farmer pays a price for the machine that is too high when he purchases it for cash. When only a partial payment is made, and the remainder is paid in instalments, the cost of the machine is enormously increased. Under the most favorable circumstances, the machine costs more money than it is actually worth.

"The immense profits of general agents and sub-agents have been added to the manufacturers' prices; then, when cash has not been paid, the prices have been still further increased to cover interest at twenty per cent. Finally, as Mr. McCormick testified in Washington that from one-fourth to one-third of his sales were bad debts, an average of 33 per cent. must be added to cover these. The grand result is, that the farmers pay from 50 to 100 per cent. more for their machinery than they ought."

But farm machinery is not the only thing that the

22

farmer is compelled to pay an exorbitant price for. If he buys a sewing machine for his wife, he pays fully twenty-five or thirty per cent. more than the article is actually worth, the difference going to swell the enormous profits of the sewing machine companies. And so the list could be extended indefinitely. All kinds of machinery sell too high, and the farmer is required to pay a most extravagant profit to the dealer who supplies him. People seem to think that he is a legitimate object of plunder, and he is charged prices that no one would ever dream of asking a sharp city merchant.

The country districts are flooded with the agents of the agricultural machine manufacturers, who are using all their arts to dispose of the wares of their principals. Reapers, corn-planters, seed drills, cultivators, mowing machines, sulky corn-plows, and the like, are offered in profusion. All are costly, and many of them are worthless, for it is a lamentable fact that about one-half of the so-called labor-saving machines are so wretchedly constructed that they either break down at an early stage, or are incapable of performing the work for which they are designed. None of them are absolutely necessary, for the work of the farm can be done without them, and should be, unless the farmer is fortunate enough to have the purchase money in hand, and is able to spare it for that purpose. But too often he buys against his own judgment, yielding to the blandishments and persuasions of the agent, and when he awakes to a sense of his error, it is too late to withdraw. He has assumed a burdensome obligation, and he must meet it at any cost. One hears many sad stories of the struggles of farmers to meet the debts thus incurred. "Since I have come into this State," says the correspondent of the New York *Tribune*, writing from Iowa, "and mingled with the farmers here, I have found that it is a common occurrence, that thousands of farmers who are still struggling to keep their heads above water, and are obliged to economize in every conceivable way and deny themselves and families many comforts, first began to run behind when they purchased a reaper or mower, or some other piece of farm machinery which they could not pay for and which they could have done without."

The sad history of Farmer Green (a veritable character, although we introduce him here by a fictitious name) should be a lesson and a warning to all his brethren.

Farmer Green was a resident of Iowa, and was reputed to be a sensible and prosperous man. He was far on in life, and had cleared his farm of debt, had stocked it with many things needful to his business, and was generally counted a prosperous man. His snug farm was his pride and boast, and he looked forward to the time when he should be able to add to it by the purchase of a desirable section of land adjoining it.

It was the early summer, and Farmer Green was rejoicing in the magnificent crop of wheat that was springing up on his land, and giving the promise of a handsome return for his care and labor. Day after day he watched the superb growth, and counted over in his mind the number of bushels of golden grain it would yield when the summer sun had warmed it into maturity. Many were the plans he laid for the use of the proceeds of that glorious crop. The goodwife's wants should be all supplied this year, and none of the children should be forced to put up with the deprivations that had fallen to their lot when he was still struggling to clear the farm from its encumbrance of debt.

One day, as he stood watching the bright field of green that spread out before him, and imagining what he would do when the grain was harvested and the money received for it, he was accosted by a stranger who came driving down the road from the village.

"A beautiful crop of wheat you've got there," said the stranger, as he drew rein before the farm gate.

"Yes," said Farmer Green, "I reckon it will turn out pretty well."

"A fine farm you have, too," said the stranger, glancing admiringly around him.

"Yes," said the farmer, pleased with the compliment to his place. "There's none better in the neighborhood."

"Paid for yet?" asked the stranger.

"Every dollar, thank God," said the owner, heartily. "It's clear at last, and I hope to keep it so."

"That's right," said the stranger. "Never contract a debt you're not sure of paying, and the farm will remain yours. That's a mighty nice crop of wheat," he added, as if speaking to himself. "I never saw anything look prettier. It will be ready for cutting soon. How do you cut it? By hand?"

"Yes," replied the farmer. "We've no reapers in this part of the country, and we farm in the oldfashioned way."

"That's a pity," said the stranger. "A reaper would work beautifully on this land. Why it would be no trouble at all to get your wheat in with a good reaper."

"That's true," said Farmer Green.

"You ought to have a reaper to cut it with," said the stranger.

"Can't afford it; haven't got the money to spare," said the farmer.

"See here, now," said the stranger, in a more confidential tone. "I'm selling a patent reaper—a first-class machine, and dirt-cheap at the money asked for it. You'd better let me sell you one."

"It's no use to talk about it, my friend. I haven't the money to spare."

"I don't want your money now," said the man, temptingly. "I'll sell you one at a bargain, and wait till it has paid for itself."

And with that the agent produced pencil and paper, and went into a calculation, showing the farmer how much it would cost him to cut his crop that year, and how much the reaper would save him, as well as a calculation of the amount of grain he could cut for other farmers in the vicinity.

"So you see," added the agent, persuasively, "before the time of payment comes around you will have saved and earned enough to pay for the reaper, and will still have a fine machine capable of doing more work, equally profitable, next season."

Farmer Green's better judgment bade him refuse the terms thus offered, liberal as they seemed. He knew the evil consequences of running into debt, and his conscience bade him put the temptation behind him. He wanted a reaper, however; he had always wanted one; and here was an opportunity of purchasing one upon terms which would enable him to pay for it out of its actual earnings. There was not a reaper in the county,

FARMER GREEN TRIES HIS REAPER.

and he felt confident that he would be able to keep it busy on his neighbors' farms, all through the season, after he had cut his own crop.

The agent was a smooth tongued, plausible fellow, and he plied the farmer with every argument he was master of. The result was that the farmer bought the reaper. He had not the money to pay for it, but he gave what is called in Iowa "an iron-clad note" for it. In plainer English, he gave his note accompanied with a statement of property. By the laws of Iowa such a note is equivalent to a mortgage. And so, in order to purchase the reaper, the farmer had imperilled his property, and had placed the safety of his home upon the turn of a chance.

The machine arrived in due time, and was found to be all the agent had claimed for it. It was a capital reaper, and a very handsome machine withal. Farmer Green could not help feeling a little downhearted as he remembered the risk he had incurred in order to obtain it; but he consoled himself with the hope that he would be able to make it pay for itself. When the harvest came around, the machine proved itself a good worker. Farmer Green soon had his crop cut and stacked, and then began to look about him for engagements for cutting his neighbors' grain. Some were willing to make the trial, and a few jobs of this kind enabled him to earn something with his reaper. But the work was less in amount than he had looked forward to, for the agent who had sold him the reaper had found other customers in the vicinity, and the demand for Farmer Green's machine was very much less than he had anticipated. The reaper stood idle under its shed during the better portion of the harvest season, and the farmer was doomed to a severe disappointment.

When the crop was sold there was another disappointment. There had been a heavy decline in the price of wheat, and the farmer did not receive as much as he had expected for his grain. All this while the day upon which the note must be paid was drawing near, and the farmer's chances of meeting it were rapidly diminishing. And still another blow fell upon him. Just after the harvest his wife fell sick, and her illness was long and expensive.

Upon the appointed day, the agent of the Reaper Company presented the note of Farmer Green, and demanded its payment. With a sad heart the farmer related his troubles to him, and told him he was unable to meet his note. He had not the money. The agent's face grew very long as he listened to the woful tale, and after considerable hesitation, he said he was very sorry; that Farmer Green should have made allowance for all these risks, in making the purchase. However, the mischief was done, and there was nothing but to accept the situation. If the farmer could not pay, he supposed the time would have to be extended, but it would be necessary to charge him a fair rate of interest. Farmer Green said that that was only just. He had done his best to meet the note, but failing to do so, he was willing to pay for his failure. What, he inquired, would be a fair rate of interest?

"Twenty per cent. per annum," replied the agent, gravely.

Farmer Green's heart sank, and he said in a despairing tone, that the rate was too high.

"For ordinary interest, perhaps," replied the agent; "but, you see, we assume a serious risk in this case. I'd rather have the money down than one hundred per cent. interest. But you havn't got it. We take the risk of your failing entirely to pay us, and it is only fair that we should be paid for this risk as well as for the delay we are put to."

There was no help for it, and Farmer Green was obliged to pay the extortionate demand. He had placed himself at the mercy of the Reaper Company, and he must do their bidding. He hoped that a succession of good crops would enable him to pay the interest and take up the note; but, alas for him, this hope was destined to disappointment also. He paid the interest onee or twice, but the burden was too heavy for him, and at last, in sheer despair, he mortgaged the farm, paid the note, and got rid of the Reaper Company. But he had only shifted his burdens. The mortgage proved as

FARMER GREEN MORTGAGES HIS FARM.

troublesome as the note had been, and instead of being able to decrease it, he was obliged to increase it as time passed on. By the first false step he had placed the farm of which he was so proud in danger. He had voluntarily incurred a useless debt, and the rest of his bad luck was simply the logical consequence of a reckless and foolish act. He ran behind steadily, and at length his difficulties increased to such an extent that in order to rid himself of the debts he had no hope of paying in any other way, he sold his farm, discharged the mortgage, and bidding adieu to his old home and friends, went farther West, to a section where lands were cheaper, and there began life anew at the time he had once hoped to enjoy some rest from his labors.

And yet, Farmer Green, with all his shrewdness,

never attributed his misfortunes to their true cause. He never admitted, even to himself, that his great error had been in contracting a useless debt, and assuming an obligation he had no certainty of meeting. He never believed that it was the reaper that ruined him, yet such was the case. Had he put by the temptation held out to him by the Reaper agent, there would have been no burden resting upon him, and his short crop, and other misfortunes, would not have driven him to the expedients he was obliged to resort to. "Out of debt, out of danger" is a true maxim; the wisdom and force of which only those who have passed through the agony and humiliation of such a slavery can appreciate.

There are debts enough that the farmer cannot help assuming; burdens that fall upon him through no fault of his. They are heavy enough, God knows, and they should teach him to assume none from which he can possibly escape.

Improved machinery is useful where it is honestly made, but even the best is worth less than the farmer ordinarily pays for it. He is charged too high, and his hard earnings, instead of constituting a fund for the rearing of his children and the protection of his old age, go to make up the colossal fortunes of the manufacturers and dealers in such machinery. A reform is needed, and it is near at hand.

CHAPTER XIX.

FARMER SMITH SPEAKS HIS MIND.

The Secretary of the Illinois Farmers' State Association—Speech at Carrollton—Views of a Practical and Thinking Farmer—Sound Views for the Consideration of the Farmers of the Union—Mr. Smith's Homestead—A comfortable Western Farm—A quiet Talk with Farmer Smith—His Statement of the Farmers' Wrongs, and his Views as to the Remedy—Corn selling for less than Cost—"Sixty Bushels of Corn to buy Two Pairs of Boys' Boots"—The Mysteries of Western Coal Selling—The Farms more heavily taxed than the Railroads—The Grange offers the best Remedy, and the best Means of attaining it.

PROMINENT among the Western farmers who have devoted themselves to a practical solution of the questions we have been discussing, is Mr. Stephen M. Smith, of Illinois. He is the secretary of the Farmers' State Association of Illinois, and a man of vigorous and independent mind. As he speaks not only for himself, but for a large and influential class of farmers, we quote here at length from his public declarations. We do this not only to commend his remarks to the careful consideration of those of our readers interested in these questions, but in support of our assertions respecting the grievances and opinions of the farmers of the United States. In an address delivered to the farmers at Carrollton, Ills., on the 3d of September, 1873, he said:

"For a purpose evident enough, the report of my Winchester speech of August 9th has been published in every paper which does not happen to agree with me politically, especially all over this State and the State of Iowa. I have had hundreds of copies of papers sent to me with marked articles criticising this so-called report of my Winchester speech. Allow me to say that the words 'blood and anarchy' never came into my speech. I didn't say I would take my boys and go to the State capital and help ride the villains out on a rail, although I believe they deserved it. If I had mentioned it, I wouldn't have said more than that I believed they deserved it, but I did not say it. I did say, and this is my very language—and mark now, for I want to be particularly honest, what my remarks were—I said, in referring to the manner in which our legislators had betrayed our trust and sold out our rights and interests again and again, that they were elected to carry out certain purposes, and had failed to do so over and over again, and that the interests of the laboring and producing classes of the whole country, not merely of this State but of the whole of the United States, had been sold out over and over again to monopolists of every sort and character; and I added, there is no law on our statute books by which we can reach them and punish them for that crime of betraying the laboring people of the country. I said I was very much inclined to adopt a punishment suggested by an indignant friend of mine when our legislators adjourned last Winter after voting to come back and spend another Winter and take another half million out of our pockets which we had to pay in corn at twenty cents a bushel, which punishment was to treat the members to a coat of tar and feathers and ride them out of their counties on a rail. Now see how you can make a speech read when you take expressions that were uttered at least an hour apart and put them into a sentence. What I said about my boys was something like this: Speaking of the magnitude of this movement, how chronic the wrong was we suffered from, and how this evil—that we had no voice in fixing a price upon our labor—had come down to us from the feudal ages, I said I did not expect to live to see the full fruition of my hopes in this country, but that I would bequeath the fight to my boys with the injunction that they should never leave it until all their rights under our Constitution and laws were guaranteed to them, and they had secured the dearest of all rights to an American freeman—that of fixing the price upon their own labor. That is the connection in which I mentioned my boys and no other. I did say that when I was a boy we used to shoot crows and hang them up on a pole in the corn-field as a terror to evil-doers [laughter], and I said perhaps the time might come when, if every other remedy failed, we might hang some men about the country in that same way as a terror to evil-doers; but I believed the remedy for all the evils of which we complained was the peaceful remedy of the ballot-box. That is the language which has been so tortured, and this is enough on that point. I have, however, this comforting reflection in regard to all that has been said and all the flings at me in relation to that speech, that if you want to find where the best apples are, go into the orchard and look for the tree under which you will find the most clubs. There is no mistake about that rule.

"Whence comes this wrong—the fact that, notwithstanding all this immense production of all the necessaries of human life, our farms are still under mortgage? Take another fact and put it with this one—that a careful estimate made in 1866, when we were nearer out of debt than we have ever been in the history of this country, since the high prices of the war enabled us to pay off more mortgages and personal debts than during any previous period of our existence. The personal debt of the United States was $1,900,000,000. Three-quarters of that, or about $1,500,000,000 was borne by the agriculturists of the United States. When you consider that this latter sum has to be carried at an average rate of ten per cent. per annum, is it any wonder that we are poor? Couple with this another fact, that the annual increase of the wealth of the United States for any decade during the very best period of our existence has never exceeded three and a-half per cent., that is, the agriculturists are actually carrying $1,500,000,000 at ten per eent., while the products of industry nowhere, taking it all through, have exceeded three and a-half per cent. Is it then any wonder you are poor, and that with each year the whole agricultural population of the United States is growing poorer and poorer, while those who handle the products of our labor are growing richer and richer?

"Take Commodore Vanderbilt, for example, and suppose that twenty years ago he was worth $5,000,000, and that to-day he is worth $65,000,000, how has he accumulated $60,000,000 in twenty years? Mark it, he never earned a dollar in his life, and yet he has gotten into his hands $60,000,000 in twenty years. I might stop right here and not say another word on the subject, for here is sufficient proof that there is something wrong in this business, owing to which this man has accumulated so much. How did he do it? Has he rendered an equivalent in the service he has performed for us in transporting our productions to market, or has he not? That is the question. If he has not, then we have been wronged of just so much money. For everything beyond a fair and reasonable equivalent for the service rendered is just as much stolen from us as if he held a pistol at your head and said, 'Your money or your life;' taking it because you had no pistol and he was the stronger.

"It is useless to try to dodge this proposition, for it can't be done. When a man who has earned nothing by productive industry, but who has simply handled the products of labor, has accumulated that amount in a number of years, it is a proof that something is wrong. The whole wrong lies in this, that we are getting too little for our products, and those who handle them are getting too much.

"Colonel Coleman has shown you what it costs to get a bushel of corn or wheat to market from where he lives in Missouri, and we all know what it costs here, and that we pay three-fourths of the product of our labor to get the other fourth to market. If this is so, who fixes the price upon your labor? What have you to say in regard to its price any more than did the slave of the South in the days of his worst estate? We are in fact in a condition of slavery unless we can control the price of our own labor. If you fix the price of my labor you circumscribe my actions and fix me to one plan for my lifetime, without opportunity for rest or recreation. How do the monopolists get these prices? Take the plowmen of this State; all have their annual conventions. They come together and agree that they will have just so much for plows during that year, no matter what we may get for our products; and for the last two years they have asked one hundred per cent.

on the cost of production. Thus that combination makes a monopoly of the plow business. No matter what agent or manufacturer you buy of you have to pay the same per cent. There are, my friends, the pork-packers, from all the principal cities in this Union, who last year met and combined to fix the price of pork—you know this as well as I. St. Louis, Milwaukee, Chicago, Louisville, and Cincinnati came together just as they are going to convene next week at Cincinnati, and fixed the price of our pork for the coming year just as coolly as the master sold the slave or the products of his labor. Last year they fixed the price upon your pork at $4 a hundred live weight, and they would buy all there is in the West at the same rate. Of course they got it, and if they had fixed the price at $5 they would have had it. I only got $3.25. Probably the high price for freight made the difference in your case. What right had they to do this?

"Certain people denounce me because I use strong language. Colonel Coleman called these pork men scoundrels, and I believe it's a good word, for the man who robs me is a scoundrel. They combined to rob us. The scoundrels came together and fixed their price, and the pork began coming in. The men who were in debt, and whose notes for these reapers and mowers had matured, sold first, and when they were through the stream stopped. It costs as much to run a packing-house on half time as it does for whole time, and as the pork did not come in they put the price up forty cents a hundred, and that started the stream again; and the next set of men whose notes for reapers and corn-planters had matured sent in another lot. When that was worked up the pork-packers put up the price another forty cents, and another lot came along. I admit that there are some rich farmers in our country, but the rich ones are the exception, and the poor ones the rule. That is the difference. But did your rich men, who do not owe a dollar in the world and who have their farms stocked and paid, stop to think who fixed the price on your pork? Why, the men who owned the first notes for reapers and sold the first lot of hogs fixed the price of yours! The price of your products, be it what it may, is determined by the figure at which those who must sell dispose of theirs. Those who must sell fix the price for those who need not. Is it not then worth while to have a union of all interests, to come together and be brothers in fact as in name? We can protect each other, and while we protect our poor neighbor and assist him over a tight place, we are protecting ourselves, because if he must sell at twenty cents, that fixes the price upon our corn. Combination will beat combination. It is to your interest to come together in these farmers' clubs, granges, organizations, and combinations, for a common purpose, which is the mutual self-protection of the whole people. Did you ever stop to think, my friend, that not a single locomotive nor car can be run over these prairies of ours without the oil manufactured from the hog, and with which they grease their wheels? Shut down on your pork for one season, and you dry up every locomotive in the State. They cannot run a day without you, nor can people do without pork as an article of food, without lard, or your other products. The wheels of the world will not go unless you grease them with the products of your toil. When you come together and enter into a combination of this sort by your State, county, township, or precinct associations, one auxiliary to the others all the way up, and when the machinery is perfect, I say you have a combination that is irresistible, for the reason that you hold in your hands the breadstuffs which feed the world, and when you look at your corn-cribs and granaries and refuse to open them, you bring the world to your feet at once, because it cannot exist one day without you.

"Increase in taxation, then, is another wrong which we have brought upon ourselves, and we might as well look it squarely in the face and right it. When I went on the farm seventeen years ago it took a $10 gold piece to pay the taxes, and now it takes seven and a-half of them, which is an advance of a little over seven hundred per cent. But what corresponding benefit have I, in God's name, for that increase of taxes? Is it not time for us to look into it, and inquire of our public servants how it happens that our taxes increase, and what they give us in exchange for them? Did you know that they are multiplying officers and expenses, and voting away thousands upon thousands of dollars with a perfect disregard of the interest of the tax-payers? I am told that a county not a hundred miles from here used to be run for $300 a year, and now it costs $2500. Did you ever stop to think that it takes the product of a whole township to feed a judge? Our Circuit Judge eats up all we earn in a township in a year. Is not he a monstrous eater? Is it not time we tried to look into this matter and see if we could not get some man who would eat less? It costs too much to board that kind of cattle. Think of what you are getting for your products, and of what you are paying for his service. Is there any reason, justice, or right in all this?"

Mr. Smith then touched upon the salary grab, which he roundly denounced. Of President Grant he said: "The President who signed that bill which put $100,000 in his pocket is not one bit better in my estimation than the men who voted for it and took the steal. You will pardon me for using such harsh expressions as that, when I tell you I have been a Republican ever since the party started, and went right straight along with it. But, thank God, I did not vote for General Grant the last time. I have not that sin to answer for, at any rate. But let me say, I do not feel a bit better about it, and I do not think I degraded myself a bit more voting for him the first time than any Democrat who voted for James Buchanan. I should think there is but very little to choose between them." In pursuing this political topic, Mr. Smith spoke of the Democratic party, which, he said, died twelve years ago. Still he wanted to fix it as the boy did the dog: "'I know he is dead, but I want to make him deader.' The people have spewed it out of their mouths because of its corruption, and they will spew the Republican party out of their mouths for the same reason.

"No reform is possible within the existing parties. History has proven that no reform was ever yet worked inside the party or sect in which originated the corruptions complained of. From Martin Luther down all the great social, political and religious reforms that have ever been accomplished began clear down among the common people and worked upwards, while all oppressions, wrongs and corruptions began up yonder and worked downward. When I began this movement I said I would seek to accomplish these reforms inside of the Republican party if I can, outside of it if I must.

But that salary grab, the action of our Legislature last winter, of the Republican Legislature of Indiana upon this matter last winter, and that of the Iowa Legislature, which snubbed the Grangers, have convinced me that there is no redress for us inside any party organization except our own. I have been voted—that is the word—ever since I was twenty-one, and now I am going to vote; that is the difference. I have been led up to the polls all these years like cattle, and have been voted. Now, in God's name, let us go to voting. When you do that you will make the 'fur fly.' When you decide that you will do that, you will see more than one fellow around with hayseed in his hair and in his clothes. My advice is simply this; to vote as farmers. Did you ever stop to think that you had three-fifths and a fraction over of all the votes in this State? Yet, with all that numerical superiority, what have you done in electing men to office to protect your interests? Think what power you have. Is it any wonder that these fellows are ready to bow right down to us when they think of the power we hold as voters? Is it any wonder that they are willing to concede to us a great many things and a great many rights they have always denied us heretofore? As for myself, so help me heaven, no man who has once betrayed my trust, no man who took that salary steal, from the President down to the lowest of them, will ever get my vote for any office whatever, not even for postmaster; and I say, as I did on the Fourth of July, that if any one of these men is ever elected again outside of certain large cities, it will be agricultural votes that will elect him. And I said, and repeat the expression, that if they are thus elected you deserve to go right in and work thirty months to pay them for one in the public service. If you are wronged in the future do not complain, for you have it in your power to remedy the evil by these combinations; but let them be for good, mind you, not for evil. Let us combine. I saw a man at our national Congress who said: 'When I left southwestern Georgia I paid a dollar a bushel for your prairie corn. I come out here and I find that you are getting twenty cents a bushel for it, and that therefore somebody got eighty cents for fetching it to me. You ought to have half a dollar for that corn, and I ought to get it for seventy-five cents; and then the fellows who fetch it to us would get twenty-five cents instead of eighty, and that would equalize the thing. I would rather pay seventy-five cents than one dollar, and you never ought to raise a bushel of corn for less than half a dollar.' And so say I. You never should sell a bushel short of half a dollar, and you can have it the moment you say you will. If your poor neighbors must sell, furnish them the money; make up a purse for them, lend them the money on their cribs and enable them to hold on till the price is up. People cannot eat dry goods and nails; but we can be self-supporting on a farm; and there is where we have got the advantage, for we can make our farms support us, as we did when I was a boy, when we spun linen and muslin, and made everything we used. They must have our products, and the power to fix a price upon them is in our hands the moment we get ready for it, and that within a year, if we are wise in this matter. First begin by organizing everywhere; not for extortion, not for robbery, but to execute the first law of nature, that of self-protection. Organize, that we may be strong against the many. While segregated we are weak;

aggregated we are a power which will be irresistible for good to ourselves."

The correspondent of the New York *Tribune*, gives the following account of an interview with Mr. Smith:

"The great apostle of the farmers' movement in this State, the man who planned the organization of the farmers, and by his earnest, well-directed efforts, has done more than any other to give it shape and make it an effective power, is S. M. Smith, Secretary of the State Farmers' Association, and himself a plain, hard-working prairie farmer, residing about two miles from this village. A native of Connecticut, and descended from one of those frugal, hard-working New-England farmers who leave the marks of their enterprise wherever they go, he came to Illinois seventeen years ago, gave up his business of woollen manufacturer, and hired Willow Farm, which he now owns. The place, though situated in the midst of one of the richest prairies of Illinois, had then, after nearly twenty years' cultivation, been little improved. A very small house, such as one may see upon the poorer farms of

S. M. SMITH, SECRETARY OF THE ILLINOIS STATE FARMERS' ASSOCIATION.

the East, a few apple trees and some straw sheds to shelter the stock in the Winter, were all that the owner had been able to add to the place to promote the comfort of his family, or increase the value of his farm. Mr. Smith has enlarged the house, though it is still very unpretentious and far from modern in its appointments, planted trees, ditched the land in the low places, and placed the whole under a very high state of cultivation. His barns are still of logs thatched with straw, but his stock is in good condition, and everything about the place bespeaks the thrift as well as the good taste of its owner. Mr. Smith himself is past sixty years of age, though time has touched him rather lightly and left him all the energy and enthusiasm of a young man. A great lover of books, his well-selected library is about the only luxury in which, as a farmer, he has been able to indulge himself and his family, and, of all his books, he prizes most highly a complete bound file of Mr. Greeley's *New Yorker*, for which, as Mr. Greeley afterward assured him, he raised the first club. From that day to this the *New Yorker*, the *Jeffersonian*, the *Log Cabin*, or the New York *Tribune*, has been constantly read by Mr. Smith and his family. The truth is, I hardly find a farmer in the West, who reads or thinks for himself, and who does not speak of Mr. Greeley as having been his personal friend. You will find the portrait of the founder of the *Tribune* hanging in almost every farmer's parlor.

"In one corner of the apartment which serves as the farmer's dining-room and the family sitting-room, Mr. Smith has a table covered with letters, documents, and newspapers; and here, between the intervals of farm labor, and assisted by wife and son, a lad about 14 years of age, he has conducted the correspondence which has united the farmers of the State, and prepared the speeches that have roused them from their apathy. Mr. Smith met me at the hotel, and invited me to Willow Farm, remarking that we should have a much better opportunity to talk there, and that he had some letters and documents which he desired to show me. I think that I have before said in one of my letters, that the farmers of this State have two distinct organizations. The Grange is a secret organization extending over many of the States, non-political in its character. The initiation fee of the Grange is five dollars for men, and three dollars for women, and a tax of ten cents a month is collected from each member. Entirely distinct from the Grange is the Farmers' Club, an open society, which until the formation of the State Association, was entirely independent. In October last, Mr. Smith, who was then Secretary of the Farmers' Club of this place, called a meeting to be held in Kewanee, on the 16th and 17th of that month, and about fifty delegates from farmers' clubs and granges attended. The sessions were held in a large hall, and a few of the citizens occasionally dropped in, saw a small knot of people in one corner of the room, laughed and went out. But those few men formed the State Association, appointed a State Central Committee, and a committee of one from each county. The Executive Committee there chosen issued the call for the Bloomington Convention of last January, probably the most important farmers' meeting ever held in this State. It is of this State Association, composed of delegates from both clubs and granges, that Mr. Smith is Secretary.

"Until I met Mr. Smith, I had been unable to get any

very clear idea of the exact grounds of the farmers' complaints. 'Railroad extortion,' 'Unjust Discrimination,' 'Monopoly,' are general terms freely used by the farmers, but just what constituted them I was unable to find out. One of my first questions, therefore, was whether the farmers of this State were not prosperous. I told him that the people of the East, hearing of the wonderful fertility of Illinois soil, supposed the farmers who had been here for any length of time must be a well-to-do, comfortably situated class of people. Mr. Smith replied that this was a very erroneous idea. 'The majority of the farmers of this State,' he said, 'have hard work to support their families. Year by year new mortgages are given to pay new debts, and it is the exception rather than the rule for a farmer to be saving anything. At least one-half of the farms in this part of the State are mortgaged for money borrowed at ten per cent. interest, and the majority of them will never be redeemed. You let it be known that a man in this village has a thousand dollars to lend on first-class security, and he will have a dozen applications before night!'

"'Have these mortgages been given for balances due on farms purchased, or for money borrowed after the farms have been paid for?' I asked.

"'In some cases they represent a part of the purchase money of the farms, but in most cases the farmers have been obliged to borrow because they have been running behind. Nobody can make anything by farming here unless he has a large farm, and I'll tell you by-and-by why, with plenty of land, a man can make a little; but even then he can seldom realize ordinary interest on his investment. Now, I have here a good farm, and you can see what I have made of it. But I never could have paid for it in the sixteen years I have been here from the profits of farming. I was successful in a little speculation, and made enough to buy the land.'

"'The railroad men say that you farmers are extravagant in your living; that during your years of prosperity the silk dress got into the family, and that you have never been able to get it out—in short, that you are indulging in a style of living wholly unknown among you a few years ago.'

W. C. FLAGG, PRESIDENT OF THE ILLINOIS STATE FARMERS' ASSOCIATION.

"'It is not so,' was the reply. 'If you will notice the little boxes of houses in which the most of our farmers live, with almost nothing about them to make them attractive, and then if you could go into them and see how meanly they are furnished, and how the inmates have to economize and count every cent they expend, you would see that the assertion is not true. Take the farmers who live within ten miles west of me, and I don't believe the whole of them spend $15 a year in reading matter; and as for dress, your mechanics, who work by the day, and their families, are much better clad. All over these prairies you see magnificent fields of corn; but if it costs the farmer more to raise it than he can get for it, a good crop is no sign of prosperity.'

"'Is it true that you have to sell your corn for less than what it costs you to produce it?' I asked.

"'Certainly, and I can demonstrate it to you,' was the reply. 'I have given the matter very careful study for years, and I think I can tell just about how much it costs me to raise a bushel of corn. You may take first the labor. I think any farmer will tell you that it takes a man and a team at least five days to plow, harrow, mark out, plant, cultivate, harvest, and house an acre of corn. It can't be done in less time. Two dollars a day is no more than a fair price for the work of a man and a team. Then the first item of expense is $10. The land in this county is assessed at $33.33⅓ an acre; it is worth, unless the sale of the whole of it should be forced, $40; I refused $68 an acre for my farm last year. Interest at 10 per cent. on $40 for one year is $4. Fifty bushels of corn to an acre is more than the average of this county. To shell a bushel of corn and haul it two miles to Kewanee, with taxes, the wear and tear on farming implements, etc., costs at least five cents a bushel, or $2.50 for the crop on an acre. So you see my acre of corn has cost me $16.50. Corn is now worth in Kewanee twenty cents a bushel. My fifty bushels will, therefore, bring me only $10. That is, I barely get pay for my labor, while I lose the interest on the money invested in my farm, the wear and tear of machinery, and get nothing for shelling and carting, or with which to pay my taxes. Again, the bushel of corn has cost me 33 cents. I suppose we can raise corn at a profit for thirty cents a bushel. Our farmers don't expect to get 10 per cent. interest on their investments, though they have as good a right to it as the capitalist or merchant. Then there is one other thing which I did not take into account. The stalks have no cash value, but they furnish fodder for our stock. The only thing that has saved the Illinois farmer from complete ruin and bankruptcy has been his stock. After our corn has been husked, we turn our cattle into the fields and they will come out in the Spring fat. I said that I would tell you why a large farm might be made to pay, where a small one could not. If a man has a large farm and capital enough he can make money raising stock and hogs, because each will get a certain percentage of its growth from material on the farm that has no market, and is, therefore, really of no value unless used in this way. Beside, the corn which I feed to my stock and pigs, brings me much more than twenty cents a bushel.

"'I can give you some striking examples of the profit of raising corn and wheat in this vicinity, if you desire. One forenoon a man went past here with a load of sixty bushels of corn. He said that he had come a long distance—had started at four o'clock in the morning. As he returned in the afternoon, I asked him how much he had got for his load of corn. He held up two pairs of boys' boots, and said that his sixty bushels of corn, and $1 in cash, had just purchased them. It took at least seven days' labor of a man and team to raise that corn, and another long day to haul it to market, to say nothing of interest on the farmer's investment and other expenses. I judge that each pair of boots cost about five days' labor, or its equivalent. I knew another man who took a ton of corn to market for the purpose of buying coal. It purchased just a ton, and he spent a day with his team in hauling. One year I raised 3600 bushels of wheat, and kept a careful account of its cost. When I sold it, and balanced my books, I found that I had for my own labor, which I had not charged, and that of my wife, who had a terribly hard time of it cooking for harvesters and threshers during the hot weather of midsummer, just $300! Why, sir, $1000 would not have paid for that summer's work. Wheat is so uncertain a crop, it has so many enemies from the time it is sown until it is threshed, and it is so exacting of the farmer who must attend to it at certain time, or he will lose it, that we can't afford to raise it for less than ninety cents a bushel.

"'Now there is something wrong in all this. With our productive soil, and facilities for reaching market, the farmers of Illinois ought to be fore-handed, comfortably housed and clothed, and able to save a little every year, instead of getting deeper and deeper into debt. We are an intelligent, hard-working, economical people, and every one of us who owns his farm is to that extent a capitalist; and we ought to be able to do as well as the journeyman mechanic, with less education than we and no capital. It is not right that the Chicago, Burlington & Quincy Railroad, which only moves our crop to Chicago, a distance of 132 miles, and the trade in that city who handles it, should be growing enormously rich, while we are growing poorer. It is not worth eleven cents a bushel to take our corn from here to Chicago, and the railroad that is charging it is robbing us of a part of the fruits of our labor.'

"'Mr. Walker, the President of this road, told me that railroad property in Illinois was not profitable, and while he admitted that his road was one of the few that are paying dividends, I inferred from what he said that it was making only fair profits,' I remarked. 'He said also, that when you farmers complain that railroad corporations are getting rich, you forget that while your farms were bought for from $1.25 to $5 an acre, they are now worth on an average, from $50 to $75, and some of them even $100.'

"'We don't forget that,' said Mr. Smith; 'but let us see, now, how much it amounts to. This country was settled about forty years ago by men who came into what was then a wilderness, lived for years in log cabins, and were deprived of every comfort and luxury of civilized life. Suppose that they had remained and still hold the land, its enhancement represents all those privations, and besides the toil not only of the head of the family, but in many cases of all its members for forty years. Fifty dollars an acre is a high price for the average land in this county. The increase in value on an eighty-acre farm has been less than $4000. Beside this we have barely made a living. Perhaps Mr. Walker and his family would like to try it.

"'And now in regard to this railroad, I'll give you a few facts, and then you can see whether it is paying more than a fair percentage or not. It has paid a dividend from the very start, and in addition to that it has been constantly laying by a surplus. Sometimes this surplus has been used to water the stock, and sometimes in other ways, as I will tell you. The road has numerous branches or feeders. Each of these has been nominally built by a separate company. The towns along the line have been induced to subscribe for stock and give town or county bonds for it. These bonds have paid for grading and putting down the ties. Then the company has mortgaged the road to raise money to complete it. In a few years it has to be sold for the benefit of the mortgagees; the stock, which is mostly held by the towns and counties, and for which they issued their bonds, is wiped out, the Burlington Railroad buys the branch for perhaps one-half or two-thirds its cost, pays the mortgage out of its surplus, issues new stock and divides it among its own stockholders. These men that get the stock never paid a cent for it. In some cases, instead of dividing new stock, they divide the bonds, a dividend of this kind amounting to sixty per cent. of the capital having been recently made. I have been told of a man whose original investment of $400 in the stock of this road has increased to $20,000. I don't believe it, but I do believe that every dollar's worth of stock originally put in has been so much watered as to represent now a good many dollars.

"'You say President Walker told you that it required large sums of money to increase the efficiency of the railroads as the country becomes more densely settled. We don't object to that. If a road needs to be extended, let the company issue new stock and get the money for it. No farmer in this State will complain as long as the railroads do not make more than 10 per cent. on the money actually paid in. What we do object to is paying such exorbitant rates for freight; we don't think it right that this Burlington Road should be able, beside paying a dividend, to lay by a surplus; then, when they spend that surplus in building a new road, they ought not to expect the farmers to pay them such high rates that they can divide 10 per cent. on their surplus. In other words, they extort from us, unjustly, $1,000,-

000, and then come and expect to realize, out of our crops, 10 per cent. on the money that ought to be in our pockets. That's what we call extortion.

"'Judge Beckwith of the Alton Road tells you that millions of dollars have been sunk in the railroads of this State. I have no doubt that is true. He says that if you take the cost of each item that entered into the construction of a road, there is not a road in the State that pays a fair percentage on that amount. Well, what of it? Does Judge Beckwith expect to get 10 per cent. interest on all the money that has been sunk in establishing railroads? Suppose the man who preceded me paid $2000 for this farm and sunk $5000 more in improving it; then suppose I come and buy it for $5000, must I expect that the farm will pay me interest on $12,000? Yet this seems to me the reasoning of the railroad managers. I think I know the farmers of this State pretty well, and I tell you only the truth when I say that we don't ask the railroads to serve us for nothing, nor for any cheaper rates than they can fairly afford. Then, if we are no better off we will look elsewhere for the remedy. But I tell you, eleven cents a bushel is too much for carrying corn from here to Chicago, where it is delivered here in the elevator, is carried over only one road and delivered in the company's own warehouse, and nothing short of the figures showing the exact cost of the service will convince us that we are not robbed by the railroad of a part of the price of every bushel of corn carried over it.

"'We think the leading railroads of this State, besides charging us too high rates for the benefit of the stockholders, are not managed as economically as they might be, and that there are "rings" connected with some of them which are increasing the cost of running the roads for their own benefit,' continued Mr. Smith. 'It is generally hard to discover just where the swindle is, because when the officers of a road manage it so that the dividends come regularly, the stockholders are not often very watchful over them. Sometimes, however, we get hold of a fact. Now here is the Burlington & Quincy road, which has in Kewanee a patent shoot for coal. It takes at this point forty tons a day. The miner will sell me coal at the bank for seven cents a bushel—that is, $1.75 a ton. He carts it a mile and delivers it in the railroad shoot for eleven cents a bushel, or $2.75 a ton. He pays one cent a bushel, or twenty-five cents a ton, for carting. This leaves seventy-five cents a ton, or $30 a day, on the amount delivered, which the miner charges the railroad company buying coal at wholesale in excess of what he charges me, when I buy by the single ton. Now there must be some cat under that meal. Again, I met, recently, a gentleman who is engaged in coal mining in Indiana. He told me that a shaft had been opened a few rods from one of the branches of this Burlington road, and, having agreed with the owner to take his coal at $1.25 a ton, he went to the railroad company to arrange for its transportation. A few miles further down the road was another mine, and he thought he ought to have rates similar to those the company mining there had. The general freight agent refused to supply him with any cars at all —I suppose he pretended not to have them. An investigation showed that at the old mine the railroad company was paying $1.75 a ton for coal—fifty cents a ton more than it was worth there—and there are strong grounds for suspecting that some of the officers of the road are interested in the mine.

"'The railroad managers say that while the "fast freight" lines may conflict with the interests of the stockholders of a road, they don't injure the farmers. I contend that they do. The rates of freight and fare are fixed with reference to the dividends that it is desired to make. Anything, therefore, that reduces the profits to the stockholders increases the rates which the farmer has to pay. Suppose this Burlington road should actually make only three per cent. this year, don't you suppose they would raise their rates the next? At the same time the "ring," comprising some of the officers of the road, may be making enough on "fast freight" lines, "palace car" lines, and other inside arrangements, to increase the dividends very greatly, were the profits divided among the stockholders. These are but a few of the abuses of which we complain. On some of the roads there are "construction rings" and "rings" of various kinds. All we ask is that the railroads shall be honestly managed; that the "rings" shall be broken, and then we are willing to pay for moving our crops such prices as will fairly remunerate the railroad companies.'

"'How do you propose to bring about this reform?' I asked.

"'We hold, in the first place, that the Legislature has a right to control the railroads. It has a right to know just how much the service which they render costs, and to enact that higher than certain rates are extortionate. The act that was passed really strikes only at unjust discriminations, and has increased the burdens of the farmers at all competing points as well as at many others. Now I hold that a lower rate at a competing point is not *prima facie* evidence of unjust discrimination. But that is what this law says.'

"'Suppose the courts decide that your law fixing maximum rates is unconstitutional,' I suggested.

"'Then we must change the Constitution. But first, we want a decision into which this question shall enter. We object to the repeated quotation of the Dartmouth College case, and desire to see this question taken up by itself and disposed of independently. If it is then decided that the railroad companies are superior to the people and the State, that they are sovereign powers, and that they have the right, by raising or reducing their rates at will, to fix the price we shall get for our produce, why then we must resort to the last remedy. I think now that I have explained to you pretty fully our position on the railroad question. Some enthusiastic men indulge in denunciation of the railroads as such and make unwise threats, but the great body of farmers are not unreasonable in their demands.

"'One other thing in connection with this: We think that there has been unjust discrimination in the matter of taxation. Mr. Harris, General Superintendent of the Burlington road, testified before a legislative committee, last winter, that the road cost $47,000 a mile. But they have only been taxed on from two to five thousand dollars a mile for road, rolling-stock, depots, side-track, and everything, while the farmers have been taxed for nearly the cash value of their property. This we don't consider fair.'

"Referring again to the low price of grain, I suggested that the low price may be partially due to overproduction.

"Mr. Smith admitted that the amount of the crop greatly affected the price of grain, but said that there was another reason why the farmers got so little for their crops—they are so constantly in need of money that they are obliged to sell for what they can get, instead of fixing their own price. 'Suppose,' he said, 'that each of our grocers in town should load up a wagon every morning and send it into the country with instructions to the driver to sell it out for whatever it would fetch, and bring back corn; don't you suppose groceries would be cheap and corn high? Reverse the picture, and you have the state of affairs actually existing. The farmer must have tea and coffee and sugar. He loads up his wagon with corn and takes it to town. The grocer says: "This tea cost me a dollar a pound; I will sell it to you for a dollar and fifteen cents; the coffee cost me forty cents; I will sell it to you for fifty." The farmer is obliged to take them at these prices or not at all. He has no money and must sell his corn in order to purchase them. But he is not allowed to say to the grocer: "This corn cost me twenty-five cents a bushel; I shall have to ask you thirty." The grocer, on the other hand, says to him: "I'll pay you twenty cents." Now, this will continue just as long as the farmers are no more prosperous than they are now. I hold that there never has been an over-production; that there is a demand for all that is raised, if it could only be got cheaply to market. What is needed is for the farmers to be able to hold on to their grain when the price is low and sell when the market is favorable.'

"Mr. Smith, like many other farmers whom I have met, complained of the enormous profits made by middle-men on almost every manufactured article the farmers buy. Many of the Granges have employed purchasing agents, who are now buying farming and domestic machinery at wholesale prices. The wholesale dealers in Chicago at first refused to sell except at retail prices to the agents of the Granges, saying that by so doing they would lose the trade of agents and retail dealers. Most of them have, however, reconsidered this determination, and the saving made by the farmers by purchasing through the Grange is from ten to fifty per cent. A farmer in Bureau county gave me some figures that will illustrate the working of this system. A good farm wagon, complete, retails at $100; the Grange purchases it for $70. A plow for which the farmers have been paying $22, a Granger gets for $16. A $50 sewing machine is purchased for $30, and a $65 one for $39. From $40 to $60 is saved on the price of a parlor organ. A rapidly growing competition seems to have sprung up among the wholesale dealers for the custom of the Granges, and whatever other result the movement may have, the farmers will be greatly benefited by the coöperative business."

CHAPTER XX.

VIEWS OF A WISCONSIN FARMER.

The Master of the Wisconsin State Grange—A Model Farmer and his Farm—Colonel Cochrane's Views of the Situation—Conflict between the Railroads and the Farms—The Roads first built with the Farmers' Savings—How the Farmer was induced to buy Railroad Stock—How they are robbed by the Roads—Position of the Middle-men—The Cost of Western Farming—Through and Local Shipments—How the Grange helps the Cheese Makers—Farming in Wisconsin; what it costs and what it pays—The Farmers unable to fix their Prices.

WE commend to the reader's attention the following account of an interview between the correspondent of the New York *Tribune* and Colonel John Cochrane, one of the most influential farmers of Wisconsin, and the Master of the Wisconsin State Grange.

"Colonel Cochrane is one of the oldest settlers of the State, having come here at an early day, and is a practical farmer, whose ability and success are attested not only by his broad acres of well-cultivated land, but by the reputation for good farming which he has all over the State. He has never been a politician, though he has, of course, always taken an intelligent interest in political questions, and he enters now into this farmers' movement at a time of life when men of his habits and pursuit generally find retirement more attractive, from a strong conviction of duty and a desire to raise the farmers of Wisconsin out of the slough of despond into

THE GRANGER'S HOME.

which they have fast been sinking. I have not been able to visit Colonel Cochrane at his home, and am therefore unable to speak of his farm from personal knowledge, but some idea of the magnitude of his operations may be obtained from the fact that he has under the highest state of cultivation nearly 1000 acres of excellent land, and finished, this recently, the threshing of his crop of 3200 bushels of as good wheat as was ever sent to market from the State of Wisconsin. His crop in a single year has been 6000 bushels.

"I met Colonel Cochrane on the cars between Waupun and this place and had a good opportunity to converse with him about the Grange, the work it has accomplished and that which it proposes, and, as he expressed more fully than any one else I have met, the sentiments of the farmers on the transportation and other questions, I shall repeat in part the substance of our conversation. Very naturally our discussion turned first to the railroads and their relations to the farmers.

"I had just been examining the last report of the Wisconsin Secretary of State in regard to the operations of the railroads of the State for the year 1871, and remarked that I found no evidence that the railroad companies of Wisconsin were making any money. The dividends declared were less than three per cent. on the cost of construction and equipment; only two roads in the State paid anything to their stockholders, and the net earnings of the roads before either paying interest on their bonds, laying aside ten per cent. to keep up the road, or paying any dividends, amounted to only little more than one per cent. on their cost. The following are the figures, although I did not quote them at length to Colonel Cochrane:

"'Total length of road reported, 2518$\frac{91}{100}$ miles; operated in Wisconsin, 1485$\frac{41}{100}$ miles; operated elsewhere, 1033$\frac{5}{10}$ miles; total cost of roads and equipments, $95,190,374.01; capital stock subscribed, $24,712,098; capital stock paid, $57,879,836.82; indebtedness of roads, $45,896,647.54; total receipts and amount due companies, $22,260,085.67; of which there was earned in Wisconsin, $7,623,904.60; expenditures less interest, new construction and dividends, $10,832,545.98; total net earnings, $11,427,537.69; cost of road and equipment per mile, $37,790.30; net earnings per mile, $453.67; roads earned on their cost and equipment 1¼ per cent.'

"It will be seen that while more than half of the mileage of roads reported on is in Wisconsin, only about one-third of the gross earnings were obtained in this State, so that the showing for this State alone, were it given entirely separate, might have been still more unfavorable.

"Colonel Cochrane replied that if the companies were not making money, many of their managers were. 'Besides,' he said, 'the men who now control these roads have, in many instances, put into them comparatively little money. Do you know how these roads have, most of them, been built? In the first place, they got land grants that in some counties are worth almost as much as the roads cost. Then they sent agents to the counties through which the road was to be built, who induced them to vote bonds to the companies and take stock; in some instances they promised them a first mortgage on the road when it should be completed. Then they got bonds from cities, and, in some cases, even from townships. Of course the interest on these all has to be raised by taxing us farmers. But that is not all. Their agents went through the country, and wherever they found a farmer who had a few hundred dollars laid by, they persuaded him to buy the stock of the road, and pay cash for it. Their argument was a very plausible one; they told us that we were paying from twenty-five to fifty cents a bushel to get our wheat hauled to Milwaukee, and that when the road was completed they would carry it for from five to ten cents. The difference would be added to the value of every bushel of wheat that we raised. Then the road would belong to the farmers, and, of course, we were all going to get rich. I took $2000 worth of stock in this (Chicago & Northwestern) road, and paid cash for it. But a great many of the farmers had no money to spare, and they were induced to give the railroads mortgages on their farms. They were assured that before the time came to pay the mortgages the road would be built and the value of their crops would be enough increased to enable them to pay up, and at any rate their farms would be enough enhanced in value more than to make up for the mortgage. In some cases the farmers took stock for these mortgages, and in some the promise of a first mortgage on the road as soon as it should be built.

"'Well, now, how do you suppose they treated us? They didn't give the farmers the mortgage on the road that they promised, but when they gave them anything it was a second mortgage on long time. The first was a short one given to capitalists. That was, of course, foreclosed, the road sold out, and the value of the stock and bonds held by counties, towns, and individual farmers destroyed. I could find you stacks of certificates and bonds for which our farmers paid cash or mortgaged their places, and which are now not worth as much as the blank paper on which they are printed. Thousands of farmers were absolutely ruined. So you see that these roads have been built, in large part, by our money, and it is the height of impudence for these men who now manage them to expect to realize a large percentage out of us on the money they have stolen from us.'

"'Do you consider the charges for freight on these roads as exorbitant?' I asked.

"'Of course I do,' was the reply. 'They are a great deal higher than when the roads were first built, and they had far less business then than they now have. Their whole policy is to take every cent that we farmers can pay. They want us to go on raising wheat, because if we should stop they would of course lose business, but they are not willing that we should make any profit. If the crop is an abundant one they put up the rates of freight. You've noticed the reports in the papers this week that it is proposed to raise the rates on all the Wisconsin roads. At the old rates, with the amount of wheat to be moved, they would make more money than during any previous year. But they don't propose to let us farmers have any of the benefit of the good crop if they can help it.'

"'How much do they charge for taking wheat from your place to Milwaukee?'

"'Ten cents a bushel for a distance of about sixty miles. When the roads were first built they used to carry it for six. But we have other grounds of complaint besides high charges. Our grain is handled too much between the producer and the consumer, and everybody who touches it takes four or five cents a bushel—he expects to make a living from it. It ought to go directly from the farmer to the seaboard, and then we should save ten or twelve cents a bushel. Again, it makes very little difference how good the quality of the grain we send to Milwaukee or Chicago is, we never get credit for anything but "number two." A neighbor of mine shipped seven car loads of wheat to Milwaukee this week, not a bushel of which weighed less than sixty pounds, and much of it went sixty-one and sixty-one and a half; it was sound and clean, and was "number one" wheat, if there was ever any raised in this State; but it was inspected as number two, and that man was actually robbed of four or five cents on every bushel of that grain. Those Milwaukee dealers will mix that good wheat with a lot that is dirty or light, and sell the whole for a better price. That is one reason why I say that our grain ought to go directly to the consumer without so much handling by those fellows who steal a part of the price of our produce and charge us for the service.

"'Now we don't ask anything unfair or unreasonable of the railroads. We say, let them be economically operated; let them stop paying such enormous salaries to their officers and give us fair treatment, and we are willing to pay them reasonably for their services. But they need not expect that we are going to allow them to earn interest on the money that we ourselves have put into the roads; we are determined on that point.'

"Colonel Cochrane says that the cost of farming in this State has been greatly increased by the building of railroads, or at least since their introduction. Before the war he could hire men who were capable of conducting his farm, without his supervision, for less money than he now has to pay for hands who hardly know enough to hitch up a team and go into the field to work unless somebody tells them how to do it. The only hands he can now hire are Dutch (I suppose he meant Germans), and they do only about half as much work as the Americans he used to get. The same is true of work in the houses; no matter how able and willing the farmer may be to hire servants, his wife must be a drudge. It is almost impossible to find, he said, a girl who knows enough to cook a meal, and who will hire out to do housework. The servants are of the most inferior kind, and even they can't be depended on to stay at the very time they are most wanted. If you make a contract with them in the Spring for the season, and agree to pay them $2 a week and board, when the hot weather comes they begin to grumble, and when the harvest begins they 'can't stand the work any longer,' and the next thing you hear they are binding in the harvest field for a dollar a day while the farmer's wife is left alone with from a dozen to twenty-five harvesters to provide for. These hardships the Colonel attributed largely to the railroads—they paid unskilled laborers better wages than the farmers could afford to, and they opened up new country for homesteads for the better class of men who formerly worked out by the month. Though these might be calamities to the large farmers in the older parts of the State, it seemed to me that society had, on the whole, been benefited by the improvement in the condition of these workingmen, and that if this was the only ground of complaint against the railroad companies, the farmers would get little sympathy. But I am convinced that it is not.

"Colonel Cochrane's idea in regard to the necessity of making through shipments is a very important one, and in almost every case where I have heard of its having been tried, it has resulted in great saving to the farmers. A striking instance came to my knowledge recently. I met on the train the owner of the largest cheese factory in the State. During a part of the Spring and early Summer he used 20,000 pounds of milk a day and made 2000 pounds of cheese. That cheese he shipped to New York, paying $1 a hundred pounds for freight, so that he received here at his factory in Wisconsin, within one cent a pound of the market value of his cheese in New York. The cheese he makes is as good as that made in New York State, sells for the same price, and, I believe, is sent East unbranded. Poor cheeses made in New York and Pennsylvania, I am told, are marked 'Western,' while good Wisconsin cheeses are branded 'New York.' I suppose four or five cents a pound would be about the difference between the farmer's selling price and the New York market price of cheese, if it was sold first at the factory, then shipped to Chicago and sold again, and finally re-shipped to New York and sold a third time. Of course every farmer who sells his milk at that factory is benefited by this direct dealing between the producers and the New York merchant, for they get about twice as much for their milk. The gentleman to whom I refer is a Granger, and pays to patrons one dollar a hundred pounds for their milk, or takes the milk, makes it into cheese, boxes it, ships and sells it, paying all charges and returning the money to the farmers for a commission of four and a half cents a pound. The price of good butter in the country in this State is about fifteen cents a pound. The rates above given net the farmer about four times as much for his milk as he gets by making butter at that price."

And while upon Wisconsin matters, the following, from the same correspondent, will be found of interest:

"I came to Oshkosh for the purpose of meeting Mr. J. Brainerd, the Secretary of the State Grange, and Mr. Osborn, its purchasing agent, both representative farmers and actively engaged in the organization of the Farmers' Movement. Mr. Brainerd, in company with his brother, manages a nursery, market, and fruit garden, and Mr. Osborn is also engaged in the cultivation of trees. Both of them are old settlers here, although both are but little past the prime of life; both have been successful in their business, and both have an extensive acquaintance among the farmers throughout the State. Neither is or has been, as far as I can learn, a politician. In my conversations with these men my aim was chiefly to learn as much as possible about the condition of the farmers of this State, socially and pecuniarily, the causes of their want of prosperity, if they fail to make farming profitable, and the remedy which the Grange and the farmers generally propose. As I have already treated at some length, in a former letter, of the relations of the Wisconsin farmers to the railroads of the State, I shall now confine myself to other topics of discussion.

"The staple crop of Wisconsin is wheat, though, from my observation in travelling through the State, and without any statistics before me, I should judge that other crops constitute a considerable per centage of the total value of its agricultural products. The dairy interest is already a large and rapidly growing one; stock-raising and feeding has been found profitable by many of the farmers; wool-growing is engaged in by some of the best farmers of the State; I think that pork enough is made to supply the demand in the lumber regions of the State with a surplus for export, and the hop and tobacco crops will bring in considerable money. Of corn and potatoes I think the surplus is not large. The wheat crop this year is the largest since that of 1860, and the grain is of excellent quality. The season was in every way favorable. After the wheat once began to grow, it came forward with unusual rapidity, and just when the farmers began to fear that the straw would be too stout and their crops lodged, dry weather set in, and, while it checked the growth of the straw, it ripened the grain finely. The weather during harvest was never better, and the result has been that the entire crop has been saved in good order. The average throughout the State is thought by the best informed farmers to be about eighteen bushels to the acre, or about four bushels greater than for several years past. On some large farms, well cultivated, the average for several hundred acres is nearly twenty-five bushels to the acre, and on single fields as high as thirty bushels. Of course the farmers, who sowed a great breadth of wheat in the Spring, are in the best of spirits now.

"But the wheat is almost the only crop that is really good. The corn has been everywhere light, and not more than one-half or two-thirds of a crop has been gathered; potatoes have not done well; the dairies of the State promised well early in the season, but the dry weather of the summer greatly reduced the amount of milk. The manager of one cheese factory told me that while he received 20,000 pounds of milk a day during a part of May and June, he was only receiving 13,000 pounds now. The hops and tobacco that I have seen look well, but I have no means of comparing the crops with those of former years. It will be seen therefore that the farmers will need all that they get for their wheat this year.

"At the close of the war, and for a year or two afterward, the farmers of Wisconsin were generally out of debt and a little 'forehanded.' The high prices that they had received for their produce of every kind had enabled them to pay off what they were before owing, to purchase farming machinery in abundance, and to indulge in many little luxuries which before had been unknown among them. But for several years past the wheat crop has not been very great, and prices have often been down; at the same time the farmers have found it difficult to go back to that system of the most rigid economy which they once practised, and they have been getting gradually deeper and deeper in debt. In the purchase of agricultural implements and machinery many of the farmers have been extravagant, and in the care of them more have been negligent. 'Go out into the country almost anywhere,' said Mr. Brainerd to me, 'and you will see the plow, the harrow, and the cultivator standing out in the weather where they were last used, and in hundreds of instances the reaper and mower will lie where the last grain or hay was cut. They may have been new this year, and cost as much as the owner will get for a great many acres of wheat. Next year they will be rusty, and the third or fourth year unfit for use, while good farmers who house their machinery make it last seven or eight years. It is the object of the Grange to teach its members to make the most of what they have, as well as to help them to purchase cheaply.'

"A great many Wisconsin farmers, like those of other States, fail because they never know on which crops they are making a profit, and on which they are losing money. I suppose I asked at least a dozen farmers in this State how much a bushel or an acre it cost to raise wheat before I found one who could give me an intelligent answer. Some thought that seventy-five cents a bushel would pay, while others thought there was no money in wheat at less than $1 a bushel. Mr.

Stillson, whose farm of 960 acres adjoins this city, and who has been one of the most successful farmers in the State, keeps a careful account of every cent that is expended in the production of each of his crops, and how much he receives for it. He told me that to pay all expenses and 7 per cent. interest on the value of the land takes about $15 an acre. If the crop is fifteen bushels, at $1 a bushel, the farmer makes nothing. This year, counting the average crop at eighteen bushels per acre, and the price at the farm at $1, the average profit will be $3 an acre. Ordinary years, when the average crop is only about fourteen bushels, unless the price is more than $1, there is no profit, and a part of the interest on the investment is lost The majority of the farmers know nothing about this, but go on year after year raising crops that don't pay, while there are others that would bring them a good profit.

"There are large sections of Wisconsin where the farmers make butter and sell it at fifteen cents a pound. By erecting a cheese factory, and turning in their milk at eighty-five cents per 100 pounds, the lowest rate paid, they would get about forty cents for a given quantity of milk, which now, after deducting the labor of making butter, the salt, packing, etc., brings them only ten cents. A few farmers like Mr. Stillson, Colonel Cochrane, and others who might be named, have always managed their places with the same business tact that a merchant displays, and, with no more capital to start with than others, and no superior advantages, they have become rich. It is not to be expected that the Grange will make Stillsons or Cochranes of all the farmers, but it may, by bringing them together for discussion and consultation, by encouraging them to read more the best agricultural journals, and by assisting them to co-operate with each other, teach them to conduct their business more intelligently and, therefore, more profitably.

"'Another reason,' said Mr. Brainerd, 'why our farmers are not prosperous, is because they have no control over the market for their produce. As soon as the crop is harvested, no matter how the market is, all the farmers of this whole country begin to rush forward their grain to Milwaukee, Chicago, and the Eastern markets. A similar policy would ruin any merchant in the country. Suppose a man in the city with a large stock of valuable goods on hand has paper becoming due and has no ready money with which to pay: does he put his goods up at auction and sell them for anything they will bring without regard to cost? No, sir; he goes and makes a new loan, and before that becomes due he has probably sold a part of his goods at a good profit. Now, why shouldn't the farmer adopt the same policy? Why shouldn't he say: "My wheat, to yield me a good profit, should fetch $1.10 a bushel: if anybody wants it for that he can take it, but it isn't for sale at a less price." Of course it would do no good for one farmer to adopt this course; but suppose a majority of the farmers of Wisconsin and Minnesota were to do it, and the price set was a fair one, don't you suppose we would get it? Talk about the price of wheat in Liverpool regulating the price here in Wisconsin. It may fix the price in the Eastern States, and if it was inflexible there and we held out too, why the transportation companies would have to move our crops for the difference. The fact is, the price would either go up in New York or freights would come down. Don't you suppose if I raise 1200 bushels of wheat that I had rather get $1200 for 1000 bushels of it, and keep the other 200 bushels over, than to sell the whole for $1200? This is a lesson we hope to teach the farmers in the Grange. I and my brothers raise vegetables and small fruits for the market here in Oshkosh, and do a profitable business, but we don't go to market in the morning and ask people how much they will give us for what we have to sell; if we did we should have failed long ago. We fix our price, and we don't sell for less than that price.'"

CHAPTER XXI.

HOW THE GOVERNMENT ROBS THE FARMERS.

Relative strength of the Farming and Manufacturing Classes—Estimate of the Number of Employers and Working People—The Farmers at the Mercy of the Manufacturers—Need of a free and cheap Market—How the Tariff works—The Government protects the Manufacturers in their Extortions from the Farmers—The Farmer requires a cheap Market—What the Farmer pays for Staple Articles of Consumption—The Farmers making the Fortunes of the Manufacturers—A Tax upon Agriculture—What a Dose of Quinine costs—Necessities taxed more heavily than Luxuries—The Interests of the Farmer opposed to those of the Manufacturer—The Government hostile to the Farmers—Food for wholesome Reflection—How the Farmer can be benefited by a Free Market—How to bring it about.

We come now to consider another manner in which the farmers and people of the country are robbed.

According to the Census of 1870 there were 2,707,421 persons engaged in manufactures, mechanical and mining industries, and 5,922,471 persons engaged in agricultural pursuits. There is no means of obtaining the exact number of employers in the manufacturing class, but, making all due allowance for the great number of small establishments in this country, we may safely assert that the number of persons engaged in manufacturing operations, and employing the labor of others, is limited to a few hundred thousand, scarcely half a million in number, if so many, the remainder of the number given above being made up by the persons receiving wages and termed operatives.

The industry of the small class of manufacturers is

divided among numerous branches of trade. Immense sums are invested in these pursuits, and large fortunes amassed. We export very little; the bulk of our manufactures being consumed at home, so that the money paid for them comes out of the pockets of the American people. The cost of manufacturing is generally high in this country, and our products of this kind cannot compete in foreign markets with those of the great manufacturing nations of Europe, which are able to produce at a much lower cost, and consequently to undersell us. Our manufacturers are therefore driven upon a home market, and must sell their goods to the people of this country alone.

It is very right and proper that the manufacturing industry of this country should be encouraged, and that so powerful an element of our prosperity and strength should be fostered by all legitimate and reasonable means. It was with difficulty that our manufactures were built up. The cheap producing nations of Europe proved themselves powerful rivals. They could deliver their goods in the American market at prices below those at which our own productions could be sold, and in order to enable American manufacturers to hold their own, the General Government imposed a tariff of duties upon imported goods. This was at first limited to a few articles, and was intended to enable the American manufacturer to meet his European rivals upon equal terms. Under the protection of the tariff the manufacturing interest of the country improved rapidly. But with its growth, its demands increased. Protection was extended to article after article, until at the present day the list of protected articles makes up a good-sized volume. Not only was the extent of protection enlarged, but the degree was also increased, until at length the tariff, which was meant to be merely protective and fostering as regards our own productions, has become prohibitory as regards importations from foreign countries.

We do not propose to discuss here the history of the tariff, or the merits of that measure; but simply to show its workings, and to call the attention of the reader to its effect upon the community.

The principal effect of the tariff, as it is arranged at present, is to close the foreign market to the average American buyer, and to compel him to purchase of the native manufacturer. He has no choice in the matter. He has no opportunity of comparing foreign with native goods. He is compelled to purchase the wares of the American manufacturer, the law operating so as to exclude the foreign dealer.

Under the operations of the tariff, the manufacturers of the United States enjoy a certain and uncontested market for their wares. They are not subjected to the competition that is necessary to the protection of the buyer, and they can put on the market whatever pleases them.

In a free market, the purchaser enjoys the liberty of buying just what he wants, and is not compelled to purchase either an inferior article, or one that does not suit him. More than this, he pays only the value of the article purchased, and the manufacturer is forced to rely upon his legitimate profit. Matters are very different under the system at present existing in this country, and this state of affairs constitutes one of the most serious evils from which the farmers of the United States are suffering.

Let us suppose that English spool cotton can be sold in the American market for four cents a spool, and that American spool cotton can be sold for the same price. The purchaser, with the prices of the two articles equal, will be guided in his selection by the quality of the goods, and will buy the best. But the American manufacturer is not willing to compete with the English manufacturer upon equal terms. Congress is petitioned, and a duty is levied upon English spool cotton, which duty at present amounts to 85 per cent. By the operation of this law the English manufacturer is compelled to add the duty to his price, and English spool cotton is advanced to seven and four-tenths cents a spool. The American manufacturer at once advances his price to, let us say, six cents a spool, and at this price sells an article which is worth only four cents, and which he could afford to sell at that figure. But Congress puts it in his power to exact 50 per cent. additional from the people of the country, and they must either purchase at this figure or buy English cotton at seven and four-tenths cents a spool. But for the tariff he would be compelled to sell his cotton at four cents, its actual value, but by the aid of that measure he is enabled to wring 50 per cent. additional from the people. The entire nation is thus taxed for the benefit of a few manufacturers of spool cotton. The interest of the nation is directly opposed to that of the manufacturers. Free trade, or a low tariff for revenue only, would result in a saving to the people, and would deprive the manufacturers of their power to plunder them.

It is the same way with almost every article of consumption that can be named. Woollen goods pay a duty of 70 per cent.; cotton goods from 35 to 52 per cent.; silks, 60 per cent.; wool hats, 72 per cent.; and blankets 90 per cent. This list could be extended almost indefinitely, but these instances are enough for our purpose. By these infamous duties the American manufacturer is enabled to keep his prices at an enormous figure, and to pocket extortionate profits at the expense of the entire people.

Let us suppose that, without the duty, English blankets could be sold in the United States for $5 a pair, and that American blankets could be sold at the same figure, with a fair profit to the manufacturer. The tariff adds a duty of 90 per cent., or $4.50 to the English article, and raises the price to $9.50 a pair. The American manufacturer is thus enabled to advance his price to $8.50, and still to undersell the English manufacturer. The people are thus compelled to pay $3.50 additional for every pair of American blankets they purchase. No wonder our woollen and cotton factors amass such immense fortunes.

There are but two or three manufacturers of quinine in the United States, but, as this medicine is the special antidote to the most common disease of this country, large quantities of quinine are used annually. The bark from which the powder is made, is admitted free of duty, but the powder itself must pay a duty of 50 per cent. The American manufacturer is thus "protected" to an extent which enables him to demand 50 per cent. more for his quinine than he could obtain in a free market, and the excess is a burdensome tax exacted from the entire nation. The reader can easily form an idea of the amount of the annual tribute paid by the nation to the two or three manufacturers of quinine.

Castor oil is about doubled in price by this system.

The heaviest duties are levied upon articles of necessity, upon the clothing, medicines and other necessities of the poor. Upon articles of luxury the duties are proportionately light. It is 90 per cent. on blankets, and but 60 per cent. on silks. On flannels of the quality used by working men for their underclothing, the duty is almost prohibitory, while on the finer grades it is comparatively light. On salt it has been within a very recent period, 107 per cent.; while on diamonds it is only 10 per cent.

Thus the burden falls upon the middle and poorer classes, upon those who cannot afford to use luxuries. The rich pay comparatively little; it is the great mass of toiling, struggling people that pay the enormous tribute to the manufacturers.

The burden falls upon the farmer with very great force. Almost every article needed in the exercise of his calling, besides those required for domestic consumption, is taxed enormously. It is useless to go over the whole list. A few instances are as follows:

	Subject to a duty or tax ad valorem in average. Per cent.
Bar iron for a wagon	40
Trace chains	60
Iron for plows	42
Steel for plows	30
Iron for threshing machines and any other machines	42
Steel for the same	32
Horse-shoes and nails	35
Horse-rugs or blankets	140
Rope	35
Harness	35
Curry-combs	35
Horse brushes	40
Nails for his barn, duty on iron	42

	Subject to a duty or tax ad valorem in average. Per cent.
Wood screws	67 to 120
Iron tacks	45
Scythes	45
Spades	45
Shovels	45
Pitchforks	45
Buts and hinges for the barn and stable doors	53
Paint for the barn and stables	51
Hand-saws, twenty-four inches	39
Cross-cut saws	26
Files and rasps	50
Hammers	38
Hoes	45
Whips	35
Saddles	35
Saddle blankets	140
Rochelle and Epsom salts to doctor his horses	73

It is asserted by the friends of the tariff that the high duties serve only to afford a revenue to the General Government, and thus supply it with the means of meeting its enormous expenses. But the burden upon the people consists not so much in the taxes they pay upon the foreign goods they purchase, as in the enormous tribute they pay to the American manufacturer. The bulk of the manufactured goods sold in the country are made here, and the high prices kept up by the operations of the tariff compel the people to pay to the American manufacturer an undue share of their earnings for his wares. Let us suppose that the duty on blankets was 25 per cent. instead of 90 per cent. Sixty-five per cent. on every pair of blankets would be saved to the purchaser. Instead of being forced to pay $8.50 for American blankets, and $9.50 for English blankets, we should pay for English blankets $6.25, and if the American manufacturer wished to undersell his English rival, he would be able to charge but from $5 to $6 for his blankets. The General Government would receive a duty of 25 per cent. on all foreign blankets, and the people would be freed from their tribute to the American manufacturer.

It is useless to talk of the injury to the revenue which would result from a lower tariff. The tariff as at present arranged, is not designed for the benefit of the Government. It is arranged for the benefit of a small class of capitalists, who are allowed by Congress to plunder the nation for their individual profit. The Government receives comparatively little benefit from the heavy burdens placed upon the nation. The vast sums thus exacted flow into the pockets of the manufacturers.

It is to the interest, not only of the farmer, but of the entire nation, that the market be thrown open to a fair and full competition, which will of necessity result in a decline of prices. The saving to the purchasing class of the country will be counted by millions. The vast majority of the American people will be immensely benefited, and people of moderate means, and especially the poor, will be relieved of one of their heaviest burdens.

As a matter of course, the manufacturers, who have fattened upon the plunder they have so long enjoyed, will resist any effort to deprive them of their immense profits. Money will be freely spent to defeat the effort of the people to secure their independence, and it will require a decided and persistent stand on the part of the people to accomplish a change. But a change is needed.

There is no good reason why the entire nation should

be compelled to pay prices out of all proportion to the value of the articles purchased, in order that a few manufacturers may amass large fortunes in an unnaturally short period. The people have the right to arrange their industrial system so that they shall pay a fair price for their purchases, and no more. They have a right to protect themselves from robbers and plunderers, under whatever guise these enemies may assail them.

Theoretically, the people are the source of power under our system of government, and Presidents and Congressmen are but their servants, charged with the execution of their wishes. But is this so in practice? Is the Congress of the United States the true exponent of the popular will? Do the people really sanction the "land grab," the "salary steal," the Crédit Mobilier swindle, the numerous jobs and schemes which plunder the people, and enrich a few unscrupulous individuals, and which bear the stamp of Congressional approval? Are the people really engaged in robbing themselves? It is absurd to ask the question.

The people, thank God, *are* the source of power, and, under our wise and beneficent system, they hold in their own hands the remedy for the betrayal of their trust. Their first duty is to exercise this remedy; to remove from power and consign to official perdition the unfaithful servants who have so wronged and misrepresented them. This should be no party work. Republicans and Democrats should join hands in it. It is not a party question. It passes the limits of ordinary politics, and comes home to each individual voter. If he values his rights as a citizen, if he wishes to save himself and his countrymen from being plundered, if he would put an end to the evils which we have pointed out to him, let him see that no one is placed in power who is not pledged to respect the popular will, which is the supreme law of the land. Let him vote for honest men only. Let him combine with his fellow-citizens in demanding the necessary reforms, and refuse to support any candidate for office who will not pledge himself to carry out the wishes of the people.

Members of Congress are very powerful individuals, no doubt; but a popular breath can unmake as well as make them; and he who values his official life should take warning in time. Whatever the remainder of the people may do, or intend, the farmers of this country have come to the conclusion that they have suffered long enough at the hands of the honorable gentlemen who make our laws. They are conscious of the fact that they are a very numerous and powerful body, constituting a very large portion of the voting class of the country, and that they are able to enforce their wishes by their ballots. They are resolved that their rights shall be respected, and they are fully aware that the Honorable Members have never recognized those rights. They are aware that they have been sacrificed to enrich railroad directors; and that they are plundered in order that a small class of manufacturers may grow rich at their expense. They are fully conscious of their wrongs, and of the responsibility of the Congress of the United States for their sufferings. They are resolved that there shall be a change, and that they shall obtain their just share of the benefits intended by the founders of our Government. If Honorable Members, either as individuals or as a body, stand in the way of their efforts to obtain redress for their wrongs, or fail to respond, and that promptly, to their demands upon them, they must be content to abide the consequences. When the issue is fairly joined, as it will be very soon, the people at large will not fail to make common cause with the farmers against the monopolists, and it will be an evil day for Congress should it continue to champion the cause of the monopolist against the sovereign people, whose power is irresistible.

CHAPTER XXII.

THE REMEDY.

Review of the Wrongs suffered by the Agricultural Classes—A Minor Evil—The Remedy—The Farmer to receive a fair Return for his Industry—The Farmer's Interest that of the Nation—The Duty of the Country to protect the Farmer—The Kind of Laws needed—The Monopolists the Enemies of the Whole People—A Free and Cheap Market demanded—Power of the Farmers of the United States—The Extent to which they control the Popular Vote—Number of voting Farmers—The People in Sympathy with the Agricultural Class—What the Farmers can accomplish—Necessity of Union—A great and glorious Revolution at Hand.

At the risk of being tedious to the reader, we have enumerated some of the principal evils from which the farming class of this country is at present suffering. We did not hope at the outset to recount all the grievances of this, the most useful, industrial class. We have aimed only at pointing out merely the leading troubles against which the farmer is struggling. We trust we have done so.

We have seen how he is robbed by the Railroad Monopoly; how in order to swell the dividends of the roads upon which he is forced to depend to reach his market, he is compelled to submit to an iniquitously high rate of transportation, which renders it impossible for him to receive his fair share of the proceeds of his crop. We have seen how utterly regardless of his rights and interests are these corporations, whose only care is to earn the largest dividend upon their stocks that can possibly be wrung from the community.

We have seen how the citizens of the Eastern States are plundered by the Coal Ring, and how this is but another phase of Railroad Extortion and Tyranny. In this unhappy state of affairs, the farmer suffers in common with the rest of the community.

We have seen also how the middle-men grow rich at the farmer's expense, and how he is compelled to pay an exorbitant price for almost everything he uses. The fact is, the farmer is charged too high for nearly every article he purchases. "In this State," says a recent letter from Iowa, "the farmers are also overcharged for the groceries and dry goods that they buy at the village store. I don't mean to say that the traders make too much money; it is the system that is at fault rather than the men, and those of the farmers who think most are beginning to see that they are in a great measure responsible for the system. The majority of the farmers of this State buy their goods of the local traders on credit, paying when they sell their crops. These traders have, therefore, in fixing their prices, to make allowance for bad debts and for interest. But as they don't receive cash, they of course cannot buy for cash, and the New York and Philadelphia merchants who 'carry' the local traders have to be paid for their risks and loss of interest. And besides all this, there is hardly a town in Iowa in which there are not about twice as many stores as there ought to be. It is a mistake to suppose that by the over-crowding of the business in the small Western towns the people get the benefit of competition. Where there are two stores and only trade enough for one, their owners combine and arrange the prices between them, being sure to put them high enough so that both can live."

Did the farmer receive a fair price for his crop, there would be less need of his making his purchases on credit, and he would be relieved of many of the debtor's burdens; but as long as he is compelled to accept an unfair and inadequate return for his labors, so long will he be compelled to bear these burdens without hope of relief.

Let him receive a fair price for his products; let him be freed from the exactions of those who control his means of reaching a market, and a very different state of affairs will ensue. These exactions swallow up so large a portion of his earnings, that he is compelled to enter the market upon the most disadvantageous terms. He is not master of his own actions, but is compelled to submit to whatever conditions the railroad and the middle-man may choose to impose upon him.

It is certainly a very unfortunate and unnatural condition of society that dooms the principal and the most useful portion of the producing class to the greatest amount of oppression; but such is the evil that is upon us. A remedy is needed, and one which can be applied without unnecessary delay. To continue in the path along which we are now moving, is to commit a most serious error. We cannot afford to oppress the farming class of this country. Both the Government and the people should foster and encourage it by every means in their power. It is our chief element of strength and stability, and a wrong inflicted upon it must react upon the whole nation. If the farmer suffers, the country at large must suffer with him, so that our own interest should lead us to shield him from injury, and sustain him in his efforts to redress his grievances.

The remedy for the evils from which the agricultural

classes are suffering lies in their own hands, and it consists in the enactment and enforcement of a series of just and liberal laws by the General and State Governments, which shall assign to each class of the community the rights to which it is fairly entitled, protect the farmer in the enjoyments of the privileges and rights which are his due, and punish any attempt of one class to prey upon another; laws which shall put an end to the practice of building useless and unnecessary railroads; which shall check the enormous power now lodged in the hands of the railroad officials of the Union, and compel them to manage their roads so that they shall be a benefit and not a curse to the community; which shall inaugurate a system of fair charges for transportation, and render it possible for the produce of the farm to reach the markets of the country at rates which do not involve the ruin of the producer; which shall give us cheap coal in this land of plenty; which shall no longer make the manufacturer richer and the consumer poorer; which shall have for their object the protection and encouragement of all classes of our industry.

The war of the monopolists upon the community has been going on long enough, and it must cease, and the General Government must cease to aid the capitalist, and to oppress the community. The interests of the entire people of the United States are opposed to those of the monopolies we have been considering. We want cheap coal, cheap bread, cheap transportation, cheap clothing. We want the price of every necessary article of consumption or daily use lowered, and whatever man, or combination of men, who seek to prevent the realization of this demand, is the enemy of the public.

The people, and especially the farmers, demand a free market, into which they can go and obtain what they need for a fair price without paying a tax to any one.

This demand they have the power to enforce, for are they not the source of power, and can they not enforce their demand, and assert their power by their ballots at the polls ?

The farmers of the United States hold in their grasp a vast power, and they are beginning to see that they must use it for their own protection. We have seen that out of a total population of 38,558,371 in the United States in 1870, 5,922,471 persons, or more than one-seventh of the entire population of the Union, were engaged in agricultural pursuits. In the same year, the total male population of the Republic was 19,493,565. Of the number of persons given as engaged in agricultural pursuits, 5,525,503 were males, so that the males of the farming classes number more than one-fourth of the entire male population of the country. Of the 19,493,565 males given above, 8,425,941 were twenty-one years of age and over; or almost one-half of the male population constitute the voting class. Assuming the same proportion for the agricultural class, we shall find that about 2,000,000 of this class are voters. The truth is, however, that the proportion is greater. Of the 5,525,503 agriculturalists, 426,381 were between the ages of ten and sixteen years; 4,650,191 were from sixteen to fifty-nine; and 448,931 were sixty and over. By the census from which these figures are taken, the number of farmers and planters is placed at 2,955,030 males, none of whom are under sixteen years of age, very few can be minors, and it seems clear that the great majority of them must be voters. We think, then, that we are warranted in asserting that the number of persons engaged in agricultural pursuits and possessing the right of suffrage, is over 4,000,000, or fully one-half of the entire voting population of the Republic. We see no reason to distrust this estimate. It appears fair, and though we have no definite returns to offer upon this head, we are sure that we are very near the actual truth in placing the voting strength of the farming classes at one-half of that of the entire country.

Now if this be true, it needs no argument to prove to the farmers of this country that they possess the power as well as the right to remedy the grievances of which they complain. Four millions of voters united in a common cause, and seeking the triumph of a common principle, are capable of accomplishing anything. But they must be united. There must be no divisions among them; no quarrelling over petty issues. The great objects for which they strive must be first achieved, and minor differences settled afterwards.

What can be more important to the farmer than the cause of his own independence, his redemption from his slavery to the monopolies that have wronged him so deeply, and robbed him so thoroughly? State and Federal legislation can be so thoroughly controlled by this powerful army of voters, that no unjust or burdensome law can be enacted to their disadvantage, the repeal of those of which they complain can be effected, and the passage of such as are necessary to the inauguration of an era of justice and equality secured.

It will be a great and a glorious revolution, and it will be peaceful. There will be no strife, no bloodshed, no ruined homes, no starving widows and orphans to cast their reproaches upon the men who undertake the change. On the contrary, the enemies of the people will be thoroughly defeated in a quiet and almost invisible manner, and the power of the monopolists will be so thoroughly broken that they will no longer be able to oppress and grind the poor, or those less fortunate than themselves.

The power of bringing about a different condition of affairs being thus secured to the farmers, it becomes a solemn duty upon their part to use it. In doing so, they can benefit not only themselves, but the entire nation. The power of the monopolists to oppress must be broken, and they can break it. They can rid the country of the great curse that has been vexing it for so long. But, in order to accomplish anything, the farmers' vote must be cast as a unit in favor of the measures desired, and of the men chosen to carry into execution these measures. Hitherto this vote has been divided, and the monopolists have taken advantage of this division to fasten their yoke upon the nation. Let the farmers now combine for the accomplishment of the ends they have in view, and there will be thousands of votes thrown with them by the outside public, who have a common interest with them, and success is certain. By presenting a solid front all over the Union upon questions vital to them, and by acting as one man in the hour of conflict with the enemy, the success of the farmers' movement will be as certain as the rising of the sun.

The best opportunity ever presented to the American farmers, of combined and energetic action in behalf of their rights, is held out to them at present by the "Order of the Patrons of Husbandry," an organization which we shall now proceed to investigate.

PART IV.

THE ORDER OF PATRONS OF HUSBANDRY.

CHAPTER XXIII.

ORGANIZATION OF THE ORDER.

Mission of Mr. O. H. Kelley to the Southern States—He discovers a Remedy for the Farmers' Grievances—Conferences at Washington—Formation of the Order of Patrons of Husbandry—Organization of the National Grange—Subsequent History of the Order—Increase of the Granges—The Grange in Iowa—Strength of the Order—The Weekly Bulletin—A Wonderful History—Unprecedentedly Rapid Growth of the Order—Comments of the "Tribune" on the Increase of the Granges—The Order in Canada—List of Canadian Granges.

In the month of January, 1866, Andrew Johnson, President of the United States, directed Mr. O. H. Kelley, of the Bureau of Agriculture at Washington, to make a tour of the Southern States, and report upon their agricultural and mineral resources. Mr. Kelley was well qualified for the task assigned him, and executed it in a manner which won the high commendation of the Department. He visited all the Southern States, and inquired minutely into their condition, conversing freely with the farmers and planters, and acquainting himself with their wants, plans, actual condition, and hopes for the future.

One of the results of this tour was the awakening of Mr. Kelley to the utterly helpless condition of the farming interest, not only of the South, but of the whole country. There were evils from which this class was suffering, and which all acknowledged, but there was no remedy to be found for them. The farmers were scattered, divided in opinions, almost indifferent to their condition, and without any means of expressing or enforcing their views as a body.

It seemed clear to Mr. Kelley that if a remedy was to be found for the evils that he encountered, it must be in the associated and harmonious action of the farming class. In order, however, to bring about such action, the farmers must be given an opportunity for association, and he conceived the plan of bringing them together through the medium of an order devoted to their interests, and affording them the means of taking the best measures for furthering those interests. He did not propose to limit the new order to the Southern States, but his plan embraced the union of the farmers of the entire country for social and educational purposes, as well as for the protection of their interests.

Mr. Kelley returned to Washington in November, 1866, and mentioned his scheme to several friends, prominent among whom was Mr. William Saunders, then and now, the Superintendent of the Gardens and Grounds of the Department of Agriculture. He also communicated it to Mr. William M. Ireland, Chief Clerk of the Finance Office of the Post Office Department; Mr. John R. Thompson, of the Treasury Department; Rev. Dr. John Trimble, also of the Treasury Department; and the Rev. A. B. Grosh, of the Department of Agriculture.

The matter was discussed by these gentlemen, and

various suggestions were offered by them respecting the proposed organization. At length, Messrs. Kelley and Ireland set to work to embody the results of these deliberations, and on the evening of the 5th of August, 1867, compiled the first degree of the Order of Patrons of Husbandry.

On the 12th of August, Mr. Saunders left Washington for the West, on business for the Department of Agriculture. He took with him the first degree of the Order, and upon reaching St. Louis began his efforts to establish it in the West. He was entirely successful, and on the evening of December 4th, 1867, the National Grange was established at Washington, at the residence of Mr. Saunders. The following officers were elected:

Master.—WILLIAM SAUNDERS, of District of Columbia.
Lecturer.—J. R. THOMPSON, of Vermont.
Overseer.—ANSON BARTLETT, of Ohio.
Steward.—WILLIAM MURI, of Pennsylvania.
Assistant Steward.—A. S. MOSS, of New York.
Chaplain.—REV. A. B. GROSH, of Pennsylvania.
Treasurer.—WILLIAM M. IRELAND, of Pennsylvania.
Secretary.—O. H. KELLEY, of Minnesota.
Gate Keeper.—EDWARD P. FARRIS, of Illinois.

Soon after this, a subordinate Grange was established in Washington, together with a school of instruction to test the efficiency of the ritual. This Grange numbered about sixty members.

The first dispensation granted by the National Grange was issued to a subordinate lodge at Harrisburg, Pennsylvania; the second was to a Grange at Fredonia, New York; the third to a Grange at Columbus, Ohio; and the fourth to a Grange at Chicago.

In April, 1868, Mr. Kelley left Washington for the purpose of establishing subordinate Granges throughout the country. His efforts were directed mainly to the Western States, and were very successful. During the first month after leaving Washington, he organized six Granges in Minnesota. From this the order spread rapidly. The farmers were a little shy of the Order at first, and the fact that it was a secret society rather inclined them to distrust it; but when it was fairly presented to them, and its objects stated and explained, it became apparent to them at once that it was a necessity, and its success was assured from the moment that this conviction entered the minds of the agricultural community. The most remarkable growth was manifested in the State of Iowa, in which as many as eighty Granges per week were organized at one period of the present year. Says a letter from Iowa:

"The Grange is stronger in Iowa than in any other State. The number of subordinate Granges is about 1800, and the number of members or 'patrons' nearly 100,000. The Order was planted here soon after its formation in Washington, nearly four years ago, but for various reasons was not widely extended until within the past twelve months. During the Spring of this year it grew with astonishing rapidity, increasing until the beginning of the harvest at the rate of from sixty to eighty Granges a week. Its great strength is in the country. When the Order was first introduced it was proposed to plant it first in the towns, with the expectation that it would naturally spread from them into the country. But this was found to be impossible, for the town Granges seemed to have very little cohesive power. There was not life enough in them to preserve their own existence, to say nothing of their inability to propagate their kind. It was not until the missionaries went out into the prairies and began to work among those who gave their whole time and attention to the farm that the Order took root.

"Of many attempts to organize the farmers of Iowa for any purpose, the Grange is the first that has been successful. Farmers' clubs, established under the most favorable circumstances and fostered by the State Board of Agriculture, have always proved short-lived. Not more than eighteen or twenty could ever be kept in active operation at once, and even that small number would soon dwindle away unless the promise of rare seeds or agricultural documents was constantly held out as an inducement to the farmers to continue their meetings, and new clubs were frequently organized to fill the places of those that died. But the Grange seems to have exactly supplied a want of the people. Perhaps the 'hard times' experienced during the past two or three years by the farmers of Iowa, in common with those of other Western States, aroused them to the necessity of united, intelligent coöperation, and prepared them to welcome the Grange as the first channel opened through which they might hope for relief. At all events, the Order has taken kindly to the soil of Iowa, and has accomplished more here than elsewhere, while the farmers' clubs have long since disappeared."

The number of Granges in the United States is increasing so rapidly that it is hard to give an accurate statement concerning them. The Secretary of the National Grange formerly issued from his office at Washington a monthly bulletin giving the strength of the Order in each State. The bulletin of January 1st, 1874, places the total number of subordinate Granges at 10,015, being an increase of 1225 in a single month. As these bulletins are of interest to all concerned in the movement, we give that of the 1st of January as a specimen. It is as follows:

PATRONS OF HUSBANDRY.

NATIONAL GRANGE,

SECRETARY'S OFFICE, 612 LOUISIANA AVENUE,

WASHINGTON, D. C., *January* 1, 1874.

Alabama........ SUB. GR'S, **354** Deputies, 9
Master—W. H. CHAMBERS................Oswichee.
Sec'y E. M. LAW..........................Tuskegee.

Arkansas........ SUB. GR'S, **134** Deputies, 40
Master—JOHN T. JONES...................Helena.
Sec'y JOHN S. WILLIAMS..............Duvall's Bluff.

California...... SUB. GR'S, **129** Deputies, 25
Master—J. M. HAMILTON.................Guenoc.
Sec'y W. H. BAXTER.....................Napa City.

Delaware.......
Master—State Grange not organized.
Sec'y ....

Florida........... SUB. GR'S, **20** Deputies, 6
Master—B. T. WARDLOW..................Madison.
Sec'y W. A. BRINSON...................Live Oak.

Georgia......... SUB. GR'S, **437**
Master—T. J. SMITH.........................Oconee, C. R. R.
Sec'y E. TAYLOR...........................Colaparchee.

Illinois........... SUB. GR'S, **821** Deputies, 117
Master—ALONZO GOLDER.................Rock Falls.
Sec'y O. E. FANNING.....................Galt.

Indiana.......... SUB. GR'S, **862** Deputies, 88
Master—HENLEY JAMES..................Marion.
Sec'y MARTIN M. MOODY..............Muncie.

Iowa............... SUB. GR'S, **1845** Deputies, 111
Master—A. B. SMEDLEY...................Cresco.
Sec'y N. W. GARRETSON...............Winterset.

State		Office	Name	Post-office
Kansas	SUB. GR'S, 779; Deputies, 44	*Master*—	M. E. HUDSON	Mapleton.
		Sec'y	GEO. W. SPURGEON	Jacksonville.
Kentucky	SUB. GR'S, 172; Deputies, 28	*Master*—	M. D. DAVIE	Beverly.
		Sec'y	J. EUGENE BARNES	Georgetown.
Louisiana	SUB. GR'S, 50; Deputies, 3	*Master*—	H. W. L. LEWIS	Osyka, Miss.
		Sec'y	N. D. WETMORE	Ponchatoula.
Maine	SUB. GR'S, 1; Deputies, 1	*Master*—	State Grange not organized.	
		Sec'y	....	
Maryland	SUB. GR'S, 14; Deputies, 1	*Master*—	State Grange not organized.	
		Sec'y	....	
Mass	SUB. GR'S, 19; Deputies, 6	*Master*—	T. L. ALLIS	Conway.
		Sec'y	BENJ. DAVIS	Ware.
Michigan	SUB. GR'S, 169; Deputies, 2	*Master*—	S. F. BROWN	Schoolcraft.
		Sec'y	J. T. COBB	Schoolcraft.
Minnesota	SUB. GR'S, 406; Deputies, 37	*Master*—	GEO. J. PARSONS	Winona.
		Sec'y	WM. PAIST	St. Paul.
Mississippi	SUB. GR'S, 516; Deputies, 57	*Master*—	A. J. VAUGHAN	Early Grove.
		Sec'y	W. L. WILLIAMS	Rienzi.
Missouri	SUB. GR'S, 1302; Deputies, 94	*Master*—	T. R. ALLEN	Allenton.
		Sec'y	A. M. COFFEY	Knob Noster.
Nebraska	SUB. GR'S, 370; Deputies, 33	*Master*—	WM. B. PORTER	Platsmouth.
		Sec'y	WM. MCCAIG	Elmwood.
New Hamp	SUB. GR'S, 17; Deputies, 1	*Master*—	DUDLEY T. CHASE	Claremont.
		Sec'y	C. C. SHAW	Milford.
New Jersey	SUB. GR'S, 30; Deputies, 14	*Master*—	EDWARD HOWLAND	Hammonton.
		Sec'y	R. W. PRATT	Newfield.
New York	SUB. GR'S, 29; Deputies, 24	*Master*—	GEO. D. HINCKLEY	Fredonia.
		Sec'y	W. A. ARMSTRONG	Elmira.

State		Officer	Name	Address
N. Carolina	Sub. Gr's, **139**; Deputies, 10	*Master*—	Columbus Mills	Concord.
		Sec'y	G. W. Lawrence	Fayetteville.
Ohio	Sub. Gr's, **293**; Deputies, 48	*Master*—	S. H. Ellis	Springboro'.
		Sec'y	D. M. Stewart	Xenia.
Oregon	Sub. Gr's, **58**; Deputies, 12	*Master*—	Daniel Clark	Salem.
		Sec'y	J. H. Smith	Harrisburg.
Penna	Sub. Gr's, **56**; Deputies, 34	*Master*—	D. B. Mauger	Douglassville.
		Sec'y	R. H. Thomas	Mechanicsburg.
S. Carolina	Sub. Gr's, **206**; Deputies, 1	*Master*—	Thomas Taylor	Columbia.
		Sec'y	D. Wyatt Aiken	Cokesbury.
Tennessee	Sub. Gr's, **302**; Deputies, 34	*Master*—	William Maxwell	Maxville.
		Sec'y	J. P. McMurray	Trenton.
Texas	Sub. Gr's, **51**; Deputies, 11	*Master*—	J. B. Johnson	Fairfield.
		Sec'y	H. H. Parker	Salado.
Vermont	Sub. Gr's, **42**; Deputies, 16	*Master*—	E. P. Colton	Irasburg.
		Sec'y	E. L. Hovey	St. Johnsbury.
Virginia	Sub. Gr's, **18**; Deputies, 1	*Master*—	J. W. White	Charlotte.
		Sec'y	M. W. Hazelwood	Richmond.
W. Virginia	Sub. Gr's, **27**; Deputies, 9	*Master*—	B. M. Kitchen	Shanghai.
		Sec'y	J. W. Curtis	Martinsburg.
Wisconsin	Sub. Gr's, **289**; Deputies, 47	*Master*—	John Cochrane	Waupun.
		Sec'y	H. E. Huxley	Neenah.
Colorado	Sub. Gr's, **4**; Deputies, 1	*Master*—	State Grange not organized.	
		Sec'y	....	
Dakota	Sub. Gr's, **39**; Deputies, 6	*Master*—	E. B. Crew	Lodi.
		Sec'y	O. F. Stevens	Jefferson.
Washington	Sub. Gr's, **6**; Deputies, 3	*Master*—	United with Oregon.	
		Sec'y	....	

Canada.......... { *Master*—State Grange not organized.
SUB. GR'S, 8 { *Sec'y* ....

Monthly Statement of Subordinate Granges organized during 1872 and 1873:

	January.	Feburary.	March.	April.	May.	June.
1872......	54	61	96	98	65	86
1873......	158	347	666	571	696	625

	July.	August.	September.	October.	November.	December.
1872......	115	79	79	91	109	120
1873......	612	829	919	1050	974	1225

Total number of Subordinate Granges, 10,015; reported membership, 751,125.

Such has been the rapid and unprecedented growth of the Order. During the year 1873, its increase was most marked, and it is now spreading at a rate which will soon carry it to every county in the United States. Hitherto its chief growth has been in the Western States, but it is now increasing rapidly in the Middle, Eastern, and Southern States. It has become in all respects a national movement, and is already a power which by its immense strength and great importance is exerting a tremendous influence in every part of the Union.

"It has grown by the process of nature," says the New York *Tribune*, commenting upon its wonderful history, "out of the pressing wants of the time; and it has spread and waxed strong steadily and rapidly ever since it came into existence. Within a few weeks it has menaced the political equilibrium of the most steadfast States. It has upset the calculations of veteran campaigners, and put professional office-seekers to more embarrassment than even the Back Pay. The Grange was not the origin of the Farmers' Movement; it was only its outgrowth. Yet in calculating the moral and physical forces of the movement, the Grange must be the principal factor. It took its origin a few years ago in a Washington office. Its founders were not farmers, but Government clerks. They understood, however, the temper and the wants of the agricultural class, and they devised, with aid, perhaps, from the prairies, one of the most ingenious and effective organizations ever invented in so short a time. From Washington the Grange spread all over the great grain region, and back again to the far East, and Southward into the country of cotton and tobacco. Everywhere it found enthusiastic adherents. Everywhere it found farmers who needed its help, farmers' wives and daughters, who picked up a new life and a fresh spirit under its social and intellectual influence, and gave it in return the attraction of a refined and cheerful membership. Business and pleasure surely were never so profitably combined before. It was the old principle of the husking-frolic and the quilting-bee, applied to loftier objects and practised with a sterner eye to the main chance. The women and the young people, who met at evening to go through the little ceremonies of the Grangers' ritual, and pass an hour or so in decorous amusements and conversation, and song, and reading, may have fancied that they were only breaking the monotony of toil by a bit of harmless entertainment; but the Grange knew better. They were learning, and teaching others, to be better farmers, to be thrifty, to buy cheaper, to sell better, to rid themselves of creditors, to keep out of debt, and finally to check the enormous power of the railroads which have so long been driving the farmers to the wall. In fact, the Grange succeeded so well because it had the art to take the average men and women of the West and make them work without knowing it, and accomplish what they hardly dreamed of. That it did succeed is a sufficient evidence that it was founded upon a genuine and general popular need, and directed toward a really important object. Even if it should be worsted at last in the struggle against monopoly, it will still have done ample good to justify its existence.

"What the end of the railroad question is to be no one can yet predict. The difficulties of the case are perplexing, and they vary in different States, nay, in different towns. Thus far we cannot see that the Patrons of Husbandry have made any distinct impression upon the defences of their chief adversary. Rates have been raised rather than lowered on the transportation lines since this agitation began. Legislation where it has been attempted has done little good. The decisions of the courts have settled no great principles, and it is not clear that the courts themselves are not in danger of being unsettled by the popular passion. But it is a great thing to have brought the farmers' cause so prominently before the public that their complaints are now familiar to every intelligent man in the United States. It is a great thing to have created almost universal sympathy for them. It is a great thing to have made the farmers formidable in the eyes of the politicians, started Committees of Congress and the Legislatures travelling about the country to find out what can be done for them, and terrified caucuses into crav-

27

ing their friendship. Above all, perhaps, it is a great thing to have brought the honest agricultural element prominently into politics; it looked a year ago as if the art of government was gradually falling into the hands of an exclusive profession,—the hands being the dirtiest and the profession the meanest in all the United States."

No official estimate is given of the number of members of the various Granges, but with the number of lodges at seven thousand, or over, the number of members may be safely estimated at about half a million. A year hence, if the present rate of growth continues, as there is every reason to believe it will, the membership of the Order will be nearly three millions.

Such an Order must necesarily exert a powerful influence upon the community. If it be for good, the Order must be regarded as a blessing to the entire country; and that it is for good we hope to show in the remaining chapters of this work.

The Order has passed beyond the limits of the United States, and has spread into Canada, where there are now seven subordinate Granges. They are as follows:

No. 1. INTERNATIONAL—*Master*, Albert P. Ball, Stanstead, Prov. Quebec; *Secretary*, J. G. Field, Derby Line, Vermont.

No. 2. BARNSTON—*Master*, Geo. C. Hanson, Barnston, Prov. Quebec; *Secretary*, Ed'd A. Cushing, Barnston, Prov. Quebec.

No. 3. GOLDEN—*Master*, Albert E. Damon; *Secretary*, Wm. Major, Drew's Mills, Prov. Quebec.

No. 4. SHIPTON—*Master*, Timothy Leet; *Secretary*, R. M. J. Bernard, Danville, Prov. Quebec.

No. 5. AYLMER—*Master*, Lord Aylmer, Melbourne, Prov. Quebec; *Secretary*, John Main, Melbourne, Prov. Quebec.

No. 6. FRELIGHSBURG—*Master*, S. R. Whitman, Frelighsburg, Prov. Quebec; *Secretary*, E. E. Spencer, Frelighsburg, Prov. Quebec.

No. 7. DUNHAM—*Master*, R. L. Galer, Dunham, Prov. Quebec; *Secretary*, C. E. C. Brown, Dunham, Prov. Quebec.

CHAPTER XXIV.

COMPOSITION OF THE GRANGES.

Objects of the Order—Male and Female Members—Division into National, State, and Local Granges—Officers of the National Grange—Membership limited to Agriculturalists—Organization of the Grange—Qualifications of Members—Secrecy required—The Degrees of the Order—The Ritual Degrees of the Subordinate Grange—Degrees of the State Grange—Degrees of the National Grange—Financial Matters—How the Grange is organized—Description of the Working System of the Order—How the Expenses of the Grange are paid—The Secret Feature considered—Necessity for and Advantage of Secrecy—Advantages of Female Members—Woman's Work in the Grange—Objects of the Order discussed.

THE Order of Patrons of Husbandry is a secret society devoted to the interests of the agricultural classes. Its objects are stated in a general way by the Secretary of the National Grange, as follows:

"NATIONAL GRANGE, WASHINGTON, D. C.

"It is evident to all intelligent minds that the time has come when those engaged in rural pursuits should have an organization devoted entirely to their interests. Such it is intended to make the Order of Patrons. It was instituted in 1867; its growth is unprecedented in the history of secret associations, and it is acknowledged one of the most useful and powerful organizations in the United States. Its grand objects are not only general improvement in husbandry, but to increase the general happiness, wealth, and prosperity of the country. It is founded upon the axioms that the products of the soil comprise the basis of all wealth; that individual happiness depends upon general prosperity, and that the wealth of a country depends upon the general intelligence and mental culture of the producing classes.

"In the meetings of this Order all but members are excluded, and there is in its proceedings a symbolized ritual, pleasing, beautiful, and appropriate, which is designed not only to charm the fancy, but to cultivate and enlarge the mind and purify the heart, having at the same time strict adaptation to rural pursuits.

"The secrecy of the ritual and proceedings of the Order have been adopted chiefly for the purpose of accomplishing desired efficiency, extension, and unity, and to secure among its members, in the internal working of the Order, confidence, harmony, and security.

"Women are admitted to full membership, and we solicit the co-operation of women because of a conviction that without her aid success will be less certain and decided. Much might be said in this connection, but every husband and brother knows that where he can be accompanied by his wife or sister no lessons will be learned but those of purity and truth.

"The Order of the Patrons of Husbandry will accomplish a thorough and systematic organization among farmers and horticulturists throughout the United States, and will secure among them intimate social relations and acquaintance with each other, for the advancement and elevation of their pursuits, *with an appreciation and protection of their true interests.* By such means may be accomplished that which exists throughout the country in all other avocations and among all other classes—*combined co-operative association for individual improvement and common benefit.*

"Among the advantages which may be derived from the Order are systematic arrangements for procuring and disseminating, in the most expeditious manner, information relative to crops, demand and supply, prices, markets, and transportation throughout the country; also for the purchase and exchange of stock, seeds, and desired varieties of plants and trees, and for the purpose of procuring help at home or from abroad, and situations for persons seeking employment; also, for ascertaining and testing the merits of newly-discovered farming implements and those not in general use, and for *detecting and exposing those that are unworthy, and for protecting, by all available means, the farming interests from fraud and deception, and combinations of every kind.*

"We ignore all political or religious discussions in the Order; we do not solicit the patronage of any sect, association, or individual, upon any grounds whatever, except upon the intrinsic merits of the Order.

"The better to secure greater benefits to our members, we desire to establish Granges in every city, town, and village in the United States."

The Order consists of a National Grange with its headquarters at Washington, D. C., as many State Granges as there are States in the Union, and an indefinite number of Subordinate Granges. The names and addresses of the officers of the various State Granges have been given. The officers of the National Grange are as follows:

OFFICERS OF NATIONAL GRANGE.

Elected at Sixth Annual Session.

Master—DUDLEY W. ADAMS, Waukon, Iowa.
Overseer—THOMAS TAYLOR, Columbia, South Carolina.
Lecturer—T. A. THOMPSON, Plainvew, Wabasha Co., Minn.
Steward—A. J. VAUGHAN, Early Grove, Marshall Co., Miss.
Assistant Steward—G. W. THOMPSON, New Brunswick, N. J.
Chaplain—REV. A. B. GROSH, Washington, D. C.
Treasurer—F. M. MCDOWELL, Corning, N. Y.
Secretary—O. H. KELLEY, Washington, D. C.
Gate-keeper—O. DINWIDDIE, Orchard Grove, Lake Co., Ind.
Ceres—MRS. D. W. ADAMS, Waukon, Iowa.
Pomona—MRS. O. H. KELLEY, Washington, D. C.
Flora—MRS. J. C. ABBOTT, Clarkesville, Butler Co., Iowa.
Lady Assistant Steward—MISS C. A. HALL, Washington, D. C.

EXECUTIVE COMMITTEE.

WILLIAM SAUNDERS, Washington, D. C.
D. WYATT AIKEN, Cokesbury, Abbeville Co., S. C.
E. R. SHANKLAND, Dubuque, Iowa.

The membership of the Order is confined to persons engaged in agricultural pursuits. This limitation is necessary as the success of the Order depends upon the unity of interests existing among its members. There must be a common object, and a common incentive to attain the fulfilment of that object.

The term Grange, which is applied to the various societies composing the Order, is particularly appropriate. It is derived from the Latin *granium*, and means simply a farm.

The membership of the Order includes both males and females. The former must be over eighteen years of age, and the latter over sixteen. These constitute a secret society, which is divided into local or subordinate Granges, State Granges, and a National Grange.

According to the Constitution of the Order, the officers

of a local Grange are thirteen in number, and are elected annually. Nine are men, and four women. The male officers are a Master, Overseer, Lecturer, Steward, Assistant Steward, Chaplain, Treasurer, Secretary, and Gate-keeper. The female officers are Ceres, Pomona, Flora, and Lady Assistant Steward. The officers of the State and National Grange are the same in number and title.

The State Grange consists of the Masters and Past Masters (or Masters who have passed out of office) of the local Granges. As there must be twelve officers in the State Grange, the Constitution requires that there shall be fifteen local Granges in each State before a State Grange can be organized.

The National Grange is composed of the Masters and Past Masters of the State Granges.

The local Granges are required to meet once a month, but are permitted to hold as many intermediate meetings as they may deem necessary. The State Granges and the National Grange meet annually, at such times and places as they may appoint.

No one but a person of good moral character can become a member of the Order. Idlers and disreputable persons have no place in it. No religious or political test is demanded. Individual opinions are respected and tolerated to the fullest degree; and no discussions upon religious or political questions are allowed at the meetings of the Granges.

The Order is a secret society, as has been stated. Its proceedings are secret, and none but members are admitted to its meetings. Several degrees are included in the organization, for each of which there is a beautiful and appropriate ritual.

The subordinate or local Grange is empowered to confer four degrees. The first degree is that of Laborer. When conferred upon women it is the degree of Maid. The second is that of Cultivator, for men; and Shepherdess for women. The third, that of Harvester, for men; and Gleaner, for women. The fourth, that of Husbandman, for men; and Matron for women.

The State Grange alone confers the fifth degree, that of Pomona (Hope). The wives of Masters of the local Granges are also members of the State Grange.

The last two degrees are conferred by the National Grange only. They are the sixth degree, or that of Flora (Charity); conferred upon Masters of the State Granges and their wives who have taken the degree of Pomona; and the seventh degree, or that of Ceres (Faith), which is conferred upon any members of the National Grange who have served one year therein. It has charge of the secret work of the Order, and tries all cases of impeachment of officers of the National Grange.

Each person becoming a member of the Order is required to pay an initiation fee, which is fixed at $5 for men, and $2 for women. This covers the four degrees of the subordinate Grange. The monthly dues are regulated by each Grange, but they cannot be less than ten cents for each member.

The Treasurer of the subordinate Grange is required to pay to the State Grange the sum of $1 for each man, and fifty cents for each woman initiated into the Grange such payments being made quarterly. He is also required to pay a quarterly due of six cents for each member.

Each State Grange is required to pay to the National Grange in quarterly instalments, the annual due of ten cents for each member of the Order within its jurisdiction. The funds of the Order are guarded by a series of judicious regulations, and their proper administration is thus guaranteed.

The necessity for these payments, as well as some of the leading objects of the Order, is set forth in the following communication from a leading Southern Patron (Mr. Wm. E. Simmons, of Charleston, S. C.):

"Not quite a year ago, I called one day to see a friend on some business; this was soon arranged to our mutual satisfaction, and, after chatting a while, I got up to leave. As we shook hands my friend handed me a small pamphlet, at the same time requesting me to read it. 'What is this?' I asked. 'Read it and judge for yourself,' he replied. By reference to the title page I was informed that this mysterious little book was the 'Constitution of the *Patrons of Husbandry*.' 'Who are the Patrons of Husbandry? I never heard of them,' I soliloquized. Upon reading a little further I found that this was a name applied to a secret organization for the promotion and protection of agricultural interests, which existed throughout various portions of the United States, in the form of clubs, or, according to the Patrons' nomenclature, 'Granges.' I also discovered that there was a head centre or 'National Grange,' located at Washington, to which all other Granges were subservient, and from which emanated all authority, information, and plans of work of any importance. Now, the idea of applying for authority to organize a club of farmers in South Carolina, to a body of men nearly five hundred miles away, and of submitting to them, for their sanction, every plan of work that we devise down here, for the benefit of our immediate neighborhood, seemed to me to imply a degree of subjection to the will of others altogether at variance with my conception of republican free principles. This, however, was only my side of the question, and being unwilling to incur the odium attached to the two knights of olden time, who, having regarded a shield from different stand-points, and seen different colors, contended, each, that the color he had seen was that of the whole shield, and neither having the candor to go over to the other side and see for himself, both preferred to settle the question at the point of the sword. I, therefore, determined just to step over and see how things looked from the opposite stand-point. And I must say that a good deal of the one-sided coloring I had at first seen was lost by this little manœuvre. For instance, being subject to the will of the National Grange, which had, at first, seemed to be so great an objection, began now to look somewhat like an advantage, certainly like a necessity. For any body of men to be effective must be organized, and every organization to be perfect must have a head, with an able corps of subaltern officers. Just in proportion as an organization is deficient in these respects will it be deficient in strength, and *vice versa*, the same is equally true. The Patrons of Husbandry is simply a grand combination of societies, of which the subordinate Granges are the individual members, the State Granges the corps of subaltern officers, and the National Grange, composed of none but those who are distinguished for pre-eminent merit and ability, the great head. Now, each subordinate Grange, being only one of a thousand, like individual members of one great body, it is necessary for the good of the whole that their several workings be in harmony with each other, and consistent with the objects of the Order. Therefore, to secure this general harmony and consistency, it is necessary that each subordinate and State Grange should submit its plan of work to the approval of the National Grange, otherwise it might enter upon a field of work totally foreign to either of the foregoing principles. Again, the exaction by the National Grange, from other Granges, of an annual due for each of their members, and each degree conferred on members during the year, seemed to me to be an imposition—a well-planned scheme for extracting money from the unsuspecting farmer. Why are we not allowed to keep all our money in our own Grange? Surely we can use it to better advantage than any other body of men! When, however, I began to think of the functions of this Grange, its portion of the work of this immense body of societies, of the vast amount of information, covering every subject of interest to the Order, daily being collected by it all over the country, to be handed over to the printer and afterward re-distributed, in a printed form, to the Granges in every section, it occurred to me that to do all this requires at least one office, and one or more secretaries and correspondents. How is the rental of that office to be paid? These secretaries and correspondents must be paid for their time. How, also, are their salaries to be paid? And, more important than all, how can it pay for this large amount of printing? Besides, the postage on all this material must foot up handsomely at the end of the year. A new member of a subordinate Grange, after having taken four degrees, has paid into the treasury of the Grange five dollars, besides his regular monthly dues. Of this sum, the treasurer of his Grange pays the secretary of the State Grange twenty-five cents for each degree the new member has taken; also, an annual due of twenty-five cents for said member, making a total of one dollar and twenty-five cents for the year. The treasurer of the State Grange then pays to the secretary of the National Grange ten cents for each degree conferred upon this member, together with an additional ten cents, as an annual due for him, or a total of fifty cents for the year. Thus, of five dollars and over, paid by a new member into the treasury of his Grange during the first year of his membership, only fifty cents is claimed by the National Grange, and, after he has taken all the degrees, it claims only ten cents as his annual due.

"The woman membership feature, likewise, appeared to be a very objectionable one. 'Woman's proper sphere of action,' I repeated, 'is the fireside; when she leaves that to join societies, etc., she takes the first step towards woman's rights.' But there is no more danger of her becoming a woman's rights woman at the Grange than there is of her becoming one at the fireside, for at each place she is in company with her husband and brother. . . . .

"As to the importance of woman's aid, I thought of the numerous instances afforded by history of the powerful influence she has always exercised over the destinies of mankind, but a stronger proof of that importance exists in the mind of every man in the United States who is blessed with a faithful and intelligent wife. But where is the necessity for secrecy? Why cannot the workings of the Order be open to the gaze of all men? Men who do good only are never afraid to have their actions scrutinized. Certainly not. But wise men keep their own counsel, and of what they do the world knows nothing until it is done. The general who conceives a great strategic movement confides his plans only to a few trusty followers, and when any business of great moment comes before Congress it sits in secret session. The general does not conceal his plans, nor Congress its deliberations, through fear of the world's scrutiny, but because the safety of the interests involved demands that secrecy be observed. In like manner, then, the Patrons conceal their deliberations, because by so doing they insure greater security and efficiency in their workings. Thus, also, are bad men prevented availing themselves of the advantages of the Order to impose upon the credulity of mankind. Secret societies, too, have always been more permanent than others, and will flourish where the latter die out. When, however, I had got this far, I suddenly remembered that we had an agricultural society in our county, and I began asking myself why it would not answer all the purposes of this secret Order of farmers. After a little reflection upon the objects of such organizations, I found that agricultural societies are limited in their application to furnishing information on the *practice* of agriculture, horticulture, etc., on the nature of soils and manures, to the establishment of shows for produce, stock, etc., and to the promotion of agricultural education. Here, then, were the most important objects of agricultural societies; unless the Patrons proposed to do more, it was useless to think further on the subject of joining them. But, on turning my attention to the objects of the Patrons, it soon became evident that not only did they propose to do all of the above, but also a great deal more. Besides teaching the farmer how to practise agriculture after the most improved methods, they, likewise, protect him in the act. They are ever on the watch to detect and warn him of impositions, to prevent his intrusting his produce to fraudulent agents, and to bring about a reduction of high freights for his benefit. They enable him to purchase his supplies cheaper, and his tools and machinery at from ten to twenty-five per cent. less than he can by any other means. They prevent cruelty to animals, nurse the sick, assist the poor, instruct the youth, establish libraries and reading-rooms, and aim at elevating all classes, both socially and morally. And while agricultural societies, in general, possess no common bond of union, each being wholly independent of the other, the Granges are but so many 'parts of one stupendous whole,' which whole is a body firmly united in substance and intent, guided by one head, striving for the achievement of one end, namely, THE GENERAL GOOD OF THE AGRICULTURIST AT LARGE."

CHAPTER XXV.

THE LAWS OF THE ORDER.

Constitution and By-Laws of the Order—Preamble—Organization—Degrees—Officers—Meetings—Laws—Ritual—Membership—Fees for Membership—Dues — Requirements — Charters and Dispensations — Duties of Officers—Treasurers—Restrictions—Amendments—Birth-day of Ceres to be observed—By-Laws.

CONSTITUTION OF THE ORDER OF PATRONS OF HUSBANDRY, AND BY-LAWS OF THE NATIONAL GRANGE,

Adopted at the Sixth Annual Session of the National Grange, January 1873.

PREAMBLE.

HUMAN happiness is the acme of earthly ambition. Individual happiness depends upon general prosperity.

The prosperity of a nation is in proportion to the value of its productions.

The soil is the source from whence we derive all that constitutes wealth; without it we would have no agriculture, no manufactures, no commerce. Of all the material gifts of the Creator, the various productions of the vegetable world are of the first importance. The art of agriculture is the parent and precursor of all arts, and its products the foundation of all wealth.

The productions of the earth are subject to the influence of natural laws, invariable and indisputable; the amount produced will consequently be in proportion to the intelligence of the producer, and success will depend upon his knowledge of the action of these laws, and the proper application of their principles.

Hence, knowledge is the foundation of happiness.

The ultimate object of this organization is for mutual instruction and protection, to lighten labor by diffusing a knowledge of its aims and purposes, expand the mind by tracing the beautiful laws the Great Creator has established in the Universe, and to enlarge our views of Creative wisdom and power.

To those who read aright, history proves that in all ages society is fragmentary, and successful results of general welfare can be secured only by general effort. Unity of action cannot be acquired without discipline, and discipline cannot be enforced without significant organization; hence we have a ceremony of initiation which binds us in mutual fraternity as with a band of iron; but although its influence is so powerful, its application is as gentle as that of the silken thread that binds a wreath of flowers.

The Patrons of Husbandry consist of the following:

ORGANIZATION.

Subordinate Granges.

First Degree: Laborer, (man,) Maid, (woman.)
Second Degree: Cultivator, (man,) Shepherdess, (woman.)
Third Degree: Harvester, (man,) Gleaner, (woman.)
Fourth Degree: Husbandman, (man,) Matron, (woman.)

State Grange.

Fifth Degree: Pomona, (Hope.)

Composed of the Masters of Subordinate Granges and their wives who are Matrons. Past Masters and their wives who are Matrons shall be honorary members and eligible to office, but not entitled to vote.

National Grange.

Sixth Degree: Flora, (Charity.)

Composed of Masters of State Granges and their wives who have taken the degree of Pomona. Past Masters of State Granges, and their wives who have taken said degree of Pomona, shall be honorary members and eligible to office, but not entitled to vote.

Seventh Degree: Ceres, (Faith.)

Members of the National Grange who have served one year therein may become members of this degree upon application and election. It shall have charge of the secret work of the Order, and shall be a court of impeachment of all officers of the National Grange.

Members of this degree are honorary members of the National Grange, and are eligible to office therein, but not entitled to vote.

CONSTITUTION.

ARTICLE I.—*Officers.*

SECTION 1. The officers of a Grange, either National, State, or Subordinate, consist of and rank as follows: Master, Overseer, Lecturer, Steward, Assistant Steward, Chaplain, Treasurer, Secretary, Gate-keeper, Ceres, Pomona, Flora, and Lady Assistant Steward. It is their duty to see that the laws of the Order are carried out.

SEC. 2. *How Chosen.*—In the Subordinate Granges they shall be chosen annually; in the State Granges once in two years; and in the National Grange once in three years. All elections to be by ballot.

Vacancies by death or resignation to be filled at a special election at the next regular meeting thereof—officers so chosen to serve until the annual meeting.

SEC. 3. The Master of the National Grange may appoint members of the Order as deputies to organize Granges where no State Grange exists.

SEC. 4. There shall be an Executive Committee of the National Grange, consisting of three members, whose terms of office shall be three years, one of whom shall be elected each year.

SEC. 5. The officers of the respective Granges shall be addressed as "WORTHY."

ARTICLE II.—*Meetings.*

SECTION 1. *Subordinate Granges* shall meet each month, and may hold intermediate meetings as may be deemed necessary for the good of the Order. All business meetings are confined to the Fourth Degree.

SEC. 2. *State Granges* shall meet annually at such time ana place as the Grange shall from year to year determine.

SEC. 3. The *National Grange* shall meet annually on the first Wednesday in February, at such place as the Grange may from year to year determine. Should the National Grange adjourn without selecting the place of meeting, the Executive Committee shall appoint the place and notify the Secretary of the National Grange and the Masters of State Granges, at least thirty days before the day appointed.

ARTICLE III.—*Laws.*

The National Grange, at its annual session, shall frame, amend, or repeal such laws as the good of the Order may require. All laws of State and Subordinate Granges must conform to this Constitution and the laws adopted by the National Grange.

ARTICLE IV.—*Ritual.*

The Ritual adopted by the National Grange shall be used in all Subordinate Granges, and any desired alteration in the same must be submitted to, and receive the sanction of, the National Grange.

ARTICLE V.—*Membership.*

Any person interested in Agricultural pursuits, of the age of sixteen years, (female,) and eighteen years, (male,) duly proposed, elected, and complying with the rules and regulations of the Order, is entitled to membership and the benefit of the degrees taken. Every application must be accompanied by the fee of membership. If rejected, the money will be refunded. Applications must be certified by members, and balloted for at a subsequent meeting. It shall require three negative votes to reject an applicant.

ARTICLE VI.—*Fees for Membership.*

The minimum fee for membership in a Subordinate Grange shall be, for men five dollars, and for women two dollars, for the four degrees, except charter members, who shall pay—men, three dollars, and women fifty cents.

ARTICLE VII. *Dues.*

SECTION 1. The minimum of regular monthly dues shall be ten cents from each member, and each Grange may otherwise regulate its own dues.

SEC. 2. The Secretary of each Subordinate Grange shall report quarterly to the Secretary of the State Grange the names of all persons initiated or passed to higher degrees.

SEC. 3. The Treasurer of each Subordinate Grange shall report quarterly, and pay to the Treasurer of his State Grange the sum of one dollar for each man, and fifty cents for each woman initiated during that quarter; also a quarterly due of six cents for each member.

SEC. 4. The Secretary of each State Grange shall report quarterly to the Secretary of the National Grange the membership in his State, and the degrees conferred during the quarter.

SEC. 5. The Treasurer of each State Grange shall deposit to the credit of the National Grange of Patrons of Husbandry with some Banking or Trust company in New York, (to be selected by the Executive Committee,) in quarterly instalments, the annual due of ten cents for each member in his State, and forward the receipts for the same to the Treasurer of the National Grange.

SEC. 6. All moneys deposited with said company shall be paid out only upon the drafts of the Treasurer, signed by the Master, and countersigned by the Secretary.

SEC. 7. No State Grange shall be entitled to representation in the National Grange whose dues are unpaid for more than one quarter.

ARTICLE VIII.—*Requirements.*

SECTION 1. Reports from Subordinate Granges relative to crops, implements, stock, or any other matters called for by the National Grange, must be certified to by the Master and Secretary, and under seal of the Grange giving the same.

SEC. 2. All printed matter on whatever subject, and all information issued by the National or State to Subordinate Granges, shall be made known to the members without unnecessary delay.

SEC. 3. If any brothers or sisters of the Order are sick, it shall be the duty of the Patrons to visit them, and see that they are well provided with all things needful.

SEC. 4. Any member found guilty of wanton cruelty to animals shall be expelled from the Order.

SEC. 5. The officers of Subordinate Granges shall be on the alert in devising means by which the interests of the whole Order may be advanced; but no plan of work shall be adopted by State or Subordinate Granges without first submitting it to, and receiving the sanction of, the National Grange.

ARTICLE IX.—*Charters and Dispensations.*

SECTION 1. All charters and dispensations issue directly from the National Grange.

SEC. 2. Nine men and four women, having received the four Subordinate Degrees, may receive a dispensation to organize a Subordinate Grange.

SEC. 3. Applications for dispensations shall be made to the Secretary of the National Grange, and be signed by the persons applying for the same, and be accompanied by a fee of fifteen dollars.

SEC. 4. Charter members are those persons *only* whose names are upon the application, and whose fees were paid at the time of organization. Their number shall not be less than nine men and four women, nor more than twenty men and ten women.

SEC. 5. Fifteen Subordinate Granges working in a State can apply for authority to organize a State Grange.

SEC. 6. When State Granges are organized, dispensations will be replaced by charters, issued without further fee.

SEC. 7. All charters must pass through the State Granges for record, and receive the seal and official signatures of the same.

SEC. 8. No Grange shall confer more than one degree (either *First*, *Second*, *Third*, or *Fourth*) at the same meeting.

SEC. 9. After a State Grange is organized, all applications for charters must pass through the same and be approved by the Master and Secretary.

ARTICLE X.—*Duties of Officers.*

The duties of the officers of the National, State, and Subordinate Granges shall be prescribed by the laws of the same.

ARTICLE XI.—*Treasurers.*

SECTION 1. The Treasurers of the National, State, and Subordinate Granges shall give bonds, to be approved by the officers of their respective Granges.

SEC. 2. In all Granges bills must be approved by the Master, and countersigned by the Secretary, before the Treasurer can pay the same.

ARTICLE XII.—*Restrictions.*

Religious or political questions will not be tolerated as subjects of discussion in the work of the Order, and no political or religious tests for membership shall be applied.

ARTICLE XIII.—*Amendments.*

This Constitution can be altered or amended by a two-thirds vote of the National Grange at any annual meeting, and when such alteration or amendment shall have been ratified by three-fourths of the State Granges, and the same reported to the Secretary of the National Grange, it shall be of full force.

BY-LAWS.

ARTICLE 1.

The fourth day of December, the birthday of the Patrons of Husbandry, shall be celebrated as the anniversary of the Order.

ARTICLE 2.

Not less than the representation of ten States present at any meeting of the National Grange, shall constitute a quorum for the transaction of business.

ARTICLE 3.

At the annual meeting of each State Grange it may elect a proxy to represent the State Grange in the National Grange in case of the inability of the Master to attend, but such proxy shall not thereby be entitled to the Sixth Degree.

ARTICLE 4.

Questions of administration and jurisprudence, arising in and between State Granges, and appeal from the action and decision thereof, shall be referred to the Master and Executive Committee of the National Grange, whose decision shall be respected and obeyed until overruled by action of the National Grange.

ARTICLE 5.

It shall be the duty of the Master to preside at meetings of the National Grange; to see that all officers and members of committees properly perform their respective duties; to see that the Constitution, By-laws, and Resolutions of the National Grange and the usages of the Order are observed and obeyed; to sign all drafts drawn upon the treasury, and generally to perform all duties pertaining to such office.

ARTICLE 6.

It shall be the duty of the Secretary to keep a record of all proceedings of the National Grange, to keep a just and true account of all moneys received and paid out by him, to countersign all drafts upon the treasury, to conduct the correspondence of the National Grange, and generally to act as the administrative officer of the National Grange, under the direction of the Master and the Executive Committee.

It shall be his duty at least once each month, to deposit with the Fiscal Agency holding the funds of the National Grange all moneys that may have come into his hands, and forward a duplicate receipt therefor to the Treasurer, and to make a full report of all transactions to the National Grange at each annual session.

It shall be his further duty to procure a monthly report from the Fiscal Agency with whom the funds of the National Grange are deposited of all moneys received and paid out by them during each month, and send a copy of such report to the Executive Committee and the Master of the National Grange.

Article 7.

Section 1. It shall be the duty of the Treasurer to issue all drafts upon the Fiscal Agency of the Order, said drafts having been previously signed by the Master and countersigned by the Secretary of the National Grange.

Sec. 2. He shall report monthly to the Master of the National Grange, through the office of the Secretary, a statement of all receipts of deposits made by him, and of all drafts or checks signed by him during the previous month.

Sec. 3. He shall report to the National Grange at each annual session a statement of all receipts of deposits made by him and of all drafts or checks signed by him since his last annual report.

Article 8.

It shall be the duty of the Lecturer to visit, for the good of the Order, such portions of the United States as the Executive Committee may direct, for which services he shall receive compensation.

Article 9.

It shall be the duty of the Executive Committee to exercise a general supervision of the affairs of the Order during the recess of the National Grange; to instruct the Secretary in regard to printing and disbursements, and to place in his hands a contingent fund; to decide all questions and appeals referred to them by the officers and members of State Granges; and to lay before the National Grange at each session a report of all such questions and appeals and their decisions thereon.

ARTICLE 10.

SECTION 1. Such compensation for time and service shall be given the Master, Lecturer, Secretary, Treasurer, and Executive Committee, as the National Grange may, from time to time, determine.

SEC. 2. Whenever General Deputies are appointed by the Master of the National Grange, said deputies shall receive such compensation for time and services as may be determined by the Executive Committee: *Provided*, In no case shall pay from the National Grange be given General Deputies in any State after the formation of its State Grange.

ARTICLE 11.

SECTION 1. The financial existence of Subordinate Granges shall date from the first day of January, first day of April, first day of July, and first day of October subsequent to the day of their organization, from which date their first quarter shall commence.

SEC. 2. State Granges shall date their financial existence three months after the first day of January, first of April, first of July, and first of October immediately following their organization.

ARTICLE 12.

Each State Grange shall be entitled to send one representative, who shall be a Master thereof, or his proxy, to all meetings of the National Grange. He shall receive mileage at the rate of five cents per mile both ways, computed by the nearest practicable route, to be paid as follows: The Master and Secretary of the National Grange shall give such representative an order for the amount on the Treasurer of the State Grange which he represents, and this order shall be receivable by the National Grange in payment of State dues.

ARTICLE 13.

Special meetings of the National Grange shall be called by the Master upon the application of the Masters of ten State Granges, one month's notice of such meeting being given to all members of the National Grange. No alterations or amendments to the By-laws or Ritual shall be made at any special meeting.

ARTICLE 14.

These By-laws may be altered or amended at any annual meeting of the National Grange by a two-thirds vote of the members present.

CHAPTER XXVI.

THE GRANGE AS A MEANS OF PROTECTION.

Advantages to the Farmers of the Grange—A Means of Combination afforded them—Good Results of Combination—Harmonious Action secured—The Grange intended as a Means of resisting the Farmers' Enemies—How it proposes to correct Abuses—The War against the Railroads—The Grange pledged to secure Measures just to all Parties—The Entire Order working for the Accomplishment of One Object—The Order the Protector of the Farmer—Plan of Action—How Measures are devised and carried out—Position of the Grange towards the Railroads—The Grange not a Political Institution—The Power of the Order, and how it is exerted—Individual Opinions respected by the Order—Prospects for the Future—Its Work—Membership confined to Agriculturalists.

WE propose, now that we have stated the general character of the Order of Patrons of Husbandry, to examine into some of its avowed objects, and to discuss some of its claims to the confidence and favor of the class to which it appeals.

In the first place, the Grange offers to the farmers of the United States a means of combination, of harmony of action such as they have never before possessed. It offers them the means of expressing their views and wishes as a body, and of enforcing them. Its objects and mode of accomplishing them are distinctly stated, and are such as to commend themselves to every member of the Order. The weakest local Grange pursues a policy and seeks the furtherance of ideas and interests that are the objects of the efforts of every Grange in the Union. There is no division. Individual differences are cheerfully surrendered to the common good. An opportunity is afforded to each member to give expression to his views, and the general discussion which follows such expression subjects the ideas advanced to a test which proves either their excellence or the reverse.

It has been generally admitted that "the farmer is a power in the land," but this admission has been made in a vague sort of way. The farmer has hitherto been conscious of his power, but it has been as a voter that he has regarded himself. He has known the extent to which his power has been rendered useless by the divisions that have existed among his class. Farmer Smith may have had one idea of his grievances, and the proper mode of righting them, and the view held by Farmer Brown may have been entirely different. But in the Grange these two men meet upon a common platform. The Order recognizes as evils certain things upon which the whole farming class have long been united, and it proposes a method of redress which men of the most extreme views can accept. Its primary object is to bring about a union among the farmers of the Republic, for it is its cardinal maxim that only in union can the agricultural class show its strength and make it felt.

The Grange recognizes the plain fact that the American farmer is the victim of certain evils, and it proposes to correct these. In order to accomplish this, it organizes the farmers into one harmonious body; makes them a unit, and then exerts their combined strength for their protection. Conscious that the farmer needs protection against his foes, and that a single man can accomplish nothing, the Grange exerts the strength of a powerful combination to protect the farmer against the evils from which he suffers.

The Grange recognizes that the farmer is robbed of his fair reward by the extortions of the railroad companies, and it seeks to bring about the existence of a more liberal state of affairs in which the railroads, while earning a just return on their investments, shall allow the farmer a more generous and less ruinous rate of freights. It proposes to accomplish this by the united action of the farmers of the entire country; by placing in power men of tried integrity, and who will secure the passage of a series of laws which shall protect the farmer and at the same time do justice to the railroad companies. This is a task of great magnitude, for, as we have shown, the power of the Corporations is immense, and they will not easily surrender it. But the Grange has taken a lesson from them. It has observed that they have achieved their power to plunder by combination and unity of action, and it proposes to fight them with the same weapons. It proposes to combine the farmers of each and every State against them. No single man, no single local or State organization of men, could accomplish such a work. But with the machinery at the command of the Order, the work is simple enough.

In each local Grange the programme is the same as in the National Grange. The same evil is recognized, and the same method of remedying it is sought to be enforced. The measures necessary to this end are arranged, and have received the sanction of the National Grange. They are recommended to the Order at large, and are communicated by the National Grange to the various State Granges, each of which, in its turn, communicates them to its respective subordinate or local Granges. The local Granges, through their individual members, carry these measures into effect, and thus we have the whole Order working for the accomplishment of one object. It is easy to see that when the Order shall embrace the entire agricultural class, as it now bids fair to do, its power to accomplish its objects will be irresistible.

The Grange then constitutes itself the protector of the farmers and their interests, and thus at the outset appeals to their sympathy and secures their co-operation. Its acts being the results of the combined wisdom of its members, it is clear that the protection it offers will be enlightened and efficient. Its deliberations ensure the avoidance of rash and hot-headed action. Nothing is done until all means are discussed, and the best and most suitable secured. The farmer is conscious that he has powerful and unscrupulous enemies, and that he urgently needs protection against them; as a member of the Order he can secure the accomplishment of the object nearest his heart, and self-interest prompts him to be a Patron of Husbandry.

The Order has, as we have said, clear and well-defined views of the evils from which the farmers are suffering, and its chief object is to remedy them. It claims to be the best judge of the wisdom and efficiency of these measures, and declines to allow the farmer's enemies to decide the question for him. Such opposition as it has met, has come from the monopolies and their supporters; but as one of the objects of the Order is the destruction of these gigantic frauds upon the people, this opposition is natural, and was to have been expected.

It has been charged that the Order is hostile to the railroads of the country, and is bent upon their destruction. This charge is absurd; but as it has been made, we may as well meet it. The Grange is not hostile to the railroads as a means of transportation, for it recognizes the necessity of this establishment to our system of society, but it is bitterly hostile to the corrupt management of this great industry. It is entirely opposed to the system of building railroads at the cost of the nation for the benefit of a few stockholders. It is opposed to the system of watering the stocks of railroad corporations, and of over-charging the people who are compelled to use the roads, in order to extort from them the means of paying large dividends on this fictitious increase of stock. It is opposed to the tyranny and corruption of the railroads, to their disregard of the rights of individuals and communities; and it is in favor of subjecting them to a series of laws which shall place them on a footing with other industries, and compel them to respect the rights of others.

These things it proposes to change, and we have shown that it has the power to accomplish its object. It is an object which appeals to the sympathy and demands the co-operation of the people at large, and there cannot be a doubt that the Order will receive all the outside aid that it needs.

The Order seeks no affiliation with either or any of the political parties of the present day. It has nothing to do with what men usually call politics. It is devoted to the interests of the farmers, and leaves political questions to its individual members, respecting every man's right of opinion and action in these matters.

Its views upon the political questions of the day

differ in different States. Attempts have been made by politicians to ally it with one or the other of the great parties that divide the nation, but they have not been successful. The Order has kept aloof from all such parties, and aims only at protecting the farming classes from the wrongs from which they are suffering, and devising and carrying out measures for their relief. It does not seek to interfere with or supersede either party, but, undertaking a different work from that of either, draws its recruits from both. The work with which it is charged is enough for it; and will occupy its attention and engage its energies to the exclusion of all other questions.

Nor has the Grange yet perfected its system of operations. The evils it seeks to combat are so old and great that there can be no difference respecting them; but the remedies are yet under discussion, and no definite plan of action will be resolved upon until a thorough investigation of its merits and a free interchange of opinions can be had.

Says a writer who has spent much time among the farmers of the West for the especial purpose of learning their views:

"The Grange makes no war upon railroads as such. Its members generally recognize the fact that without railroads their rich farms would soon be deserted except along the rivers, and become once more the homes of wolves and wild fowl, and they are willing that men who put their money into railroads shall receive fair returns on the capital they invest. But they believe that the people have some rights which even railroad corporations are bound to respect, and they are not willing that railroad charges shall be put so high as to pay ten per cent. on stock which the present owners never paid anything for, nor on stock that has been issued as a dividend. Many of the roads have been partially built with money subscribed by the farmers themselves, or by the towns and counties through which they extend, and the people are unwilling that men who have since got possession of these roads, often by the payment of comparatively little money, shall make large dividends until they have low rates. Above all, they are unwilling that the price of their crops shall be fixed by a ring of railroad men.

"The remedy proposed is different in almost every State. Some propose a *pro rata* law; some desire a fixed rate of maximum tariffs for freight and passengers; some desire that the question shall be regulated by the State, and some by the United States. In some States the present controversy is over the power of the Legislatures to control the railroads; in others that power is conceded either in the charters of the companies or the constitutions of the States, and then the question is, how shall the power be exercised? Some hold that the right of eminent domain exercised by a State in condemning private property for the use of railroads is a right pertaining only to the State in its sovereign capacity, and one of which it cannot in any way divest itself. Railroad property, they say, is no more sacred or exempt from the exercise of this right, when the interests of the people demand it, than any other. Should a railroad company now existing, therefore, become so oppressive in its charges as to make it for the public interest that a new company should be formed under greater restrictions, the State has the power to charter a new company to operate a road over the same line, and, in its exercise of the right of eminent domain, to appoint a commission to appraise and condemn the property belonging to the old company. Nowhere are violent or illegal measures proposed. No tracks have been torn up, no buildings burned; the motto of the Grange is, equal justice to all; and as the farmers have the power, by united action, to carry any measure they propose, they feel confident of ultimate success.

"The Grange is not a political organization; politics and religion are forbidden topics of discussion in the Grange-room. But it strives to educate men to think for themselves and not to follow the dictates of party leaders and packed caucuses unless their own judgment approves. A majority of the people in the West, as is well known, have been Republicans, and a majority of the Grangers voted for General Grant last year. The Democracy has been their *bête noir*, and though the faith of many of them may have been shaken in the infallibility of the Republican party, they would never go into any other of which the Democrats formed an influential part. But the Grange makes the farmers a power within themselves and outside of any political party, and now, in the States where they are strongest, should they step out of the ranks of the party with which they have hertofore acted, it would not be necessary for them to seek shelter in the camp of their long time political enemy. They might leave the old ship that served them so long and bore them safely through so many a glorious fight, but which is now strained and worm-eaten, not to go on board the Democratic ship, but to launch a new one of their own. How wisely they may build remains yet to be seen. Just now, the influence of the Grange is little more than to loosen the bands that bind men to old parties and to make them free to choose their future places.

"The Grange, although organized several years ago, did not become a formidable body until within the past twelvemonth. Immense crops of corn which had to be sold for less than the cost of production; short crops of wheat, with no corresponding increase of price; railroad combinations to prevent competition and reasonable rates of freight; wheat and corn rings, formed to control the price along many of the great railroad lines, and to prevent the farmers from receiving any advantage from favorable markets; the insatiable greed of some implement makers and agents; the accumulating mortgages on farms—these and many other circumstances have at length aroused the long-suffering farmers, and the Grange, already instituted, gave them the means to make their demands effective. This explains the astonishing growth of the Order since October, 1872.

"I have said that none but farmers and their families may be members of the Grange. I see it reported that a number of grain-dealers and others in Boston, not practical agriculturists, have obtained a charter and organized a Grange. I don't know by what authority Mr. Abbott, the State Deputy of Massachusetts, has initiated men who were not farmers into the Order, but every prominent Patron with whom I have spoken on the subject disapproves of this extension of the Order, and the matter will probably come before the National Grange at its next session. Hundreds of men in every State I have visited have, for personal ends, attempted to obtain admission to the Grange. Some have been politicians, who have desired to promote their political prospects; some have been commercial agents, who have had an eye to business; and some have been editors, who have desired to make the Order their constituents. Grangers are ready to clasp hands with any one for the purpose of promoting reform, but they do it outside of the Grange room."

CHAPTER XXVII.

SOCIAL ASPECTS OF THE GRANGE.

Dull and Monotonous Life of Farmers and their Families—The need of the Farmer for Social Intercourse—Hard lot of Farmers' Wives and Daughters—Scarcity of Amusements—"All Work and no Play"—Demand for a Change—The Work of the Grange—The Grange a Means of Social Enjoyment—Advantages of the Social System of the Grange—Farmers' Wives and Daughters in the Grange—The Lesson of Innocent Enjoyment taught—Festivals and Pleasures of the Grange—How the Order promotes Sociability and Friendship among the Farmers—Interesting Details—Barbecues—Sociables—Public Meetings—The Lesson of Courtesy—What the Grange has done for the Happiness of the Agricultural Class—A Great and Good Work.

THERE is another feature of the Grange that, alone, would make it invaluable to the farmers of America. It is the best means that has yet been devised of cultivating social relations among them, and in its social aspects, it is a perfect success.

Few who have not been residents of the country, can rightly understand the monotony of a farmer's life. Day after day the farmer and his family pursue the same appointed round of toil. There is no change save the regular recurrence of the Sabbath, and attendance upon religious services where such privileges are accessible. During the busy season constant toil leaves little leisure on the hands of any member of the household; but when the long winters set in, and several months of forced inactivity are upon them, the monotony is often very hard to bear. It is always felt, even by the dullest.

Visiting is rare, and as a rule is not encouraged. Strange to say, the farmer does not value social intercourse, and yet no one needs it more. He lives a lonely and secluded life, rarely caring to go beyond the limits of his farm, except to visit the village or the country store on business. Occasionally a circus, or some travelling show, or some political meeting would draw the farmers out of their seclusion, but with this exception, the monotony was unbroken. No wonder, then, that with constant toil and unbroken solitude as his only companions, the farmer should be a careworn, prematurely old man. No human being can exist without a certain amount of recreation and change. If these be denied, the whole mental and moral nature must suffer. The indifference of the farmer to social pleasures and relaxations was, perhaps, the worst feature of the case.

Now, if this was the condition of the farmers, what shall we say of their wives and daughters? Women are much more dependent upon society than men. Monotony affects them quicker and more powerfully, and they need relaxation and amusement to a greater degree than men. Yet how inexpressibly dreary is the lot of the farmer's wife and daughter. Theirs is a life of constant toil—the same routine day after day, week after week—with scarcely a break in it. A funeral or a wedding, or a county fair, are great events in their existence, as they bring them together with their neighbors and afford them some little society. But as a rule the loneliness of their lives is unbroken. They are confined to the limits of the farm, and there they must remain.

Who that has attended a country fair has failed to

mark the noisy and, at the first glance, unnecessary mirth of the farmers' wives and daughters? To city people, with scores of pleasures and amusements within reach, these outbursts may seem ridiculous; but they are natural. They are the assertion of the protest of nature against the long and dreary restraint that has been put upon them, and the mirth of these women is as natural and irresistible as the song of the long imprisoned bird escaping from its cage. They laugh and are boisterous because they have been silent and subdued so long. Such occasions, such opportunities for enjoyment come rarely to them, and they are quick to take advantage of them. Their time for pleasure is brief, and they make the most of it. Then they go back to their dreary monotony at home; for no matter how comfortable the home, how liberal the provision of the husband and father, there is a monotony and a loneliness about it which the most loving wife and dutiful daughter feels most keenly.

"Time was when young American women born and bred in the country were glad to 'go out to do housework,' and a woman's 'help' in the house was intelligent and capable. That time has passed; intelligent American girls, if their services are not needed at home, and they are obliged wholly or partially to earn their own living, become teachers or seek employment in the cities and villages, while the only household 'help' that can be obtained is of the raw Irish or German variety, which requires a generation in which to be educated, and which when educated ceases to be obtainable. The farmer's wife, therefore, though she may be able and willing to pay for good assistance, cannot get it, and is obliged to make a slave of herself, working from sunrise to sunset through the long Summer days until nature itself fairly gives way. I do not exaggerate; I have seen the haggard looks and heard the weary sighs of overworked farmers' wives in the East and in the West. I have seen broad acres of highly cultivated land groaning under the abundant crops, good houses and barns, fine stock and money to the farmer's credit in the bank, but the order and cleanliness that reigned in-doors in harvest time, when twenty hungry men sat around the farmer's board, as well as when the family only were there, were too often purchased at the price of the premature old age of the wife. Anything that will break in upon this tread-mill life which, though not quite universal, is altogether too common, should be hailed with joy by the farmer and his family."

Now the Grange proposes to change this state of affairs, and render to the farmers and their families one of the greatest services that can possibly be done for them. It offers them the means of improving their social intercourse, of adding to their pleasures, and of improving their condition mentally as well as socially.

"The social feature of the Order," says J. C. Abbott, Deputy of the National Grange, "consists in being associated with ladies who are admitted as members of the Order, and upon whom the four degrees of a subordinate Grange are conferred. Other Orders close their doors against women, and shut her out from their councils. But believing that she is the help-meet of man, and that we need her counsel, as well as her aid, we open the doors of the Grange and bid her welcome.

"At the regular monthly meetings of the Grange a feast is held, the ladies supplying a bountiful repast of the good things of this life, and this is made the happy medium whereby introductions are made and many pleasing and lasting acquaintances are formed.

"Mankind, in their natures, are social beings, and when in solitude all pine for social and friendly intercourse. This being an organization designed especially for farmers, we say that its social features are particularly pleasing, and well adapted to meet the necessity which exists for some method to bring them and their wives and families together, so that they may know each other better, and be brought into a closer connection and sympathy than now exists.

"In fact, the aid already rendered to our Order by woman is invaluable, and her services could not well be dispensed with. To divest the Order of this feature would be to go far toward despoiling it, and detract greatly from the enjoyment now felt in all our meetings. We say, then, that woman's presence is indispensable in all places where good conduct and moral and religious principles are sought to be inculcated."

"Socially," says Captain E. L. Hovey, in an address delivered at St. Johnsbury, Vermont, February 22d, 1872, "it is the right thing in the right place, for it is a Farmers' Society. If there is anything that tends to break up the humdrum life they have been living, and are living, it should be fostered with every possible means. Of all the evils that fetter and hamper this class of our people, there is nothing so destructive of that happiness human beings were permanently destined to enjoy as the seclusion in which they drag out their lives. Isolated from the arena of business life, with nothing to stimulate thought, they too often live and die strangers to any of those finer and ennobling feelings that are so readily nurtured by commingling of society. They are becoming more and more unsocial, and have been tending in this direction since the first settling in this country. A half century ago and more, when general poverty and insecurity rendered mutual protection a necessity, there was a more genial feeling among the inhabitants. They went long distances on foot for an evening's enjoyment of social intercourse; but since those good old days a competence has come to the majority of farmers, and they stick to the homestead with a tenacity that fosters every social evil. They go through with about the same routine of duties from sunrise to sundown, from one year's end to another, through the whole active part of life, never unloosing the mind from the drudgery of farm life. The human being alone was created with the faculty of social intercourse, and he who fails to improve it scarcely rises above the level of the brute creation.

"One of the principal objects of this Society is to enlarge this God-given faculty. It calls the laborious worker of the soil from his duties and places him side by side with those engaged in the same occupation. A thousand questions are discussed that interest and benefit its members.

"Place a person in solitary confinement before any indications of intelligence are manifest, and actual experiment proves that the appearance, the shape of head, the features, suffer from such treatment, and the actual knowledge is excluded. Since these things are so, farmers who enslave themselves, who are semi-imprisoned, cannot expect to wear a very prepossessing personal appearance.

"You all know the value of a social home; you

know the difference between it and one continuously darkened with silence, wrangling, or brutal violence, it may be. What tends more to enlighten the mind and fill it with principles that will shed their lustre down through the whole course of life than a family gathered after the work of the day is completed, engaged in healthy, mind-invigorating, social intercourse? Any one who has paid any attention to the positions of families reared in these different ways cannot fail to bid God-speed to one institution that will improve the social condition of the farmer.

"Some who are inclined to see a humbug in every new move assert that this is a 'woman's rights' movement; others that it is a cover for political intrigues. Nothing could be further from the truth. The fact that women are admitted to full membership in the Order I regard as one of its most worthy features. I do not believe in woman suffrage, nor never can. I do not believe in making a plow-point of a gold watch; but the condition of a people, its customs, its manners, its morals, its social standing, its educational status, depend more upon its women than upon man. Is there not as wide a field for improvement in woman's sphere as in man's? Besides, when men are assembled for mental culture or social chat, what more stimulates them to high-minded action than the presence of woman? But there is no need of my dilating upon this important theme. The solution of a mathematical problem decides the matter. If great good comes from a meeting of only two—provided that *both* sexes are represented—how much advantage will result from a gathering of a hundred?"

It may be a slight thing to the ordinary reader of a

newspaper to see such paragraphs as the following, which we clip from an exchange:

"Delegates of the several Granges of Dubuque county, Iowa, met at Rockdale on the 8th, and arranged for a monster basket pic-nic at Ebworth on the 17th."

"The Granges of Randolph county, Ind., held a pic-nic at the fair grounds at Winchester on the 9th."

These items, which are now very frequent in the newspapers of the day, may mean nothing at all to the ordinary reader, but to the farmer, or to one who is familiar with the old regime of country life, they are eloquent indeed, for they tell of a different era that has dawned upon the agricultural community, and to the "Granger" they are apparent as the work of his Order. Who ever heard of farmers taking the trouble to organize themselves for enjoyment until the Grange taught them that pleasure is a duty as well as labor?

In the monthly meetings of the subordinate Grange, the farmers of a community are brought together twelve times a year if no oftener, and are accompanied by their wives and daughters. The ordinary proceedings of each meeting are such as to interest them, and to place them in a happy frame of mind for the cultivation and promotion of social relations. Acquaintances are made, new friendships are formed, and old ones strengthened. The farmer is taught that the world does not end for him at the boundaries of his farm; that there are hopes, fears, joys and sorrows beyond his domain in which it is his duty to take an interest. The entire farming community is bound together by the bonds which unite men working for a common cause. A few hours are spent in pleasant intercourse.

The week or the month has one bright spot in it for those who have taken part in the meeting.

The farmer is taught that social relaxation and pleasure are a necessity of human existence, and the duty of granting these to his family and dependents is made an obligation which he is bound to comply with. Leading members of the Grange arrange for gatherings of pleasure and social intercourse apart from the regular meetings of the Order. Pic-nics, barbecues, sociables, processions, public meetings are arranged and carried out at such times as will not interfere with the work of the farm, and the whole power of the Order is exerted to break up the dullness of farm life and enliven it with innocent social pleasures, which shall lighten the cares of the farmers and their families and increase their happiness.

In all the meetings of the Order, in all its gatherings for pleasure, the two sexes are brought together, and placed upon an equality, and the farmer is thus quietly and forcibly reminded that his wife and daughters are ladies entitled to all the courtesies and attentions of polite society, and not mere drudges charged with the performance of household work; something he has been too apt to forget.

Courtesy and high-toned feelings and deportment in all things are the lessons taught by the Grange, which thus becomes the instructor and guide of its community. Coarse and improper pleasures, rude and unmanly or unwomanly conduct, are not tolerated by the Order. Its pleasures are innocent and healthful, and it aims at the elevation and improvement of its members in every respect.

A letter from Iowa, in which State the Order has

attained its greatest strength, thus refers to the great change which the Grange has already brought about there:

"The social condition of a majority of the farmers in this State before the organization of Granges is described to me by leading members as anything but satisfactory. The country is comparatively new, having been settled only 10 to 25 years, and the people are still very much isolated. The dull monotony of their lives has only been broken in upon by an occasional wedding or funeral, and they have plodded on year after year, working from sunrise to sunset, taking very few holidays, rarely meeting each other except at the cross-roads store, church, or town-meeting, reading very little, and, in fact, transforming themselves into corn and wheat-producing machines. Of business methods they have known almost nothing. It was rare that a farmer was able to tell how much it cost him to make a bushel of corn or of wheat, a pound of beef or of butter, or to bale a ton of hay. The condition of the farmer's wife was even worse. Her work began earlier and ended later than that of her husband. It was a slavish life, with almost nothing to give it variety or to lift the woman out of the deep rut of her daily drudgery. Perhaps the most of these people have never known any different kind of life; perhaps they have had better food and a greater abundance of it, more comfortable homes and better clothing than before they became Iowa farmers, but their enjoyment of life has been of a low order, and any one who will give them broader ideas will be hailed as a benefactor. I have not been describing the average farmer of this State from personal observation; that would be impossible for a stranger spending only a few days, or at most weeks, in the State to obtain. I am obliged to take the picture as it is painted for me by those who have beem familiar with it for years, and who have often sat at the farmer's table and slept in his 'spare room.' At the same time no one can ride across this State without observing even from the window of a railway car a painful contrast between the richness of the fields and the poverty of many of the homes. I don't believe there is a better farming country in the world than the Des Moines Valley, with its beautiful rolling prairie lands, and it every where shows evidence of good culture. And yet the owners of those farms live too often in little cramped-up houses, unattractive, in many cases uncomfortable. If there are good houses, they are generally owned not by the men who get their living from the farms, but by the men who buy wheat and ship it, or who have other means of support, and farm for pleasure, not for profit.

"Such a state of affairs as I have described cannot be corrected in a month or a year, and yet, I am assured, the influence of the Grange in elevating the farmers socially is already very apparent. In the first place, it brings together the farmers of a neighborhood, old and young, men and women, and if it did nothing more it would not have been established in vain; for the people of a town cannot spend an hour a week in informal conversation, even, without gaining new ideas and carrying away something to think about during the days that follow. But the meetings of the Grange are not entirely informal. A portion of the time is spent in the discussion of topics that are of especial interest to the farmers. The best crops for particular lands, the best methods of cultivation, the experience of the different members, the cost of different kinds of crops—any questions the solution or discussion of which tend to make better farmers are considered. The women read essays on the various duties of their departments, and thus learn to be better housewives. Sometimes the Grange considers social and moral questions, and sometimes its exercises are of a literary character. The discussion of political and religious questions is strictly forbidden by the constitution of the Order.

"Another custom which originated in the Grange is that of holding festivals at short intervals during the season. It is impossible for the farmer to leave his business for a month in the Summer and spend the time in recreation, and, if his work could be left, his purse is rarely long enough to pay the expense of such a vacation. But he can spend a day between planting and cultivating, another before harvest, and a third when the grain is stacked; and the Grange taking advantage of this, either invites those of neighboring townships to a basket pic-nic, or accepts an invitation itself. A day is spent in some pleasant grove; there is speaking and music, and perhaps a little dancing, and the farmer goes back to his field better prepared for his work, some of the marks of care are smoothed out of his wife's face, and the business of both field and house go on with less of fret and worry for the day's innocent recreation. I was once a farmer's boy myself, and know from experience that those who till the soil work too many hours and have too few holidays. There is nothing that makes the work on a farm go easy like a holiday, and if it is rightly spent it puts new life into the work for a long time after."

CHAPTER XXVIII.

THE LESSONS OF THE GRANGE.

The Grange as a Means of disseminating Agricultural Information—Grange Tracts—How they are circulated—Efforts of the Order to improve the Farmer's Condition—The Grange as a School of Reform—It makes Better Farmers—How it spreads Information—Advice as to Improvements—The Grange the Enemy of Careless and Improvident Farming—It encourages Good and Careful Work—The Stacks of Wheat—Only Virtuous and Industrious Members admitted into the Order—The Grange making Intelligent Farmers—Beneficial Effects of the Discussions of the Grange—The Grange teaches Habits of Thrift and Economy—Discountenances Debt—The Grange the Enemy of Selfishness—Encourages Education—The Friend of the Schools—The Grange making Better Men as well as Better Farmers—Claims of the Order upon the Sympathy of the Country.

SAYS a recent number of *Frank Leslie's Illustrated Newspaper*, "Over 500,000 tracts were issued to farmers by the National Grange last year."

This little paragaph, in an out of the way corner of the journal from which it is taken, is full of meaning to the members of the Grange. It shows that the Order is honestly and efficiently performing one of the principal portions of its work, for one of its main objects is the collection and dissemination of information relating to agricultural interests and of value to the farmer. Half a million little tracts sanctioned and sent out from the Central Grange find their way to millions of readers. They are couched in plain and simple language, and abound in practical information. *They are read, also.* The farmer receiving such a tract reads it with confidence, as he knows that it has come from men who are earnestly seeking his good as a member of the great interest for which they are working. Being small and convenient in form, and, above all, not too long, he can carry it about with him. He can read it by his own fireside "when all the house is asleep," and he is keeping watch with his own thoughts; or as he sits under the shade of some wide-spreading tree to rest from the heat and the toil of the day. The little tract deals with questions which are of vital importance to him, and it sets him to thinking, and to thinking in the right direction, too.

The Grange does not regard the farmer as a mere machine, a mere drudge. It looks upon him as a reasonable, responsible being, and seeks to elevate and improve him. It stretches over him the shield of its protection against the enemies that assail him and seek to rob him of the rewards of his industry; it offers him the means of social enjoyment and teaches him the duty of healthful recreation and pleasure; it recognizes the right of woman to share the pleasures as well as the cares of man, and secures her pure and ennobling influence and co-operation in its work; and it teaches and enforces the lesson that the most intelligent and thoughtful farmer is sure to be the most successful.

In short, the Grange seeks to make better farmers of the agricultural class. It claims no authority to coerce them into any course of action, no right to command. It is merely an advisory body, and it seeks the improvement of its members only by its moral power. It is to the farmer a wise and judicious friend, and it advises only that which his own good sense tells him is the best course for him.

Its mode of operation is very simple. Reports are

sent in to the National Grange by the State and subordinate Granges of the condition and prospects of the agricultural interests in their respective localities. The authorities of the National Grange are in this way enabled to exercise an intelligent supervision of the farming interest of the whole country. Reports are regularly sent to the State Granges, and by them distributed to the subordinate Granges. The National Grange is thus a National Intelligence Office for the benefit of its members, and the humblest farmer can, through the system thus adopted, keep himself informed as to the actual condition of the interest to which he is attached. The state of the crops, the probable amount of the harvest, upon which anticipations of value are based, can be thus ascertained. The farmer can be kept advised of the ruling prices in the various markets of the country, and freed from his dependence upon the grain dealers and their organs for this information. In this way he is better prepared to go into the market and dispose of his wares.

But not only does the Grange keep the farmer advised upon these points. It recognizes that agriculture is susceptible of great improvement, and it makes it its business to ascertain and make known the best means of attaining a high state of excellence in this pursuit. From the National and State Granges information is sent out concerning the latest improvements, and the needs of the system. Nothing is ordered. The subordinate Grange is informed that such and such improvements have been used with profit in certain parts of the country, and is invited to discuss the question whether they are needed, and may not be advantageously introduced into its own community. The local organization is free to adopt or reject them; but the discussion is of infinite value to all concerned in it. It brings out the opinions of the entire farming community, and many useful and admirable suggestions, which are often carried into practice by the members, spring from it.

The Grange is the uncompromising enemy of carelessness and disorder. Its golden rule is, "What is worth doing at all, is worth doing well," and it enforces this law by every means in its power. Careless and untidy farming are discountenanced. The farmer is taught that he must keep his place in order, and make it look its best every day in the week.

"While riding over the country the other day with a leading Granger," says a letter from Minnesota, "he called my attention to the wheat stacks that we saw. 'There,' he said, pointing to a row of stacks that stood up straight and trim, were well capped, and in every way calculated to allow the grain to dry while it was protected from the weather, 'there are the stacks of a Granger. Those over yonder, in which the bundles are thrown together haphazard, belong to a man who is not a Patron. He thinks that he will thresh out his grain soon, and that it is therefore unnecessary to take pains in stacking. But something may happen to prevent this, and the stacks may remain for months, and the grain be injured. The Grange teaches the farmers that it always pays to do their work well, and it is making better farmers than we ever had before.'"

This is important testimony in favor of the Grange, and if the organization did no more than make "better farmers than we ever had before," it would still be doing a great and beneficent work, and a work which would lay the whole country under obligations to it.

An Order which can thus improve and elevate over five millions of human beings is certainly deserving of the encouragement of the nation.

The Grange is the bitter and uncompromising foe of idleness. It has no room for a lazy man. While it recognizes the fact that frequent relaxation and indulgence in innocent pleasures are a necessity of mankind, it teaches that industry is the only sure foundation of success, and that the idle or lazy man can be neither prosperous, virtuous, nor useful. The Order aims to accomplish a great work in the community, and it accepts only workingmen and women. Each one has a part to play in the execution of its great designs, and it will tolerate no idlers, no mere lookers-on. A powerful stimulus is thus given to its members, and it will not be long before it will be as easy to recognize a Granger by his prosperity as it is now by his habits of neatness and order.

As it is the enemy of idleness, so it is of vice. The Grange will have no dealings with drunkards, swindlers, or immoral men and women. It demands a good moral character as the first requisite for membership. A drunken or dissolute farmer can have no sympathy with an Order which teaches that all men should be temperate and pure minded. A swindler can have no fellowship with men who believe in honesty as the basis of every relation in life, and who act upon this belief. And so the Grange keeps unworthy men and women at a distance, and its influence in and upon the community is entirely in favor of virtue.

Says a letter from Ohio:

"This is what the Grange aims to do. Once in two weeks (sometimes every week) its members meet in some convenient hall which they either hire or own, each family bringing its basket of food. Many hands make light work; cooking utensils, dishes and tables are owned by the Grange; a bountiful feast is soon prepared, and the afternoon is spent in social pleasures or in discussions upon subjects in which they are mutually interested. Who can doubt that an occasional breaking away from work by the farmer and his family, even though he should get no new ideas, will improve them all in health and make them better able to perform their routine of duties?

"But the Grange strives directly to make better farmers, and of this there is certainly need. Many of the agriculturists of the West and North-West left Eastern farms where high cultivation and intelligent management were necessary to insure a living; and if they were fair farmers there, they have generally been abundantly successful in the West. But there is a large class of men who have gone upon the wild lands of the West—Irish, German, Scandinavian immigrants gathered in Europe by railroad and emigration agents—whose knowledge of agriculture is of the most limited kind, and who have everything against them but the strength of their arms, their ability to endure privation, and the wonderful fertility of the soil when its tough sod has once been broken. They put lots of muscle into their business, but very little brains. Nor are all of the bad farmers of the West of foreign birth. Thousands of men reared in cities have been induced, by the promise of cheap land and rich crops, to forsake the life in which they were reared, for the reaper and the plow. Some of these men have done well; others have naturally failed. Another fact I have noticed is that the very men who are most in need of advice such as a good agricultural journal would give them, are the ones who don't take it—probably their failure to read such a paper explains their need of it. To all farmers, good or bad, the Grange offers opportunities of improvement never before within the reach of the country people except in Farmers' Clubs, and in them only to a limited extent. Experienced, successful men tell in the Grange room how they have made good crops, or why they failed to do so; agricultural newspapers are taken, read and exchanged; important advice is given to young and inexperienced farmers, and each member, no matter how well he understands his business, is sure to obtain some item of useful information.

"The Grange teaches the farmer to contract habits of thrift and economy. The man who buys on credit always buys in the highest market, and of no class in the community is this remark more strikingly true than of the farmers. It is no uncommon thing for a bill at the village store to make a veritable slave of a farmer. A partial failure of his crops, sickness in his family, or other unforeseen occurrence, makes it impossible for him to settle when pay-day comes around, and a mortgage on his farm, at fifteen per cent. interest, is the result. Other men may offer to sell goods to him cheaper, but it may be impossible for him to transfer his trade when such transfer might involve a foreclosure of a mortgage. The Grange advises all of its members to buy and sell for cash, and to demand such favors as cash purchasers are justly entitled to. If ten per cent. of a man's sales on credit become bad debts, the increase in prices to make up for such loss ought to be charged against those who buy on credit, and not against those who buy for cash, and on whose purchases there is, therefore, no risk. The Grange also assists its members to get down to a cash basis, by making contracts with local dealers to allow a discount to Grangers who pay on the spot for their purchases, by making extensive contracts to purchase agricultural implements, sewing-machines, etc., at wholesale from the manufacturers, and in a few cases by lending money at low rates of interest to enable the farmers to take advantage of these arrangements. I have spoken of this feature of the Grange movement at considerable length in one of my letters from Iowa: the Grange in that State has thus far been the model which those of other States are imitating with greater or less success.

"The Grange, I have said, teaches its members to be thrifty and economical. By this I do not mean that it teaches them to pinch and starve themselves, or to deny themselves the comforts or even the luxuries of life. On the other hand, it shows them how to acquire the means to gratify their finer tastes. Instead of leaving his plow in the last furrow, to rust and rot through the long season and wear out in four years, when it ought to last six, the Grange teaches the farmer to put it under cover, and so save enough to pay for the subscription to a good newspaper or magazine, or to purchase a good book. Instead of allowing his wheat to lie in the shock and sprout before it is threshed, the Grange tells the farmer that its value will be increased several cents on a bushel if he carefully stacks it. It shows the careless, thriftless farmer the secret of his more successful neighbor's success, and gives him a helping hand to make that secret of practical value to him."

The Grange is the foe to selfishness. It is too much

the habit of the farmer to regard himself and his own family only, and to be careless of the welfare or interests of others. The Grange teaches him that he is only a single member of the vast community of men who till the ground, and that his interests are identical with theirs, and that he must consider others as well as himself. He is thus drawn out of himself and made to entertain larger and more liberal views of life and its duties.

The material interests of the farmer are not the only ones which receive the fostering care of the Order. His intellectual improvement is also aimed at. The Grange teaches its members that education and intellectual culture are necessary to the farmer as well as to other men. It impresses upon him the duty of encouraging the growth and prosperity of the public schools, and reminds him that money saved at the cost of his children's education is saved at too high a price. It encourages the farmer to purchase and read good and useful books, and the best periodicals of the day. At its meetings discussions are encouraged which serve to keep its members informed upon the leading questions of the times, and to accustom them to express their views in an intelligent manner. In one respect the Grange may be considered as an educational club, with the very positive and definite object of achieving the intellectual improvement of its members. Certainly this is a noble work, and should win for the Order the best wishes and cordial sympathy of every citizen of the Republic. The work of the Order is just commencing. It is still in its infancy, but it gives promise of glorious results.

CHAPTER XXIX.

THE COÖPERATIVE FEATURE.

Coöperative Feature of the Grange—How the Grange saves the Farmers the Middle-man's Profit—Circular of the Secretary of the National Grange—A Means of Practical Economy—The System of Purchases adopted by the Grange—The System on Trial in Iowa—The System productive of Economy—How the Iowa Grange conducts its Operations—Bringing the Manufacturers to Terms—The Plow Trade—A Saving of Fifty Thousand Dollars on Plows—A Liberal System of Discounts—Work of the State Agent—Joint Stock Stores established—Method of Coöperative Selling—Elevators established by the Granges—Direct Shipments—Magnificent Success of the Grange in Iowa—The Granges saving more Money than they cost—Efforts to embarrass the Grange—Warning of the National Grange—Opposition of the Middle-men—A Successful Effort at Coöperation abroad—The History of the Civil Service Supply Association of London—A Lesson and an Encouragement to the Grange.

"THE Order of Patrons of Husbandry," says the Secretary of the National Grange, in a circular addressed to Manufacturers of Agricultural and Domestic Implements and Machinery, "is an organization of farmers and horticulturalists, one object of which is to secure to its members the advantages of coöperation in all things affecting their interests. . . . To enable the members of the Order to purchase implements and machinery at as low cost as possible, by saving the commission usually paid to agents, and the profits of the long line of dealers standing between the manufacturers and the farmers, the Executive Committee of the National Grange desire to publish a list of all the establishments that will *deal directly* with State and subordinate

Granges. This list will be regarded as strictly confidential, and one copy only will be furnished to each Grange.

"Large orders can thus be made up by the consolidation of the orders from Granges in the same State or vicinity, and special terms for freight, etc., arranged with transportation lines, thereby effecting another large saving to the purchaser.

"Manufacturers of all articles used by farmers, who desire to avail themselves of this means of disposing of their products directly to the consumer for cash, thereby avoiding the losses incident to the credit system, or the storing of goods in the hands of commission merchants or agents, are invited to send their catalogues and wholesale price lists to, and to correspond with

O. H. KELLEY,

"Sec'y of the National Grange, Washington, D. C."

This circular embodies one of the principal features of the Grange movement, and one which will render it indispensable to the farmers. The Grange recognizes the fact that the farmers have been charged too much for the majority of the articles they purchase, and it undertakes to save them from this loss.

The method of coöperative buying adopted by the Grange for this purpose, is very simple. A State agent is appointed, whose duty it is to correspond with manufacturers and wholesale merchants, and ascertain the most favorable terms upon which they will sell their wares to members of the Grange. If these terms are satisfactory, they are communicated to the subordinate Granges, the members of which make up their orders, and accompany them with the amount of the purchase money of the articles desired in cash. These are sent to the State agent, who forwards them to the manufacturer, with instructions for the shipment of the purchases direct to the purchasers.

The Grange thus makes it to the interest of the dealer to sell to its members direct. A large cash order is given, and the manufacturer or dealer is relieved of the ordinary risks of business, while the farmer is enabled to purchase his goods for much less than he could formerly buy them. A single farmer, or a single local organization of farmers, could effect nothing in this respect, for they would be powerless to compete with the middle-men or local dealers; but the trade of a county or State is a valuable consideration, and the manufacturer will make advantageous concessions to secure it.

Perhaps we cannot better illustrate this feature of the Grange than by presenting here an account of the successful inauguration of the coöperative system in Iowa, where it has been tried with admirable results.

"One of the first lessons which the Grange of this State seeks to teach its members is to buy for cash, to avoid as far as possible the purchase of agricultural machinery not absolutely needed, and when in rare instances the farmer must buy implements for which he cannot pay, to borrow the money outright, furnishing the required security on his place, rather than to purchase on three, four, or five years' credit, giving 'iron-clad' notes, and paying the enormous prices which are always charged under such circumstances. This has been no holiday job; machinery-bankrupt as thousands of the farmers have been, with their notes, given for implements, in the hands of every sheriff of the State for collection, and, by official figures, constituting almost half the delinquent debts, they have doubted and hesitated when relief through coöperation has been offered, and have been far more ready to advocate reforms with which they personally have nothing particular to do than to begin at home and help themselves. Happily, the Grange of the State has had among its members a few men of good business ability and sterling integrity, who have felt interest enough in the welfare and prosperity of the farmers and the success of the order of which they were members, to give much of their time and in some cases to risk their money in this much-needed home reform. While at Des Moines, and since I came here, I have found out some of these men, and, in their desire to spread a knowledge of their success, and to encourage farmers elsewhere to imitate their example or improve on it, they have given me every facility to learn the inside workings of the system of coöperative purchases and sales they have adopted, and to see what its results have thus far been.

"It is but little more than a year since prominent Grangers of Iowa were first successful in making large coöperative purchases, although previous to that, I think, they had appointed a State agent and a few county agents. They then found the manufacturers almost wholly in the power of the agents. Not only had they made their contracts with these agents for the year, giving to each a monopoly of the sales for his particular district, but, had they been disposed to disregard those contracts and sell to the agents of the Granges at wholesale rates, they did not dare to do it, because to lose the trade of the agents, who would have nothing to do with manufacturers selling at wholesale prices to other customers within their districts, would, while the present methods of doing business are adhered to, be nothing short of ruin. The farmers who were moving in the matter understood this; they knew that it was unreasonable to ask a plow maker, a sewing-machine dealer, or any other manufacturer or wholesale merchant, to abandon a business system by which he was supported, unless the Grange could offer him in exchange an equally profitable and extensive patronage. And just here may be explained the failure of all local attempts at coöperative machinery buying. When the farmers of a county have united to purchase their plows, and have sent their agent to the manufacturer, they have found that they could get no material reduction of price. The manufacturer has said: 'My trade in your county belongs to Mr. A, and I have agreed not to sell goods to persons living there below his prices, or at any rate to pay him his customary commission on all such sales that we do make. You want twenty plows; if we sell them to you at our wholesale price we shall either have to lose the agent's commission on them or lose his trade, and he takes a thousand plows a year.' No local coöperative association could command trade enough to make it an object for the manufacturers to show them any important favors.

"But the wide spread of the Grange in this State gave to the farmers the means of holding out to any manufacturer whom its members should generally patronize, an inducement to give up the trade of the agents and sell directly to them, and the managers of the Grange were not slow to avail themselves of the power they thus acquired. Having agreed to buy nothing on credit but to pay cash for all their purchases, and having received assurances from a sufficient number of Granges that their members would purchase through their own agent, application was made to three manufacturers of plows in Des Moines for wholesale rates. Two of them refused to make any terms with the Grange, but the third agreed to make a deduction of twenty per cent. on the retail price of each of his plows and twenty-five per cent. on cultivators. The result was that this man, although he made up a large stock in advance, was unable to supply the demand of the Grange, and the freight agent of one of the railroads at Des Moines remarked the other day that the paint had not been dry on a single plow that had been shipped from that man's shop this year. One of the other manufacturers very soon discovered his mistake, and got some of the orders that the first could not fill, and the third is now ready to trade with the Granges. Plows have also been bought of other manufacturers, both in this State and those adjoining. How many plows the Granges have purchased within a year at these reduced prices cannot be ascertained, as, after the contract had been made by the State agent, the orders did not necessarily come through him, and no complete record has, therefore, been kept; the county agents have forwarded many of them directly to the manufacturers, the only condition being that the cash accompany the order and that the purchaser be a Granger. It is safe to say, however, that the purchases have amounted to many thousands, and that not less than $50,000 have been saved to the farmers of the State, within a year, in the purchase of plows and cultivators alone.

"In the purchase of sewing machines the saving has been still greater and the sales very large. The retail price of sewing machines in this State has been from $50 to $95, according to variety; they are now sold to the Grangers at 40 per cent. discount from these prices, or for from $30 to $57. The demand has been so great that 1500 machines have been ordered to be delivered during the coming year. Supposing all of these to be of the cheapest variety, the saving will be $30,000. The number purchased will probably far exceed 1500. On parlor organs the discount to the Granges is from 20 to 25 per cent.; on scales, from 25 to 33⅓ per cent.; on shellers, 15 per cent.; on wagons, 20 per cent.; on hay-forks, 33⅓ per cent.; on miscellaneous implements, like feed-grinders, stalk-cutters, harrows, field-rollers, hay-rakes, grain separators, etc., 25 per cent. On mowers the discount is 25 per cent.; that is, a machine which retails at $120 is sold to the Grangers for $90. A lot of reapers which a manufacturer who was going out of the business had on hand were offered to the Grangers for $75 each, provided they would take the whole of them. They were carefully examined and tested by the State agent and others, who reported that they would be cheap at $150. A circular conveying this information was sent to the Granges of the State, and the whole lot was disposed of at once. They have given universal satisfaction. I might go on at great length quoting prices, but those I have given are sufficient to show that by intelligent coöperation the farmers of the West can save a great amount of money. General Wilson, Secretary of the State Grange, thinks that $2,000,000 has already been saved in this way. Mr. Whitman, the State agent, to whose well-directed and untiring efforts the success that has so far crowned this experiment is, in very great measure, due, thinks this figure too high, though he has no data from which to make an estimate. The Grange has not only benefited its own members by its coöperative purchases, but has caused a reduction in the prices of all kinds of farming implements, sewing machines, etc., in the stores and when sold by agents. A single example will illustrate this fact. A year ago, when the agents for the sale of a certain cultivator supposed that they had the entire control of the market, they charged $35, and threatened to raise the price. Since the Grange has been purchasing similar cultivators for $26.25, the agents have reduced their prices to $30.

"The manner of conducting this coöperative buying is very simple, although to insure success it is necessary to place it in the hands of competent and honest men. Mr. J. D. Whitman, the State agent, has his office at Des Moines, and is the principal manager. He gives a bond of $50,000 for the honest and faithful performance of his duties, and receives a small salary. In each county of the State there is a County agent, who may also be placed under bonds, if the Granges of the county think it necessary. The State agent places himself in communication with manufacturers and wholesale merchants, learns the terms on which they will sell their goods to the Granges, makes contracts with them when it is desirable, and informs the Granges by circular of the prices, etc. Orders may then be given through either the State or County agents. All orders must be accompanied by the cash to pay for the article desired, and a certificate from the Master of the Grange that the purchaser is a member of the Order.

The State agent on receiving money credits the remitter with the amount on his books, specifying the article to be purchased and sending him a receipt. He at once forwards the cash to the manufacturer or merchant, and then debits the purchaser with the amount remitted. The goods are shipped directly from the manufacturer to the purchaser, but the receipted bill is sent to the State agent, who files it away as his voucher. If the goods are imperfect, or not as good as have been contracted for, and the seller refuses to give the purchaser satisfaction, then the Grange transfers its entire trade to some other firm. A man who was furnishing the Grange with plows, last Spring, sent a few that were much inferior to the sample. A circular was sent to all the Granges informing them of this fact, and in less than a week orders for that plow stopped, and the man has not sold one to a Grange in the State since.

"The State agent always gives preference to home manufacturers. Wherever an Iowa plow-maker or manufacturer of any kind can furnish first-class goods as cheaply as they can be purchased at Chicago, St. Louis, or New York, the Grange gives him its trade, but its motto is to buy in the cheapest market which ready cash will command. In some sections of the State the members of the Grange have established joint-stock stores, and have thus been able to purchase their groceries and dry goods much cheaper than before. This has not been generally encouraged by the leading Grangers, except in cases where the local traders have refused to deal with them on what they considered fair terms. The great bulk of the home trade in this State has been done on credit, and the farmers who have remained solvent have had to pay not only a fair profit on the goods they have purchased, but something in addition for the time that has been given them, and to make up for the losses of the traders by bad debts. Now the members of the Grange who propose to pay cash for what they buy, think that they ought to have their goods cheaper than before. Some of the traders have admitted the justice of this claim, and have made satisfactory terms, but others have refused. Where no terms could be made, the Grangers have been forced to establish their own stores. Their plan has been to divide the stock into shares of $10 or $15, so that each member of the Order can afford to own one or more shares. The goods are then all bought and sold for cash, an advance of eight per cent. on the cost being charged. At Waterloo, where a store of this kind has been established, the farmers find that they obtain better articles at less prices, and that their stock pays them a good profit. The average sales in that store, since its establishment, have been $112 a day, a considerable portion of it coming from the railroad shops situated there.

"But it is not alone by coöperative purchases that the Grangers hope to save money. They have not only bought their goods on credit, and therefore in the highest market, but they have sold their crops at home to middle-men for cash, and therefore in the lowest market. They now hope, by coöperative selling, to get better prices for what they raise than they have hitherto received. Until quite recently such a thing as shipping his own grain to Chicago, or any eastern market, has been almost unknown among the farmers. Whenever any of them have attempted it they have often been swindled so badly that they have lost all confidence in commission merchants. One of the first steps that the Grange took was to select a commission house of the highest character in Chicago, and another in New York, and make them its agents. Each of these houses has given bonds to the amount of $100,000, and agrees to receive everything that is consigned to it by the Grange or any of its members, and dispose of it to the best possible advantage, taking only one per cent. for commission. Since this arrangement has been made, many of the farmers have shipped their own grain, and the Chicago agent has been able to sell it for them on the cars upon which it was originally loaded, thus avoiding altogether elevator charges and the cost of trans-shipment. The prices thus realized by the farmers have generally been several cents a bushel better for grain than those offered at home, although the railroad companies have given them no special rates.

"In order to take advantage of favorable markets, the Granges have established at several points in the State elevators and warehouses of their own. In some places these warehouses have been built by two or three prominent members of the Order; in others the stock has been divided into small shares and is owned by great numbers of the farmers. The plan of conducting the business is the same in both cases. If the farmer prefers to sell his grain outright and get the money for it when it is delivered, the managers will pay him the highest price the state of the market warrants; if he is willing to take the risk of the market, they handle his grain for him, sell it, and return him the proceeds for a commission of a cent and a half a bushel. This makes the farmers almost independent of middle-men between them and Chicago or New York markets. If the price offered for their grain is not in their judgment enough, they are not obliged to sell it at home, but can ship it themselves, feeling perfectly sure that they will be honestly dealt with, will have to pay no exorbitant commissions, and will get the best market price. At Waterloo, about 100 miles west of this city, an elevator was established by the Grange about nine months ago, the stock being held by a great number of farmers. Grain that has been shipped from that point both to New York and Chicago has brought the farmers considerably better prices than the local traders would pay, and beside this, recently a dividend of fifty per cent. was made on the stock. The Grangers' elevators now do all the work at that and other stations where they have been established, the local middle-men having gone entirely out of the business. If a Granger who does not live near one of these Grange elevators desires to make a shipment on his own account, he applies for cars at his local station, loads them, directs them to the Grange agent in Chicago or New York, and sends the receipts which the railroad company gives him to the State agent at Des Moines. By him the proper papers are forwarded to the Chicago or New York agent, and to him the returns are made. The State agent then returns to the shipper all the papers showing the charges and receipts on his shipment, with a check for the balance due.

"The farmers of this State raise every year a great number of hogs, that have always passed through the hands of at least one middle-man before they have reached the packers. The Granges in some parts of the State concluded, last year, that they might as well sell directly to the packers themselves, and appointed one of their number to collect the hogs and deliver them. A considerable saving was made in this way, and the experiment will be much more extensively tried this Fall. In some parts of the State the Grangers are already talking of establishing their own packing-houses, so that, instead of selling the hogs alive, they can sell them in the shape of bacon, hams, and lard, packed and ready for shipment. They hope to realize much larger prices than by the old system.

"Another experiment, which was tried to a limited extent last Spring, and was attended with a gratifying degree of success, was the direct shipment, by members of the Grange in this State, of provisions to planters in the South, who are members of the Grange there. This business was managed by Mr. Shankland of this city, a member of the National Executive Committee of the Grange. He received orders from Grangers in South Carolina, accompanied by the cash, and purchased flour and bacon, which he shipped directly to the consumer. The purchases were made of the farmers when it was possible to do so, and when not, of the packers and millers in this city. The shipments were made by rail, by way of Cairo, Hickman, Kentucky, Nashville, Chattanooga, and Atlanta, to Columbia, two trans-shipments being made on the route. The railroad companies made special rates at from \$1.08 to \$1.15 per hundred pounds. Bacon was thus laid down at Columbia, South Carolina, at less than eight cents a pound, while its market value there was from 12 to 14 cents a pound. One planter informed Mr. Shankland that he saved by this plan \$400 on a single car-load of flour and bacon. Only twenty-three or twenty-four car-loads of provisions were shipped in this way last Spring, for the reason that these transactions are all between members of the Order, and the Grange had not at that time become much of an institution in the Southern States. There are now in the South 597 Granges, outside of Missouri, and it is expected that the number will be increased to at least 1000 by Spring. By that time it is also hoped that the Grangers of this State will have established a number of packing-houses and perhaps a few mills, so that they will be able to ship provisions directly to Southern consumers without their passing through the hands of any middle-men. There is no doubt that a large business of this kind will be done next year.

"Coöperative buying and selling by the Granges is as yet but an experiment, but the facts which I have set forth in this letter, all of which I have gathered from official sources, show it to be a very promising one. Of course there have been, and will be, difficulties to overcome. The farmers are often timid; a sudden decline in the market causing them to lose money on a single shipment of grain sometimes alarms them, and they are prone to go back to the old grain-buyers, forgetting, perhaps, that their gains on other shipments compensate many times for the loss on one. Dishonest men may secure appointments as agents and swindle them, and a hundred other things may occur to retard the success of the system. But the leaders of it, most of whom have been life-long disciples of Mr. Greeley, believe that the principle is right, and that the Grange organization furnishes the machinery by which it can be put into practical operation.

"Several Western journals have contained articles purporting to give an estimate of the immense sums of money that the Grange is taking from the farmers in the shape of fees, dues, etc., and have hinted that it was a foolish waste of money, which should be stopped before the people could be made to believe the farmers' cry of 'hard times.' They have also criticised the Grange for its secrecy. In justice to the Order, as I have observed it, I wish to repeat what I said in a former letter—that all attempts to organize the farmers for any purpose previous to the establishment of the Grange were failures; that its constitution and by-laws are public; that, as at present organized, it cannot be converted into a secret caucus to further the ends of any men or party, though it teaches certain principles which its members will probably demand shall be recognized by the parties or the men they support; and finally, that, though the Grange does collect from its members, in the aggregate, a considerable sum of money, it has already returned to them in this State alone, through coöperation, which is yet only an experiment, more money than has been paid into it throughout the country, to say nothing of the intellectual and social benefits it has conferred upon its members. The Granger who has bought a plow only through the agent of the Grange has saved more than enough to pay his fees for a year, while those who have purchased sewing machines or other expensive machinery have saved enough to pay their fees for several years."

As was to have been supposed, reckless and irresponsible men have sought to take advantage of the business feature of the Grange, to increase their sales to farmers, and it is a common thing to find dealers claiming that their wares have received the endorsement of the Order. Farmers should beware of all such persons, and bear in mind the following letter of warning issued to the Order in general by the Secretary of the National Grange:

"TO PATRONS:—All members of the Order are hereby cautioned against noticing any circulars that may be sent them purporting to be issued on the recommendation of the National Grange, or its Officers, as no certificate in favor of any article, seed, or implement will be furnished to any individual, under any circumstances whatsoever, until the merits of such article is endorsed and properly certified by at least ten Granges, (whose endorsement will appear on the circular,) and not in any case to those who are not members of the Order; and when any such endorsement is issued by the National Grange, all and every subordinate Grange will be duly notified by the Secretary of the National Grange."

No feature of the Order has had to encounter more opposition than the coöperative system we have described. The middle-men early took the alarm, and assailed it as a mere chimera, as utterly impracticable, and ridiculed unmercifully the claims of its advocates. Even that portion of the press most favorable to the Order expressed grave doubts of the possibility of carrying out this part of the programme. The repeated and numerous failures of coöperative systems were brought forward and urged with great force in opposition to the claims of the Grangers, and for a long time the farmers themselves held aloof from this feature of the Order. But time and experience have demonstrated both the practicability and the excellence of the system. The magnificent success of the scheme in Iowa shows that it is not chimerical, but that it is based upon sound business principles, and only needs the cordial coöperation of members of the Order elsewhere to make it equally successful in other States.

In Minnesota, where no organized system of coöperative buying and selling has been adopted, a number of separate Granges have made purchases at wholesale, but they have not been able to make as favorable terms as the State agent of the Iowa Grange; but even these efforts, it was stated at the last meeting of the Minnesota State Grange, had resulted in the saving of three-quarters of a million of dollars to the farmers of the State.

The advantage of this feature of the Grange to the farmer is incalculable. Millions of dollars can and will be thus saved to the agriculturalists of the country, and the beneficial effects of this saving will soon be apparent.

One of the good results of the system came under the observation of the writer while these pages were being written. In a certain large city of the Eastern States, there is a mercantile house largely engaged in the purchase and exportation of Southern cotton. Among the debtors of this firm was a certain planter in a Southern State, whose indebtedness amounted to several thousand dollars. Wishing to close the account, and being in need of money, the firm applied to the planter for a remittance. He having business in their city, replied in person. He told them that, owing to the great stringency of the money market, he found himself unable to obtain ready money, and would be obliged to ask their indulgence for a little while longer. "I expect to pay you very soon," he said, "and to pay interest on the debt. I can only raise money now by selling my cotton at a sacrifice, and I cannot consent to do that. Cotton is worth more than the price I should get for it now."

"What would it bring you?" asked the head of the house.

The planter named a price below the market rates in New York.

"What do you think it is worth?" asked the merchant.

"A cent and a half a pound more," said the planter.

This last price was still below the market rate, and the merchant said to the planter that if he would deliver to the Southern correspondent of the firm enough cotton to cover the amount of the debt, at the rate he had named, it would be as acceptable as the money.

"I should like to do so," said the planter, hesitatingly; "but I cannot. I am a member of the Grange, and we have recently established an agency of our own for the sale of our cotton, and have pledged ourselves to sell through no other source. I must stand by my agreement, or I would accept your offer."

But for the Grange the planter would have sold his cotton at a sacrifice, even at the higher price he had named. The Order, however, gave him the means of obtaining the worth of his crop. It prevented him from acting in haste, and, although he was but poorly informed as to the state of the market, the agency through which his business was transacted possessed the necessary information and saved him from the effects of his ignorance.

The coöperative system of the Grange is yet in its infancy. That it will be generally adopted, we cannot doubt, and that the greater portion of the Granges will enjoy the benefits seems clear. That it will have to encounter much opposition is to be expected. The middle-men will oppose it wherever it is introduced, as it will free the farmers from the tax hitherto paid to this class of merchants; but we have good reason to believe that it will be ultimately carried out by the entire Order in all parts of the country.

That successful coöperation is possible, and may be carried to a very great extent, is proved not only by the experience of the Iowa Grange, but by the operations of the "Civil Service Supply Association," of London. The following account of the establishment and growth of this remarkable association, perhaps the most successful of its kind in existence, was prepared by one of the original members. It is full of encouragement and suggestion to those who are working in the Grange for a similar object, and we commend it to them. It is as follows:

"The Civil Service Supply Association is the oldest Coöperative Society in the Service, and it has been the model upon which all London Coöperative Societies have been formed. Although barely eight years old, and in its commencement most humble, it is now selling goods at the enormous rate of £780,000 a year, and is fast revolutionizing the retail trade, not only of London, but of the whole country. Surely the story of its rise and progress is worth the telling.

"The Association originated in the Post-Office. The Winter of 1864–5 (like many other Winters, and for that matter Summers too) found a good many of us

Post-Office men engaged in a rather hard struggle to make both ends meet. Some of us had ventured to ask for higher pay, and had been favored with the usual sympathetic but depressing reply, that it was regretted that the circumstances of the case would not justify any addition to our salaries, etc., etc.

"Feeling as we did sharply, the general rise in the cost of living, especially in the price of all articles of clothing consequent on the American War, one or two of us had already bethought ourselves of coöperation as a means of lessening our difficulties. I, for one, being a Liberal in politics (for there are some few Liberals in the Civil Service) had watched with interest the doings of the Rochdale Pioneers, but could not at all see how to apply their experience to our own case.

"One day, however, two office friends came to me—it was, as I well remember, a foggy, gloomy day in November, enough to make one more than usually despondent—and declared once for all that they must either have more to spend or manage to spend less. They had given up all hope of more pay, and as a last resource they proposed that we should try to spend less by means of coöperation. Their idea was that we should induce a number of Post-Office men to procure their supplies of coal from some one coal merchant, in the expectation that by the largeness of the united order, and by the payment of ready money, we should obtain a considerable abatement in price. Talking the matter over, we resolved to try buying on this plan; but we soon agreed that coal was not a good article for the experiment, and in the end we decided to make a beginning with tea. That very afternoon one of us on his way home called at a celebrated wholesale house

(I even now withhold names for fear of the wrath of retail traders) and learnt that by buying half a chest at a time, and paying for it in ready money, we should save from 6d. to 9d. a pound. We therefore invited a few other office friends to join us. Each wrote down on a list the quantity he would take, at the same time handing in the money to pay for it. Some of the most cautious limited themselves to a single pound: others boldly coöperated to the extent of two pounds, a few rash men pledged themselves to three pounds, and we promoters had to take enough to make up the full order. The tea was bought, and after office hours we weighed and divided it among the purchasers. It proved to be excellent, and soon a demand arose for more. Other men in the office, who had heard of our successful venture, wished to join, and this time there was no need for us promoters to take more than we wanted. Some one now luckily discovered an empty cupboard in the office, and here we locked up our second half chest of tea till we could divide it among ourselves.

"This cupboard was the original store of the Civil Service Supply Association.

"More tea being very shortly needed, we prepared for a third purchase, and now so many joined us that we had to buy a whole chest. It was no joke to make up 100 pounds of tea into parcels of two or three pounds apiece, but we were lucky enough to find one who, like old Trapbois, was willing, nay eager, to undertake the task for a consideration. This was a funny little fellow, since dead, whose duties were very humble, and salary yet more so. Though nominally a clerk, he was regarded as a kind of cross between a clerk and a messenger. Poor fellow! while his small salary had no prospect of increase, his large family increased but too fast. His remuneration for this piece of extra service was the surplus tea (some three or four pounds) contained in each chest, beyond the nominal amount.

"Our success in tea led us on to buy coffee; and each time that our list went round the office more and more men asked leave to join. Our poor cupboard soon became too small for our ever-increasing stocks, to which moreover we thought of adding sugar and other groceries. With no small anxiety we found ourselves constrained to hire a store-room outside the building, a step that we felt could not be safely taken unless we formed ourselves into a regular association. Hence arose the Post-Office Supply Association, which, being afterwards extended to the whole of the Civil Service, in the end took the title of the 'Civil Service Supply Association.' Our first impulse was to call ourselves the 'Post-Office Coöperative Society," but even the boldest of us shrank from so hazardous an avowal—so strong only eight short years ago was the prejudice against coöperation, regarded as it was by many as identical with socialism. In a word, we took the thing, but not the name.

"A small committee of Post-Office men was formed; and after much anxious deliberation they resolved, and a daring step they thought it, to take a little room at a rent of twelve shillings a week, in the perhaps not over-fashionable neighborhood of Bridgewater-square, Barbican.

"The following is an extract from the original prospectus of the Association, now a very scarce and highly-prized document:

"'This Association has been formed for the purpose

of supplying officers of the Post-Office and their friends with articles of all kinds, both for domestic consumption and general use, at the lowest wholesale prices.

"'The advantages of the scheme are obvious, but its full benefits can best be secured by a general combination in support of it on the part of the officers of the various departments.

"'It is intended that the articles mentioned in the accompanying price-list shall be purchased by the committee and distributed among the members. Arrangements for the supply of all other articles have been entered into with the firms named in the accompanying list.'

"Even when the Association was fairly started, and carrying on its business on its own premises, the committee did not venture to order any goods without ascertaining from the members what quantity of each article was needed. The business soon outgrew the room in Bridgewater-square, and the committee in a fit of extraordinary daring, engaged from a printer the upper floor of a small house in Bath street, on the ground floor of which the worthy typographer carried on his own business. The memorable house wherein the third store (counting the original cupboard) was carried on, has long since been pulled down to make way for the new Post-Office buildings, but those who went there to coöperate in those early days must have a vivid recollection of the narrow staircase, where one was elbowed by printer's devils, and of the dark little rooms crowded with purchasers. Here, however, we stayed but a short time, the business growing so rapidly that within a very few months the committee had again to seek larger premises, and this time, after making temporary use of some premises in Wood street, they took a really desperate leap. After many a hunt for a house big enough to meet any probable increase of business, two of our committee discovered a suitable one in Monkwell street, a very narrow, out of the way thoroughfare near Cripplegate Church, and filled with confidence by past success, they took it on their own responsibility at a rent of £400 a year. Great was the anxiety of the remainder of the committee at this bold proceeding, though the intention was to sublet the upper floor or the house to some firm that should undertake to sell goods to the members at wholesale prices. Tenants were found in certain hosiers, relatives of one of the Post-Office clerks, and the arrangement worked fairly well for a time, but as soon as it could safely do so, the committee regained possession of the floor, and undertook the sale of hosiery on its own account.

"From this point the narrative, from being one of small beginnings, becomes the story of a large and rapidly increasing business.

"First the committee obtained part of an adjoining house, then the whole of it, and after a time the other adjoining house, and part of a house on the opposite side of the street. A fresh house was taken in Villiers street, and subsequently a larger one in Long Acre, for the convenience of West End members. Before this time a great pressure had been put upon the committee to open a West End store; but they would not then make the venture, and this, among other causes, led to the establishment of the sister Association, entitled 'The Civil Service Coöperative Society,' which has its stores in the Haymarket.

"The city business of the Association will, during the present month, be removed to very large and handsome premises, near the Heralds' College, in Queen Victoria street, now building expressly for its use.

"I have not mentioned the extreme difficulty which the committee experienced in inducing wholesale houses to deal with the Association, especially when its doings found their way into print. Though ready money was always offered, together with good orders, most of the wholesale houses hung back, declaring that unless the orders were very large indeed, they should not feel warranted in encountering the fierce opposition of the retail traders. And now let us mark the consequences of this opposition. Very large orders being out of the question, so long as custom proceeded only from a limited number of persons, each of moderate income, and Civil servants generally not yet joining in the movement, the coöperators were obliged, in self-defence, to extend admission to quasi-membership beyond Civil Service bounds. Even this extraneous aid barely carried them through the struggle; the retailers having, over and over again, succeeded in deterring particular firms from supplying them with goods. These quasi-members, however, called by us 'subscribers,' were by no means admitted to any share in management, which indeed during the first year was strictly confined to a Post-Office Committee, though afterwards extended to representatives from the Civil Service generally. The exclusion of the general public from authority we have regarded as one of the chief causes of our success. Subscribers, however, by an annual payment of 5*s.*, obtain all the commercial advantages enjoyed by full members, except that their purchases are not delivered carriage-free. The full members become so by taking each a £1 share, of which, however, only 10*s.* has been called up. No one is allowed to hold more than a single share, nor are shares saleable or transferable in any way. On a member's death, his share is cancelled, and his deposit returned to his family. Until about a month ago any Civil Servant not below the rank of a clerk was eligible as a shareholder; but actual admission to the shareholding body required the approval of the committee. The number of shareholders, which has largely increased during the last three or four years, is now about 4200.

"By the rules of the Association, any profits which may be made are to be spent in reducing the prices at which the goods are sold. Even in the outset, prices were not fixed higher than is deemed needful to cover the working expenses, which now amount to only 6 or 7 per cent. on the wholesale purchase price; but, of course, the committee in its calculations has always taken good care to be well on the safe side. It is, perhaps, owing to extreme prudence in this matter, though, probably, still more to the need felt for a considerable working capital, that the Association has gradually accumulated the sum of about £75,000. The very magnitude of this capital has, however, proved a source of danger; for, without question, some persons have at different times obtained shares simply in the hope of breaking up the Association and getting a share of the spoil. Happily these unjustifiable attempts have hitherto always met with signal defeat, an overwhelming majority of the shareholders being determined to maintain the Association in honest and faithful accordance with the principles upon which it was founded.

"At the last half-yearly meeting of the Association in April, a proposal was brought forward to limit the shareholding body to the present number. After a prolonged and animated discussion, it was resolved to submit the proposal to the vote of the whole of the shareholders, which was taken by ballot. Out of the 4200 shareholders only 1200 voted, but of those who did vote there was a majority of 400 in favor of the proposal, which was accordingly carried. Of course, could the accumulated profits be divided, this limitation of the number of shareholders would give the shares a considerable value. Legal opinion, however, is entirely against the possibility of thus disposing of any past accumulations, which by the rules can only be spent in reducing the prices of articles sold. It is expected that those who have thus obtained a limitation of the shareholding body, will now endeavor to carry such an alteration in the rules as will allow future profits to be devoted to a Widow and Orphan Fund, or to some such purpose. Any change in the constitution of the Association, having for its object the benefit of the Civil servants as distinguished from their friends the subscribers, is viewed with much anxiety and disfavor by most of the earlier members of the Society.

"The number of subscribers is now limited to 15,000. While this number furnishes a *clientèle* sufficiently strong to enable wholesale houses to disregard the retail traders, some check is placed upon the enlargement of the business, and consequent increase in the labor and responsibility of management.

"The extraordinary rapidity with which the business

has grown, will best be seen from the following table showing the amount of sales at the stores during each year of the Association's existence, viz.:

Date.	Amount of Sales.	Date.	Amount of Sales.
1865	£5000	1869	£345,000
1866	21,000	1870	447,000
1867	83,000	1871	646,000
1868	218,000	1872	723,000

"During the half year ended March 31st last, the sales reached £392,000, being, therefore, at the rate of £784,000 a year, viz.: for grocery and wine, £410,000: for hosiery and clothing, £192,000; and for fancy goods, stationery, etc., £182,000. At the present time about 8100 pounds of tea and about 15 tons of sugar are sold weekly.

"The articles sold at the stores consist principally of groceries, cigars, and tobacco, wine and spirits, hosiery and drapery, stationery, books and music, watches and jewelry. But most of these articles, and, indeed, almost every other article of ordinary demand, can also be obtained by members and subscribers at low rates, though of course only for ready money, at all such warehouses and shops as have arrangements with us. The latest quarterly Price List, which, from a single small sheet has grown to be a book of more than 200 pages, shows that the covenanting firms are not less than 250, while the reduction promised in prices ranges from 15 to 25 per cent. It is believed that this additional business amounts, at least, to £800,000, and not improbably to as much as £1,000,000 a year. Contrary to what might be expected, this part of the system works satisfactorily; for, though purchasers are invited to complain to the committee if they ever have reason to suppose they do not obtain the full discount promised, few complaints are received. These, however, are all thoroughly examined, and wherever they prove to be well founded, the offending firm is struck off the list. Moreover, members soon learn from each other at what shops they are civilly and fairly treated, and act accordingly; so that some of the firms which have been connected with the Association from its early days, having gradually acquired a high reputation among us, are now doing a very large business with our members.

"The members have the advantage of a tailoring department, carried on in Bedford street, Strand, which, however, was for a long time a source of great trouble to the committee. Much difficulty was experienced in getting, and still more in keeping, good workmen, who left in a mysterious manner; and the work was frequently so badly done as to convince the committee that the workmen were being bribed to spoil the clothes intrusted to them, and thus to entail loss upon the Association. After a while, and by the exercise of great perserverance, these difficulties have all been overcome, and the tailoring department promises to be a great success.

"Notwithstanding that the retail price of the articles sold at the stores is on the average some six or seven per cent. above the wholesale price, it happens every now and then, that, owing to a rise in the market price between the publication of the quarterly price lists, the market price becomes higher than the retail price at the stores. Unless the article is one of large general consumption, such as tea, the committee adheres to its retail price until the issue of the next Quarterly Price List. This sometimes leads to an attempt by retail traders to buy up—of course through some subscriber willing to play false to the Association—all the stock in hand. During the Franco-German war an attempt was thus made to buy up all the champagne, and not many months ago a rapid rise in the market price of white pepper and of anchovies led to similar attempts with these articles. Large orders are never now executed without such inquiry as satisfies the committee of their being made in good faith.

"The Association directly employs about 400 people, and pays upward of £18,000 a year in salaries and wages. The stores in Long Acre stand at an annual rental of £600, while for the new stores in Queen Victoria street, the mere ground rent is no less than £1400. The premises themselves we are about to purchase for £15,000, while a further rent of £200 a year is paid for a warehouse at Ward's Wharf, where are kept large stocks of every article in the Price List, and where are executed all large orders for goods. Something has been said as to the causes of our well-doing, but it seems desirable to inquire further into the reason of success so unprecedented. The Association is now one of the largest buyers and sellers in England, nay, in the world; and yet it was commenced and has been carried on by a body of men who in their ordinary employment neither buy nor sell. Moreover, the *personnel* of the committee so changes that at the present time there is left upon it but one of the original members, while every fresh committee-man, of course, has to learn the very A B C of commercial business. For explanation, I believe we may fairly point first to the high sense of honor which pervades the Government service, and which always renders it easy to find abundance of men whose integrity is above suspicion; secondly, to the admirable training for business (viz., the adaptation of means to an end, as Mr. Walter Bagehot happily defines it) which the Post-Office service affords; and, thirdly, to the corporate nature of the Civil Service. In the establishment of almost every other trading company, as it seems to me, the promoters aim at some advantage for themselves and their friends beyond what is avowed, getting perhaps a larger allotment of shares, or obtaining them on more favorable terms than the general public, or at least securing appointments for their nominees. Indeed, so general is this practice, that it would, I suppose, be impossible to persuade the public that a company had been formed on such a footing as to give equal benefit to every individual shareholder. On the other hand, when the Civil Service Supply Association was formed, not only did not the originators of it obtain any special benefit for themselves, but no one ever imagined that they did. During the eight years that the Association has been in existence, though nearly £2,500,000 have passed through the committee's hands, there has arisen, so far as I know, no suspicion whatever of any dishonesty, or even of any questionable dealing.

"As I have before stated, the Association originated and was organized in the Post-Office—a department which, under the guidance and control of Sir Rowland Hill, has seen a great rise of able and energetic men. Even in earlier days, Post-Office men had, of course, taken constant part in a vast and complex business; but the introduction of penny postage had prodigiously enlarged this business in all its branches. Moreover, Sir Rowland's system of management—particularly his bold application of the principle of promotion by merit instead of by seniority—had not only advanced able men to important posts, but had brought out throughout the service powers previously latent. Mr. Scudamore, in a recent lecture, stated that the indirect results of Sir Rowland's postal reforms have been even greater than the direct. Among these indirect results, as due to the general spirit of activity and enterprise thus engendered, may, I believe, be reckoned the establishment of the Civil Service Supply Association and the kindred societies which this has called into life.

"Another main element of success is the corporate nature of the Post-Office and of the Civil Service generally. This provided a large business connection, already linked together and accessible without the aid of advertisements, so soon as the value of the Association was proved. Moreover, there was a special guarantee for integrity. Every one in the Post-Office either knows or can easily know something of every brother officer of whatever rank, and this holds good, though perhaps in a lesser degree, of every Government department. Every committee-man has felt that his reputation as a Civil servant was of far too great a value to be endangered by any unfair dealing in the affairs of the Association; the motive to rectitude being so strong, that to put men of even moderately good official standing on the committee was to render it certain that the work would be honestly and diligently done. While, however, the Association has thus far succeeded so admirably, it seems to me that its future course is not free from danger.

"The shareholding body, composed as it is of upward of 4000 Civil servants from all branches of the Service, who have been admitted to membership without any reference to their fitness for the position, has sometimes proved very unruly. Latterly, however, the introduction of the plan of voting by proxy has greatly reduced the power of the comparatively small fraction of shareholders who are disposed to be troublesome.

"The pay of the committee, too, for duties involving much sacrifice of well-earned leisure, considerable labor, and great responsibility, is very low. So long as salaries are limited to £80 or £90 a year, the committee must remain a too changeable body, since capable men cannot be permanently retained on such terms. Hitherto the Association has been mainly served by men whose chief motives were pride in its success, and a desire to benefit their fellow-officers, but of course this will not last. The time must come when the chief inducement to such service will be the desire of adding to income; nor should it be expected that the Association will be maintained in full vigor unless the payment to the committee be made sufficient to induce well qualified men to serve mainly as a matter of business.

"A reduction in the shareholding body, with a limitation of it to suitable persons, is now out of the question. Many of us Post-Office men thought, and still think, that a great mistake was made in not resolutely retaining the control of the Association in the Post-Office service; though, of course, we quite approved of admitting the remainder of the Civil Service to all the other advantages of membership. I feel no doubt that should the present Association ever collapse, the Post-Office men would rapidly and successfully organize a new society on the plan of keeping the control in the hands of a moderate number of trustworthy and reasonable men of their service.

"About two years ago, when our Association limited the number of subscribers to 15,000, a new society entitled, 'The New Supply Association,' was projected to take in those friends of Civil Servants and others who could not gain admission to the old. Several of the then members of our committee joined the direction of the new Association, which is conducted upon the same general principles as our own. I see by the first annual report that the Association, which has its stores in Long Acre, has during the past twelve months sold £20,000 worth of goods to its members, so that it has made a good commencement.

"I must mention, in conclusion, that I have never served, and certainly never intend to serve, on the Committee of Management myself, although I have had the opportunity of watching its work from the commencement to the present time."

CHAPTER XXX.

THE FUTURE OF THE GRANGE.

Retrospective—Future of the Order—What it will accomplish for the Farmer and for the Country—The Grange pledged to a Just and Liberal Course of Action—The Grange not a Destructive Order—Its Stake in the Community—Elements of Opposition—Distrust of Politicians—Political Views of the Granges—Platforms of the Farmers of Illinois, Minnesota, and Iowa—Necessity for the Order to confine itself to its Proper Work.

WE have now examined the organization, traced the history and growth, and discussed the prospects of the Order of Patrons of Husbandry. It is a remarkable order, and has a remarkble history. Its growth is unprecedented. Not even the old Know Nothing party spread with such rapidity. Although organized as far back as 1867, the growth of the Order has been confined almost exclusively to the past year, in which it has spread with a swiftness which has exceeded even the wildest hopes of its most sanguine friends. But a mere handful at the beginning of 1873, it is now a vast army, stretching over the entire Republic, with a well arranged and satisfactorily working system of government, with definite and honestly avowed aims, and ample means of attaining its ends; and it is increasing by many thousands every week.

No man can predict its future; but it seems safe to assert that at no very distant day it will embrace the entire farming community. Certainly those who have the farmers' interests at heart should strive for such an end, for the Grange has shown itself the farmer's best friend. Its spread means protection to him, encouragement to him, a greater degree of prosperity and happiness to him, for its only object is to make him a better and more prosperous farmer and man. None but farmers are admitted to its membership, for it does not concern itself with outside issues; its work is with and for the farmers only. There is every reason why the farmers of the country should work for its success, and it will be a great mistake to hold aloof from it.

When it shall have accomplished its work, the results will be such as will affect the condition of the country for remote generations. It will have broken the power of the railroad monopolies and secured to the farmer a cheap means of reaching a market. It will have rendered the financial demoralization from which we are now suffering impossible, by securing the passage of laws meting out equal justice to all men. It will have given the whole country cheap coal, and cheap bread. It will have secured the farmer a fair return for his industry. It will have relieved him of the necessity of incurring debts, and have enabled him to make cash purchases at reasonable rates. The accomplishment of all this must exercise a powerful influence upon the country, and change the entire current of its progress and history.

That this will be accomplished, we firmly believe. We do not expect to see it done in a day, or a year. It is the work of time, and the Grange must be patient; but it will be accomplished. We have shown the power of the farmers to make their wishes respected, and it is the work of the Grange to guide this power in channels which shall benefit the entire country.

The Grange is a destructive institution only so far as abuses are concerned. It seeks to eradicate these; but it also seeks to build up a better system in their place. It is pledged to destroy railroad tyranny, but it is also pledged to secure the inauguration of a system which shall be just to the roads while destroying their power to do harm. It will do nothing hastily. Its very organization and mode of procedure are guarantees against indiscreet and dangerous action. It is composed of men who have the highest stake in the welfare of the community; of honest, temperate, and industrious men and women, a class of which the Republic is especially proud. Its measures will be the result of the combined wisdom of this class, and will be discussed and examined patiently and fairly before being decided upon. The Order has nothing to gain by haste, but everything to compel it to act calmly and judiciously.

That its work will be accomplished without opposition, we cannot venture to hope. Several elements of opposition naturally array themselves against it. The railroads, whose corrupt use of power it seeks to check; the friends of the land grab system; the coal monopolists; the men who dam up the avenues of approach to a free market—the "protected class;" the middlemen, whose vast gains are directly endangered by the coöperative feature of the Order—all these are its natural and bitter enemies, and they will seek by every means which their ingenuity can devise to weaken and distract the Order and prevent the achievement of its great work. A portion of the press of the country, in sympathy with them, will take up their cause and endeavor to discredit the Order in the eyes of the public. The opposition which the Grange must encounter has, as yet, scarcely begun. The rapid and astonishing growth of the Order has taken its enemies so entirely by surprise that they have not yet recovered sufficiently to organize their opposition; but it will come.

But let the farmers remain steadfast in their purpose, and above all, let them confine their Order to their own class. The very life, the very existence of the Order depends upon the oneness of the interests of its members. Such outside support as it needs it will quickly obtain when the people see, as they soon will when the fight is fairly opened, that the Grange is battling for the rights of the whole people as well as those of the farmer. The Grange has the good of the nation at heart, and its aim is to be just and generous in the exercise of its powers.

With political parties as such it has no affiliation, and yet it must act as a political party itself in one sense. Many of its ends can be attained only by exercising the political rights of its members. It desires to break down abuses and secure the adoption of just laws. To accomplish this it must put men in power who will faithfully carry out its wishes. It proposes to do this; to see that its individual members entitled to the right of suffrage cast their votes only for men who are pledged to the accomplishment of the objects which it is working for. Patrons under these circumstances will vote for no man as a Republican or as a Democrat, but as a man pledged to the adoption of a definite reform. Already the politicians, appreciating the power of the Order, have sought to use it for their own purposes; but the Order has declined, and will continue to decline, all such overtures. What sympathy has it with the men who have aided in bringing about the very abuses it seeks to correct? What bond of union can there be between the Order and the men who have aided and encouraged wild cat railroad speculation; who have fastened the land grab system upon us; who have put it in the power of the monopolist to rob the people; and who are stained with the infamy of the "Crédit Mobilier," and the "back pay steal?" This is well understood by the farmers of the country, and the politicians have undertaken a hopeless task. The day is approaching when better men will be charged with the work of the State, and it is the task of the Grange to hasten the advent of this period.

The Order has not yet committed itself to any definite avowal of its views upon political questions. Various minor issues have been advocated by the various State organizations, but no general platform has yet been presented to the Order.

The Farmers' Convention, which met in Chicago towards the last of October of the present year, made the following declaration of principles:

Resolved, 1. That Congress be asked to pass a general law fixing a maximum rate for transportation between the States, and the Legislatures laws governing transportation within the States, and that no more subsidies to private corporations be given.

2. We demand the construction of railroads and the improvement of water communications between the interior and seaboard, the same to be owned and operated by the General Government, for the purpose of affording cheap and ample transportation, and to protect the people from the exactions of monopolies.

3. That people should create and patronize home manufactures in order to do away, in so far as is possible, with the necessity for transportation.

4. That the people should free themselves of debt in order to be better prepared for the struggle with monopoly whenever it comes.

5. That no industry can be protected save at the expense of all other industries, and that all special legislation is wrong.

6. That the farmers should organize throughout the country to secure reform of abuses and equal justice for all.

The Minnesota farmers, at their convention, at Owatonna, September 2d, 1873, adopted the following platform:

Whereas, The leading issues that have hitherto divided the people of this country in political parties have ceased to exist, and it is unwise to seek to continue the old party organization now that new and momentous questions have arisen; and

Whereas, The principal question now demanding consideration is that involving the privileges and powers of corporations as antagonizing with and operating in opposition to the well-being of the people; and

Whereas, We, the farmers, mechanics, and laborers of Minnesota, deem the triumph of the people in this contest with monopolies essential to the perpetuation of our free institutions and the promotion of our private and national prosperity; and

Whereas, In addition to this, and to the honest and economical administration of the Government, we recognize no party distinctions nor political issues now before the country as worthy of more than minor consideration; be it therefore

Resolved, First: That the purpose of all proper government is the promotion of the welfare of the entire people, and that therefore the conduct of any citizen, association, or copartnership, whether chartered or otherwise, which may operate to the prejudice of this general welfare, is antagonistic to the true objects of our Government, and violative of the fundamental principles upon which all correct law is based.

Second: That we recognize no political party nor individual aspirant for office as worthy of our support, unless it or he will unite with us in declaring that the Government cannot alienate its sovereignty either in whole or in part to any person, association, or corporation for any purpose whatever, but such are always and must forever remain subject to the sovereign authority and control of the Government.

Third: That we will not aid in elevating any man to any important public position whatever who will either deny or object to the exercise by the Legislature of the power to reverse or annul at any time any chartered privilege or so-called vested right or any privilege claimed to be involved in any charter to any corporation, railroad, or otherwise, which experience has shown is or may be exercised by such corporation or by other similar corporations to the detriment of the public welfare; and that we will demand from every candidate for a high executive, legislative, or judicial position to whom we accord our support that he shall pledge himself to recognize the maintenance of this right by the Government as a sacred duty essential for the preservation of the liberties of the people and the stability and prosperity of the Commonwealth.

Fourth: That taxes can only be rightfully levied for the purpose of raising revenues to defray the expenses of the Government in the discharges of its legitimate duties, supporting public institutions, and promoting the public welfare; and that the levying of imposts as inure to the benefit of a class or classes in the community, while being detrimental to other classes, is unjust and oppressive; and that tariffs levied on imported articles may be and are often so arranged as to become thus discriminative and injurious; and that it is therefore essential that the utmost care should be taken in framing such tariff laws, in order that the objectionable features may be avoided and that they may operate for the well being of the entire community.

Fifth: That it is contrary to the spirit and purpose of a Republican Government that its servants should be compensated for their public services to an extent that will make office holding attractive to human cupidity, and that in the late act of Congress, increasing the official and Congressional salaries, notwithstanding the pleas and excuses urged in its palliation, we recognize only a corrupt and reprehensible avarice and reckless disregard of the public weal, which deserves the severest censure; and we demand the repeal of the law at the earliest practicable moment, and declare every man who supported and approved, or aided and abetted in procuring its passage, or received benefit through its enactment, whether in the shape of back or future pay, as unworthy the confidence of his fellow-citizens and unfit for the further occupancy of any position of public trust.

Sixth: That all participants in the Crédit Mobilier and the corrupt transactions exposed by its investigation of the late Congress and by the late Treasury investigation of the State, deserve to have been punished as criminals, and that those who aided in screening them from complete exposure and consequent punishment, should likewise become objects of public scorn and contumely.

Seventh: That every public officer is amenable to the people for his conduct, and that public sentiment should demand and compel the resignation of all those who are guilty of misrepresenting their constituents, of malfeasance in office, and of neglecting to execute faithfully the duties intrusted to them.

Eighth: That the fees and salaries at present allowed to county and other officials within this State are frequently excessive, and that these should never be greater than is paid by private individuals to their employés engaged in similar duties and bearing similar responsibilities, and that we demand that the State Legislature shall at its next session remedy this evil, and reduce such salaries and fees to what will be no more than a just and reasonable compensation, and thus, by removing the inducements for holding, lessen the desire for seeking office, and obviate to a considerable extent one of the most patent causes of local political corruption.

Ninth: That our experience proves that persons elected by parties are subservient to the leaders and wire-pullers of the parties electing them in the performance of their public duties, to the neglect partially or wholly of the opinions and wishes of the mass of the people; and that therefore we, as farmers and laborers, despair of ever having our wishes complied with or our interests subserved in the administration of public affairs until we shall take upon ourselves the discharge of the duties we owe to ourselves and to each other of choosing and electing our own candidates independently of the action of all other political organizations, and we therefore earnestly recommend to the farmers and laborers of the State that we shall do all in our power to procure the nomination and election of full and complete county, district, and State tickets, embracing candidates elected in the interests of the masses of the people for all the positions in the Executive, Legislative, and Judicial branches of the Government to be elected this Fall, and that, to the end that this policy may generally obtain, we solicit the coöperation of the industrial masses of the other States in order that the influence of the movement may be extended to the administration of our national affairs.

Tenth: That we receive with satisfaction the decision of the Supreme Court of this State in the case of Blake agt. The Winona & St. Peter Railroad Company, in which the Court holds, in effect, that the railroads are simply improved highways, public roads, and that as such the right to prescribe a rate of tolls and charges is an attribute of the sovereignty of the people of which no legislation can divest them, and that the thanks of all the people of this State are due to W. P. Clough, the attorney for the plaintiff, whose skill, ability, and devotion to the cause of the people secured this judicial triumph.

Eleventh: That we have seen with alarm the startling revelations in reference to the condition of our State Treasury, the undoubted defalcation of our Treasurer of over $100,000, and the reported defalcation of his successor of nearly $40,000; the loan of the public funds to merchants and lumber dealers; the making of accounts with bogus certificates of deposit; the fact that nearly half a million of the school fund, the precious heritage of our children, was left unindorsed as required by law, and completely at the mercy of these dishonest officials; the perjuries of the State Treasurer before the Commission, and finally, the desperate efforts that were successfully made to hide the Ring of guilty parties who had used the State Treasurer as their tool.

Twelfth: That we claim that the law requiring these companies to fence the lines of their roads should be strictly enforced, and that the said companies should be compelled to pay for all loss and damage to stock caused by the absence of such fences.

Thirteenth: That we are opposed to the monopoly of wood and coal in our great cities by the Rings, as a shameful tax on the industry of the people.

Fourteenth: We are in favor of free water communication with the ocean by means of the improvement of the Mississippi and other great rivers of the State, and the improvement of the great lakes; that we are in favor of an examination by the National Government of the region between the St. Croix and Lake Superior, to ascertain whether canal communication can be made to connect the tributaries of the Mississippi with the waters of Lake Superior.

Fifteenth: We are in favor of such reasonable limitation of the hours of labor in the shops and factories of the State as will give the laboring people opportunity for moral and mental improvement.

Sixteenth: That we demand a State law that will pay out of the public funds the costs and charges of all suits brought by individuals to enforce the laws of the State against railroad corporations.

Seventeenth: That we can sympathize with all attempts for the moral improvement of the people, and that we regard the temperance societies of the land which are working by moral suasion for the advancement of the cause as deserving of the consideration of good men everywhere.

Eighteenth: That the honor of our State demands that the delegation in Congress from this State call for a thorough investigation into the equitable settlement, so-called, of the transfer of the Fort Snelling property.

Nineteenth: That the subserving of the present candidate for Governor on the Republican State ticket to the interest of railroads, shows him to be an enemy to the rights of farmers and laborers, and a friend of monopoly.

The Farmers' Anti-Monopoly Convention, which met at Des Moines, Iowa, on the 13th of August, 1873, adopted the following platform:

Whereas, Both political parties have discharged the obligations assumed at their organization, and being no longer potent as instruments for the reform of abuses which have grown up in them, we deem it inconsistent to attempt to accomplish a political reform by acting with and in such organizations; therefore,

Resolved, That we, in free Convention, do declare as a basis of our future political action, that all corporations are subject to legislative control; that those created by Congress should be restricted and controlled by Congress, and that those under State laws should be subject to the control respectively of the States creating them; that such legislative control should be an express abrogation of the theory of the inalienable nature of chartered rights, and that it should be at all times so used as to prevent moneyed corporations from becoming engines of oppression.

Resolved, That the Legislature of Iowa should by law fix the maximum rates of freight to be charged by the railroads of the State, leaving them free to compete below the rates.

Resolved, That we demand a general revision of the present Tariff law that should give us free salt, iron, lumber, and cotton and woollen fabrics, and reduce the whole system to a revenue basis only.

Resolved, That we demand the repeal of the back salary act, and the return to the United States Treasury of all money drawn therefrom by members of the last Congress, and of the members of the present Congress we demand the repeal of the law increasing salaries and the passage of a law fixing a lower and more reasonable compensation for public officers, believing that until the public debt is paid and the public burden lightened, the salaries of our public servants should be more in proportion to the rewards of labor in private life.

Resolved, That we are opposed to all future grants of land to railroad or other corporations, and believe that the public domain should be held sacred to actual settlers, and we are in favor of a law by which each honorably discharged soldier, or his heirs, may use such discharge in any Government land office in full payment for a quarter section of unappropriated public land.

Resolved, That in the corrupt Tammany steal, Crédit Mobilier frauds, Congress salary swindle, and official embezzlements, and the hundreds of other combination steals, frauds, and swindles by which Democratic and Republican legislators, Congressmen, and office-holders have enriched themselves, defrauded the country, and impoverished the people, we find the necessity of independent action and the importance of united efforts, and cordially invite all men, of whatever calling or trade, regardless of political views, to join us in removing the evils that so seriously affect us all.

From these resolutions, embracing three of the principal agricultural States of the Union, the reader can easily gain a clear conception of the views of the members of the Order upon the topics of the day.

By remaining true to itself, then; by resisting outside influences, and especially by avoiding political complications, the Grange can safely and successfully accomplish its great mission. It is a noble work that it has taken upon itself, and its success must result in benefit to the whole country. Its objects are pure and lofty, and its success can be attained only by high and ennobling means. An Order which seeks the individual elevation and material prosperity of nearly one-half of the whole nation, merits, and will receive the warm and hearty sympathy of our entire population.

CHAPTER XXXI.

LEADING GRANGERS.

DUDLEY W. ADAMS,

Master of the National Grange.

THE Master of the National Grange is the presiding officer of the Order. His duties are thus defined by the By-laws of the Organization:

It shall be the duty of the Master to preside at meetings of the National Grange; to see that all officers and members of committees properly perform their respective duties; to see that the Constitution, By-laws, and Resolutions of the National Grange and the usages of the Order are observed and obeyed; to sign all drafts drawn upon the treasury, and generally to perform all duties pertaining to such office.

He is chosen by the National Grange, by ballot, and his term of office is limited to three years. He exercises a general supervision over the Order, and his duties are by no means as light as some persons seem to think. The position is one of very great responsibility, and calls forth the exercise of executive abilities of a very high class. The Master must be a man of great firmness and force of character, fertile in resource, possessing great tact, and, above all, must be a practical farmer and thoroughly devoted to the interests of the class for whose benefit the Order is working.

These qualities are happily united in the present

Master of the National Grange, Dudley W. Adams, Esq., of Iowa.

Like most men of his calling, Mr. Adams' life has been quiet and uneventful. "My life," he says, in a letter which we have been permitted to use, "has been a rather uneventful one, and not such as will make a stirring narrative." "I was born," he continues, "in the pine woods of Massachusetts, in the town of Winchendon, in the year 1831." He is consequently forty-two years of age at present. His father was at this time "running a small saw-mill," by means of which he managed to make a modest provision for his family. When the subject of this sketch was four years of age the family moved to a small rocky farm, which the father had purchased, and on which the childhood and youth of Dudley Adams were spent. His was the life of the ordinary New England farm lad—working on the farm in the milder weather, and attending the district school in the winter. He was a bright, quick lad, and manifested a strong desire to excel in his studies, so that when but a mere youth, we find him in possession of all the erudition the district school could afford him. Then the pupil became the teacher, and for several years the young pedagogue presided over the school in which his own education had been gained.

At length the critical period in his life arrived. He was twenty-one years of age, and a free man. He was also the owner of a modest little sum that he had saved from his earnings as a teacher. He must now make a decision which would affect his whole future; he must choose the vocation of his life. New England offered few inducements to an energetic and ambitious young man, but the great West was a field in which the opportunities for advancement were without limit. Westward his heart led him, and Westward he went. "At the age of twenty-one," he writes, "I turned my face West, and took up a piece of wild Government land at Waukon, Iowa, of which I made a farm, and on which I still reside."

The young New Englander proved no drone in the new community. He set to work with a will, and from the first was recognized as one of the most energetic and intelligent farmers of the county or State. His neighbors testified their appreciation of him by electing him President of the Allamakee County Agricultural Society within a year after his settlement in the county. He was only twenty-two years of age at the time, and it was no small compliment to be chosen over the heads of older and more experienced men. He still refers to his election with pride.

Under the intelligent and vigorous management of Mr. Adams the "piece of wild Government land" became one of the prettiest and most flourishing farms in the State. Its owner was a reading and a thinking man, and devoted his leisure to an intelligent and systematic course of reading and self-culture.

In his own pursuit his attention was directed particularly to Horticulture, and he did not confine his efforts to his own farm. Recognizing the needs of the country, his labors embraced the whole Northwest. "My principal energies," he says, in a letter to the writer, "have been directed to the development of Horticulture in the Northwest, and now I have perhaps the finest orchard in that section, numbering over 4000 trees." In 1868 he was elected Secretary of the Iowa

State Horticultural Society, and has been annually re-elected since then. He has during the present year relinquished the position in consequence of his time being occupied by other and more important duties. "The achievement of my life," he says, "in which I take most pride, is the little I have done to improve the horticulture of my adopted State." A very modest way of viewing a good and useful work, and one that will keep Mr. Adams' name in grateful remembrance in Iowa long after he has been gathered to his fathers.

A man of Mr. Adams' mental capacity and activity could not help recognizing and investigating the evils from which the farming class has suffered, and upon the organization of the Order of Patrons of Husbandry, he at once identified himself with it as the best means of remedying the defects complained of. He took an active part in promoting its growth, and upon the organization of the Iowa State Grange, on the 12th of January, 1871, was elected Master of that body. In December, 1871, he was reëlected for a full term of two years. In January, 1872, he was chosen Master of the National Grange, and resigned his position in the Iowa State Grange. Upon the adoption of the Constitution of the Order, in January 1873, at the sixth annual session of the National Grange, Mr. Adams was reëlected Master for the full term of three years.

The following address, delivered by Mr. Adams before the Granges of Muscatine and Union counties, Iowa, in October, 1872, presents him in a favorable light as a speaker and thinker. We commend his vigorous and well-timed remarks to the careful consideration of the readers of these pages. He said:

"When physicians meet in convention, as they often

do, it is customary for members of the medical profession to read papers for the entertainment and instruction of the assembled M. D.'s.

"When railroad men have a convention, such persons as have had active experience in railroad business do the talking and have charge of the meeting.

"Editorial conventions are attended by editors, and they, as firmly as any other class of people, are of the opinion that they are capable of managing their own business, and they are not in the habit of imploring the members of other callings to furnish the brains to amuse or instruct them.

"Shoemakers have organized themselves into the order of St. Crispins, and consider themselves able to paddle their own canoe.

"Lawyers not only feel competent to address and properly edify conventions of their own profession, but their modesty does not forbid them from rendering valuable assistance to less favored classes by a free use of their surplus talent.

"But, when the tillers of the soil have met in an agricultural society of any kind, it has been usually customary to select a lawyer, doctor, editor, or politician to tell us what he knows about farming. The idea has very rarely occurred to the managers of such institutions that it might be possible for a farmer to have anything to say on such occasions which should be either appropriate, interesting, or instructive. When these professional oracles of our professional managers' selection open their mouths, we are edified with a rehash of such ideas as may be prevalent in the community, served up in a great variety of forms, and presented in a great many different and beautiful lights, depending for its coloring upon the business of the orator, as this is the stand-point from which we are viewed, and, of course, this view determines the nature of the picture. Lawyers and doctors in beautiful colors paint the nobleness and independence of the farmer's life. They tell us we are the most intelligent, moral, healthy, and industrious class in all the land, and all our present is calm and our future happy. Merchants tell us that no business is so sure and free from care as farming, and that in no other calling do so few men end in bankruptcy. Politicians laud in stentorian tones the 'honest yeomanry,' 'the sinews of the land,' the 'bulwarks of our nation's liberties,' 'the coarse blouse of homespun which covers the true and honest heart,' and deluges more of equally fulsome and nauseating stuff.

"Soft-handed agricultural editors give long dissertations on the necessity of saving all the spare moments, and converting them into some useful purpose. They tell us how rainy days may be laboriously used in mending old rake-handles, and winter evenings utilized by pounding oak logs into basket stuff, while our wives and daughters can nobly assist in averting bankruptcy by weaving the baskets or ingeniously making one new lamp-wick out of the remains of three old ones.

"It has never occured to these very wise instructors that farmers and farmers' families are human beings, with human feelings, human hopes and ambitions, and human desires. It will doubtless be a matter of surprise for them to learn that farmers may possibly entertain some wish to enjoy life, and have some other object in life besides everlasting hard work and accumulating a few paltry dollars by coining them from their own life-blood, and stamping them with the sighs of weary children and worn wives.

"What we want in agriculture is a new Declaration of Independence. We must do something to dispel old prejudices, and break down these old notions. That the farmer is a mere animal, to labor from morn till eve, and into the night, is an ancient but abominable heresy. We have heard enough, ten times enough, about the 'hardened hand of honest toil.' The supreme 'glory of the sweating brow,' and how magnificent the suit of coarse homespun which covers a form bent with overwork, and which has incorporated in its every thread moments of painful labor which the over-worked wife had stolen from her needed rest.

"I tell you, my brother tillers of the soil, there is something in this world worth living for besides hard work. We have heard enough of this professional blarney. Toil is not in itself necessarily glorious. To toil like a slave, raise fat steers, cultivate broad acres, pile up treasures of bonds and lands and herds, and at the same time bow and starve the godlike form, harden the hands, dwarf the immortal mind, and alienate the children from the homestead, is a damning disgrace to any man, and should stamp him as worse than a brute.

"It is not honorable to sacrifice the mind and body to gain. It is not a trait of true nobility to bring up children to thankless, unrequited labor. It is not just or good or noble to wear out the wife of your bosom in the drudgery of the farm without a just return. You have no right to make agriculture so disagreeable as to drive all young men of spirit and enterprise into other branches of business.

"I will be met right here with the thousand time repeated rejoinder, 'Oh, we farmers have to work hard. We can't get along as mechanics in town do with ten hours' work. We can't afford to hire help. We can't afford to have holidays. We can't get time to make a vegetable, flower, and fruit garden, and supply our wants with vegetables, flowers, and fruits. We can't get time to make a lawn and plant trees around the house.' You can't? You can't? Then what are you farming for? As men, as citizens, as fathers, as husbands, you have no right to engage in a business which will comdemn yourself and your dependents to a life of unrewarded toil. If the calling of agriculture will not enable you and yours to escape physical degradation, and mental and social starvation; if it does not enable you to enjoy the amenities, pleasures, comforts, and necessities of life as well as other branches of business, it is your duty to abandon it at once, and not drag down in misery your dependent family. But I do not believe we need be driven to this alternative. I *do* believe that agriculture, followed as a business, with a reasonable regard to business principles, can be made a business success. I believe that by keeping steadily in view the primary end of life—our happiness, our comfort, our bodily health, our mental improvement and growth—they can be as well attained or better than in any other calling. Right here is the great difficulty; right here with ourselves is the remedy: We *work* too much and *think* too little. We make our hands too hard, while our brains are too soft. The day is long past when muscle ruled the world. Brain is the great motive power of this age, and muscle but a feeble instrument. The locomotive, tearing along, jarring the earth below, outstripping the wind above, and bearing in its train the beauty, honor, and treasure of a State, represents brains. The dusty, sweaty footman, wearily plodding along, carrying a pack on his back, symbolizes muscle. The self-raking reaper, driven with gloved and unsoiled hands, sweeping down like a fable the golden grain, represents brains. The bowed husbandman, painfully gathering handfuls of straw and cutting them with a sickle, represents muscle. The steamboat, plowing its way with ease against the strongest current of our swift and noble rivers, is brains. The dug-out, slowly creeping along the willow-margined shore, propelled by the Indian's paddle, is muscle. The sewing-machine, which stitches faster than the eye can follow and never eats or tires, is brains. The weary, pale, and worn wife, painfully toiling over the midnight task, is muscle. How futile the attempt, then, for muscle to compete against mind in the great battle of life! A wise man once wrote, 'The wisdom of a learned man cometh with opportunity of leisure;' and in that sentence is food for reflection and thought sufficient for an entire sermon. Unless farmers devote more time to the use of the brain and the improvement of the mind, and less to wearying and exhausting muscular labor, how can they hope to successfully compete against the vigorous minds of the present age? It is not the skilful hand, the strong arm, or the watchful eye alone that will in these days bring success to the farmer. These are needful, but a cultivated, intelligent, active brain to direct them is of ten times more importance.

"Again I say, we work too much and think too little. A farmer rises at four o'clock, goes out and does the chores among the stock, chops wood for the day, mends the harness, and is very industrious. By breakfast time, he has got all ready for the day's work. All hands then pitch into severe labor till noon. Dinner is called and dispatched in haste, and labor renewed till supper. This unavoidable but neccessary hindrance to labor is hurriedly performed, work resumed until darkness compels a cessation of labor in the field, and then the laborers return to the house. A lantern is procured, by the aid of which the milking and other chores are 'done up,' and by nine or ten o'clock at night the day's work is closed, and the family, tired and stupid, retire to bed, only on the following day to repeat the same routine of slavery. And yet such men are called good, thifty, industrious farmers. It is a lie! a base slander to call such stupid slavery of body, such starvation of mind, good or thifty, or in any wise commendable.

"Go into the country, and you will find numberless cases of men with poor health, crushed energies, ruined constitutions, and stunted souls, and women the slaves of habits of excessive labor, more fatal than the pernicious and much-comdemned customs of fashionable society. You will find children prematurely old, with the bright light of happy childhood extinguished, and everywhere a lack of that life and cheerfulness which gives to life its greatest charms. Most of these evils can be traced directly to overwork. Is such work necessary or even profitable for a famer? Most certainly not. Such work is a losing business, and farmers who adopt that course of labor will find at the end of the season that themselves, their wives, and children are worn and discouraged, and have not accomplished as much as had been attempted or expected. Why? Because they have worked like oxen and not like men, and have depended on muscle alone instead of making it an auxiliary of the mind, and they treat themselves to the luxury of a good, long, hearty growl at members of all other industries for combining to oppress the poor farmer. They growl at the shoemaker; they growl at the merchant; they growl at the railroad; they growl at the commission men; they growl at everybody and everything that lives by using its wits in sponging, cheating, and oppressing the hard-working farmer. This horde of cormorants are growled at, whined at, and snarled at, because they filch from the farmer his hard-earned dollars and live in luxury and ease thereon. Speakers at agricultural and political meetings, and writers in agricultural papers repeat these complaints, and ring the same charges over and over again, in season and out of season, until themselves and most farmers really believe that the tillers of the soil are the most industrious, moral, intelligent, hardworking, abused, persecuted lambs in the world, and everybody else are wolves, seeking whom they may devour.

"Now, as one who was born on a farm, reared on a farm, has spent the flower of his days on a farm, and still earns his bread by tilling the soil, I know my brother farmers will forgive me if I do not follow in and repeat this strain, but tell plainly the naked, disagreeable truth. Many of these complaints are true, and we ought to be ashamed of ourselves that such is the disgraceful fact. Here is a class of people exceeding any other in numbers and wealth, and claiming superior industry, intelligence, and morality, complaining of being oppressed. We ought to be ashamed of ourselves, and either cease our boasting or our whining.

"Let us take a candid look at the situation, and see if we cannot discover what is the matter. Let us try and see if there is any good reason why the great majority should be governed and oppressed by a small minority.

"In human affairs effects follow causes; results are accomplished by action, even when the actors are unseen. Look at our State and national Governments, and who are the men to whom we entrust this great responsibility? Look at our boards of trade, industrial expositions, and in fact any great project for the advancement of science, art, liberty, or industry, and you will find at its head and the moving spirit thereof a lawyer, doctor, preacher, student, merchant, or, in fact, almost anything but a farmer. These men rule the nation. They shape the laws; they make the channels of trade, and place trade in its channels. They build ships, harness steam to their wagons, make lightning carry their messages; they compel rivers to turn their saws, twirl their spindles, and throw their shuttles. They use their brains, and mind governs the world.

"Just think of competing against such men by stupidly hoeing corn fifteen hours a day and selling it at twenty cents a bushel, and then laying awake nights, growling at railroad men and merchants. The dog who barks at the moon comes nearer accomplishing his purpose than such a growler. Why have not farmers taken a position of influence and power in the councils of the nation and otherwise, in proportion to their numbers and wealth? Simply because we have not used our brains.

"The world pays homage to intelligence, to intellect, and puts it in places of honor, of trust, of responsibility. The world is not partial to lawyers, ministers, and doctors, but the world wants to use brains, and accepts them wherever found, and uses them to promote its wishes; and if we farmers want to be placed in the foremost rank in the nation and in the world; if we wish to be put in positions where we can have power to aid our fellows; if we wish to have influence and make our mark on the institutions of the land; if we wish to stand where we can do something towards governing the price of our commodities; if we wish to weigh according to our size in the scale of public opinion; if we want to have farmers in demand for places of trust and honor and profit, and for husbands for beautiful, refined, and intelligent women; if we want to escape from our present vassalage, we must furnish some brains, sound in quality, liberal in quantity, polished with constant use, refined by study and thought. Show me such a farmer as that, and I will show you a man whom his fellow-men will want to use in places of trust.

"I speak it in sorrow; I admit it with deep and burning shame, that the farmers can furnish but comparatively few men whose minds are fitted to organize great enterprises. Look at the farmers in our Legislature. In numbers they are very small in proportion to the population of the State, and smaller yet in the influence they have upon the legislation. When they come in contact with men who are in the habit of close and logical reasoning, they, with a few exceptions, prove wanting. It may, and probably will be said that head-work will not hoe corn or feed the pigs. Granted. But prove to me that an intelligent man is disqualified from performing the duties of a farmer and you will prove to me that farming is a business which it is disgraceful to follow, and that it is grossly unjust to say aught to induce any young man of common sense to become a farmer.

"It is seen that thought, intelligence, mind, brains, used in other branches of business, lead to success. It is found that men with clear heads, sharp wits, sound judgment, and business habits go straight along and compel success even under adverse circumstances. Now, is it any advantage to have and use brains? Can a man with brains get, in tilling the soil, a fair compensation for their use? Can brain work be employed on the farm and return to the owner as much of comfort, wealth, happiness, honor, and general prosperity as in other branches of business at the present time? This is a knotty question, but it is one we have got to meet, and meet it now. There is no use in attempting to evade or ignore this great alternative. If there is anything in agriculture that necessarily dwarfs the mind and makes it secondary to mere physical exertion, then it is a disgrace to be a farmer, and common honesty requires that we cease talking about the honorableness of the noble yeomanry. But, on the contrary, if agriculture will give scope to thought and research; if it will cause a man to think while he works and study while he has leisure; if his business is such that talent and tact will transform his soil to gold and his house into a beautiful and happy home; if the same amount of bodily and mental labor on the farm will produce as much pleasure, wealth, and happiness as in the shops, counting-room, and mines, then we may conscientiously recommend agriculture as one of the desirable employments. Can this be done?

"Brother Patrons of Husbandry, our Order has been formed to assist in answering this great question in the affirmative. How shall we proceed?

"I do not underrate the importance of making an effort to buy our reapers a few dollars cheaper and sell our wheat a few cents higher and get our freights a little lower. What is gained in this way is certainly added to the profits of the farm, but I very much fear that many members of the Order place too high a value upon this matter of purchase and sale. This is not what ails us. It does not reach the root of the difficulty at all. It only prunes away a few slender twigs which grow again in a single night. We can never accomplish what we want, and make agriculture respectable, remunerative, and desirable; farmers intelligent, contented, and honored; farmers' wives envied and respected, and farmers' sons and daughters eagerly sought by the wise, good, learned, and beautiful of the land for husbands and wives; we cannot make beautiful homes, fertile farms, and improving flocks by saving five dollars on a plow and five cents a bushel on wheat. No! Never! When we build like that we must dig deeper, lay the foundations broader, and use *brains* as the chief stone of the corner. An ox excels us in strength, a horse in speed. The eagle has keener sight, the hare a quicker ear, the deer a finer sense of smell; but man excels them all in mind and rules above them all. So among men it is not the strong, the swift, the keen-sighted, the quick-eared or fine scented who rules the world, but the clear-headed. Human beings are like pebbles on the sea shore; by rubbing against each other they become rounded, smooth, polished, symmetrical: alone, they are rough, uncouth, repulsive.

"Farmers are too much alone. We need to meet together to rub off the rough corners and polish down into symmetry. We want to exchange views, and above all we want to learn to think. A man who has performed fourteen hours of severe physical labor is in no condition to think, and we may as well decide at once that any class of men which starts out in life by working at severe labor fourteen hours of the twenty-four, and faithfully adheres to the practice, will fill forever the position of hewers of wood and drawers of water for men who use the God-given mind, and nourish the soul with liberal and abundant mental food.

"I have already tired your patience, and in closing will only say that in my opinion the coming farmer will not toil with his hands fourteen hours out of the twenty-four and compel wife and children to the same slavery. But he will give a liberal share of his time to thought, study, and recreation. He will know of what his soil is composed, in what it abounds, in what it is deficient. He will know what elements of earth and air are needed to plant growth, and under what conditions they can be most readily assimilated. He will understand the laws of plant and animal life, that he may more successfully treat them. His house will be abundantly supplied with books and papers on agricultural and matters of general interest. Pictures and abundant amusements will make his home attractive. A beautiful lawn and flower beds, a fruit and vegetable garden, an orchard, groves, and evergreens and deciduous trees for ornament, shelter, and use, will make his home so lovely and homelike that his daughters will not be so disgusted with farm life as to marry a village dolt, or the son so worn, weary and dispirited as to leave the farm at the first opportunity and open a barber shop in some country village. Can this be done, and can the farms really be made the happy homes of refined, intelligent, honored men and women, instead of the abodes of overworked slaves? Yes! emphatically yes! But not by neglecting to rust the God-given mind, but by rousing it up and making it the compass, the sail, and the rudder in the voyage of life. The body is but the hulk. Then set your sails, stand by the rudder, steer by the compass, and start out boldly on the great journey, whose passage is pleasure and whose end is success."

Mr. Adams is married to an excellent lady, endeared to a wide circle of friends by her many virtues. She holds the post of Ceres in the National Grange.

O. H. KELLEY,

Secretary of the National Grange.

Mr. Kelley is a native of Boston, Massachusetts, and has been for some years a clerk in the Bureau of Agriculture at Washington, D. C. He is about forty-six years of age. In appearance he is a man who would be singled out of a crowd as a thinker. His high bold forehead, large earnest eyes, long white beard, and generally scholarly appearance, would seem to stamp him as a philosopher rather than a man of action. He is eminently fitted for the position he holds in the Order, a position involving an infinite amount of detail, and requiring ability of a marked and special character. His services in founding the Order have already been related in these pages, and it is useless to repeat them here.

COLONEL JOHN COCHRANE,

Master of the Wisconsin State Grange,

Is one of the most prominent members of the Order. His views upon the questions of the day have been given at length elsewhere. He is one of the old settlers of the State, resides at Waupun, and is known throughout Wisconsin as one of the best practical farmers in the West. He has never been a politician, though he has taken a deep interest in all the political questions of the day; "and he enters into the farmers' movement at a time of life when men of his habits and pursuits generally find retirement most attractive, from a strong conviction of duty, and a desire to raise the farmers of Wisconsin out of the slough of despond into which they have been fast sinking." Colonel Cochrane's farm comprises a tract of 1000 acres, which he has brought under the highest state of cultivation. He has had a crop of wheat of 6000 bushels in a single year on this land.

S. H. ELLIS,

Master of the Ohio State Grange,

Is a native of the State, is forty-three years of age, and has been a farmer all his life. His connection with the Order dates back to September, 1872, when, with fifty of his acquaintances, he succeeded in organizing the first Grange in the State. He was elected Master of this Grange, and was subsequently appointed by the National Grange a deputy for Ohio to organize new

Granges. By the 9th of April, 1873, there were in the State thirty Granges. In this month the State Grange was organized, and Mr. Ellis was chosen Master.

JOHN WEIR,

First Master of the Indiana State Grange,

Is a native of East Tennessee, but emigrated to the Wabash, in Indiana, in 1817, being then seventeen years of age. A large majority of the inhabitants were Indians. The settlers were destitute of churches, schools, mills and roads. Comparing that time with the present, he has witnessed perhaps a greater change brought about by civilization than any other man. The State Grange was organized at Terre Haute, February 28th, 1872. On the 15th of January last there were 49 organizations, since which time the number has increased nearly six-fold.

F. H. DUMBAULD,

First Master of the Kansas State Grange,

Was born in Pennsylvania, and is now 45 years of age. With his father's family he removed at an early age into Ohio, where he remained 18 years. In 1864 he settled in Kansas. He has "made" three large farms in his life. The State Grange was organized July 30th, 1872, and Mr. Dumbauld elected Master. With the assistance of George Spurgen, the Secretary, he has organized over 400 subordinates in the last nine months. The Patrons have effected quite a revolution in Kansas, having brought dealers to supply agricultural implements and other necessaries at prices varying from 30 to 40 per cent. lower than usually charged.

T. R. ALLEN,

Master of the Missouri State Grange,

Is a practical and prosperous farmer. He resides about thirty miles west of St. Louis on the Missouri Pacific Railroad. In the discharge of his official duties he has travelled through every county of the State during the past year, and under his management the Order has spread rapidly throughout that State.

The Illinois State Farmers' Association,

Is often confounded with the Order of Patrons of Husbandry, but is entirely distinct from it. It is composed of both Granges and Farmers' Clubs, and has taken a very prominent part in the politics of the State. The President of this Association is Mr. W. C. Flagg, of Mora, Madison county, and the Treasurer is Duncan M'Kay, of Mount Carroll, Carroll county. Stephen M. Smith, whose views and speeches we have given at length, is the Secretary of the Association. He is also an active member of the Grange.

DUNCAN M'KAY, TREASURER OF THE ILLINOIS STATE FARMERS' ASSOCIATION.

DECLARATION OF PURPOSES

OF THE

ORDER OF PATRONS OF HUSBANDRY.

Adopted at the Annual Session of the National Grange, at St. Louis, Mo., February 11th, 1874.

PROFOUNDLY impressed with the truth that the National Grange of the United States should definitely proclaim to the world its general objects, we hereby unanimously make this Declaration of Purposes of the Patrons of Husbandry:

1. United by the strong and faithful tie of Agriculture, we mutually resolve to labor for the good of our order, our country, and mankind.

2. We heartily indorse the motto, "In essentials, unity; in non-essentials, liberty; in all things, charity." We shall endeavor to advance our cause by laboring to accomplish the following objects;

To develop a better and higher manhood and womanhood among ourselves. To enhance the comforts and attractions of our homes, and strengthen our attachments to our pursuits. To foster mutual understanding and cooperation. To maintain inviolate our laws, and to emulate each other in labor. To hasten the good time coming. To reduce our expenses, both individual and corporate. To buy less, and produce more, in order to make our farms self-sustaining. To diversify our crops, and crop no more than we can cultivate. To condense the weight of our exports, selling less in the bushel, and more on hoof and in fleece. To systematize our work and calculate intelligently on probabilities. To discountenance the credit system, the mortgage system, the fashion system, and every other system tending to prodigality and bankruptcy.

We propose meeting together, talking together, working together, buying together, selling together, and in general acting together for our mutual protection and advancement as occasion may require. We shall avoid litigation as much as possible by arbitration in the Grange. We shall constantly strive to secure entire harmony, good will, vital brotherhood among ourselves, and to make our order perpetual. We shall earnestly endeavor to suppress personal, local, sectional, and national prejudices, all unhealthy rivalry, all selfish ambition. Faithful adherence to these principles will insure our mental, moral, social, and material advancement.

3. For our business interests we desire to bring producers and consumers, farmers and manufacturers, into the most direct and friendly relations possible. Hence we must dispense with a surplus of middlemen; not that we are unfriendly to them, but we do not need them. Their surplus and their exactions diminish our profits. We wage no aggressive warfare against any other interests whatever. On the contrary, all our acts and all our efforts, so far as business is concerned, are not only for the benefit of the producer and consumer, but also for all other interests, and tend to bring these two parties into speedy and economical contact. Hence we hold that transportation companies of every kind are necessary to our success; that their interests are intimately connected with our interests, and harmonious

action is mutually advantageous. Keeping in view the first sentence in our declaration of principles of action, that "individual happiness depends upon general prosperity," we shall therefore advocate for every State the increase in every practicable way of all facilities for transporting cheaply to the seaboard, or between home producers and consumers, all the productions of our country. We adopt it as our fixed purpose to open out the channels in Nature's great arteries, that the life-blood of commerce may flow freely. We are not enemies of railroads, navigation, and irrigating canals, nor of any corporation that will advance our industrial interests, nor of any laboring classes. In our noble order there is no communism, no agrarianism. We are opposed to such spirit and management of any corporation or enterprise as tends to oppress the people and rob them of their just profits. We are not enemies of capital, but we oppose the tyranny of monopolies. We long to see the antagonism between capital and labor removed by common consent and by enlightened statesmanship worthy of the nineteenth century. We are opposed to excessive salaries, high rates of interest, and exorbitant profits in trade. They greatly increase our burdens, and do not bear a proper proportion to the profits of producers. We desire only self-protection and the protection of every interest of our land by legitimate transactions, legitimate trade, and legitimate profits.

4. We shall advance the cause of education among ourselves and for our children by all just means within our power. We especially advocate for our agricultural and industrial colleges that practical agriculture, domestic science, and all the arts which adorn the home be taught in their courses of study.

5. We especially and sincerely assert the oft-repeated truth taught in our organic law, that the Grange, National, State, or subordinate, is not a political or party organization. No Grange, if true to its obligations, can discuss political or religious questions, nor call political conventions, nor nominate candidates, nor even discuss their merits in its meetings. Yet the principles we teach underlie all true politics, all true statesmanship, and if properly carried out will tend to purify the whole political atmosphere of our country. For we seek the greatest good to the greatest number, but we must always bear in mind that no one by becoming a Patron of Husbandry gives up that inalienable right and duty which belongs to every American citizen, to take a proper interest in the politics of his country. On the contrary, it is right for every member to do all in his power legitimately to influence for good the action of any political party to which he belongs. It is his duty to do all he can in his own party to put down bribery, corruption and trickery; to see that none but competent, faithful and honest men, who will unflinchingly stand by our industrial interests, are nominated for all positions. It should always characterize every Patron of Husbandry that the offices should seek the man and not the man the office. We acknowledge the broad principle that difference of opinion is no crime, and hold that progress towards truth is made by differences of opinion, while the fault lies in bitterness of controversy. We desire a proper equality, equity, and fairness, protection for the weak, restraint upon the strong; in short, justly distributed burdens and justly distributed power. These are American ideas, the very essence of American independence, and to advocate the contrary

is unworthy of the sons and daughters of an American republic. We cherish the belief that sectionalism is, and of right should be, dead and buried with the past. Our work is for the present and the future. In our agricultural brotherhood and its purposes, we shall recognize no North, no South, no East, no West. It is reserved by every Patron, as the right of a freeman, to affiliate with any party that will best carry out his principles.

6. Ours being peculiarly a farmers' institution, we cannot admit all to our ranks. Many are excluded by the nature of our organization, not because they are professional men, or artisans, or laborers, but because they have not a sufficient direct interest in tilling or pasturing the soil, or may have some interest in conflict with our purposes. But we appeal to all good citizens for their cordial cooperation to assist in our efforts toward reform, that we may eventually remove from our midst the last vestige of tyranny and corruption. We hail the general desire for fraternal harmony, equitable compromise, and earnest cooperation, as an omen of our future success.

7. It shall be an abiding principle with us to relieve any of our suffering brotherhood by any means at our command. Last, but not least, we proclaim it among our purposes to inculcate a proper appreciation of the abilities and sphere of woman, as is indicated by admitting her to membership and position in our Order. Imploring the continued assistance of our Divine Master to guide us in our work, we here pledge ourselves to faithful and harmonious labor for all future time to return by our united efforts to the wisdom, justice, fraternity, and political purity of our forefathers.

www.ingramcontent.com/pod-product-compliance
Lightning Source LLC
LaVergne TN
LVHW021240110826
845150LV00002B/369

9781425561444